The Evolution of Development Thinking

Politics, Economics, and Inclusive Development

Series Editors
William Ascher, Claremont McKenna College
Natalia Mirovitskaya, Duke University
Shane Joshua Barter, Soka University of America

Politics, Economics, and Inclusive Development focuses on promoting humanistic development by publishing single- or multi-authored monographs. These books use an interdisciplinary approach to understand the current challenges facing individual nations, world regions, or the entire global system in pursuing peaceful, democratic, and technically sound approaches to sustainable development.

Economic Development Strategies and the Evolution of Violence in Latin America
Edited by William Ascher and Natalia Mirovitskaya

Development Strategies, Identities, and Conflict in Asia
Edited by William Ascher and Natalia Mirovitskaya

The Economic Roots of Conflict and Cooperation in Africa
Edited by William Ascher and Natalia Mirovitskaya

Civilian Strategy in Civil War: Insights from Indonesia, Thailand, and the Philippines
Shane Joshua Barter

Smallholders and the Non-Farm Transition in Latin America
Isabel Harbaugh

Development Strategies and Inter-Group Violence: Insights on Conflict-Sensitive Development
William Ascher and Natalia Mirovitskaya

The Evolution of Development Thinking: Governance, Economics, Assistance, and Security
William Ascher, Garry D. Brewer, G. Shabbir Cheema, and John M. Heffron

The Evolution of Development Thinking

Governance, Economics, Assistance, and Security

William Ascher, Garry D. Brewer, G. Shabbir Cheema, and John M. Heffron

Prepared under the auspices of the
Pacific Basin Research Center,
Soka University of America

palgrave
macmillan

First published 2016 by
PALGRAVE MACMILLAN

Palgrave Macmillan in the UK is an imprint of Macmillan Publishers Limited, registered in England, company number 785998, of Houndmills, Basingstoke, Hampshire, RG21 6XS.

Palgrave Macmillan in the US is a division of Nature America, Inc., One New York Plaza, Suite 4500, New York, NY 10004-1562.

Palgrave Macmillan is the global academic imprint of the above companies and has companies and representatives throughout the world.

Hardback ISBN: 978–1–137–56038–4
E-PUB ISBN: 978–1–137–56040–7
E-PDF ISBN: 978–1–137–56039–1
DOI: 10.1057/9781137560391

Distribution in the UK, Europe and the rest of the world is by Palgrave Macmillan®, a division of Macmillan Publishers Limited, registered in England, company number 785998, of Houndmills, Basingstoke, Hampshire RG21 6XS.

Library of Congress Cataloging-in-Publication Data

Names: Ascher, William.
Title: The evolution of development thinking : governance, economics, assistance, and security / William Ascher, Garry Brewer, Shabbir Cheema, and John M. Heffron.
Description: New York City : Palgrave Macmillan, [2015] | Includes bibliographical references.
Identifiers: LCCN 2015022379| ISBN 9781137560384 | ISBN 113756038X
Subjects: LCSH: Economic development—Developing countries. | Economic assistance—Developing countries. | Internal security—Developing countries. | Developing countries—Economic conditions.
Classification: LCC HC59.7 .A8339 2015 | DDC 338.9009172/4—dc23 LC record available at http://lccn.loc.gov/2015022379

A catalogue record for the book is available from the British Library.

To future generations of development scholars and practitioners who will take the lead in advancing human dignity.

Contents

Illustrations

Figure

Tables

Preface

This book, the eighth in the Palgrave Macmillan's *Politics, Economics, and Inclusive Development* series, shares with the other volumes the commitment to explore the complex task of simultaneously pursuing economic growth, broad participation and equity, democratic peace, and sustainability.

The collaboration that produced this book was motivated by the recognition that progress in addressing the aspirations of people in developing countries has been very mixed, in terms of both the soundness of approaches and the results. A broad assessment, recognizing both achievements and faltering, is needed to determine why some very promising theories have not been translated into doctrines and why some sound doctrines have not been in evidence in development practice.

But rather than just rail against the slippage from theories to doctrines to practice, we try to understand the dynamics that stand in the way of sound practice. We explore, for example, how fundamental goals of enhancing human dignity in its multiple dimensions give way to lesser objectives, how development objectives are often dominated by security concerns, and how technical uncertainty allows political decisions to masquerade as simply technical.

The unusual length of this book reflects our realization, based on our combined experience of more than 200 years of studying policy and policy processes, that the specialization on each of the constituent themes (governance, economic development, foreign assistance, the roles of the military, and the roles of nongovernmental organizations) has largely neglected the interconnections among these challenges. Economic strategies have to take into account the limitations of governance; governance must take into account the economic constraints on state capacity; foreign assistance must recognize the issues of security, economic needs, and governance. Nongovernmental organizations have to navigate the governance institutions, contribute to economic development, and often collaborate with military initiatives or try to salvage people's well-being in the aftermath of military actions.

Assessing the impact of development approaches across multiple world regions does yield many insights, but it should not be taken to mean that we have discovered the one and only strategy that applies in all cases. The importance of context makes that aspiration foolish indeed. Rather, we hope that our readers will become more aware of the dynamics that have supported or thwarted progress and recognize that what may at first seem novel in development approaches would turn out not to be—this should lead to a fuller appreciation of the obstacles to mounting development approaches. We have tried to emphasize the remaining challenges: fundamental disagreements about the theoretical and methodological bases for economic development strategies; the tensions between governments and civil society; painful trade-offs among the objectives of foreign assistance; and the uncertain future of the role of the military. It is most troubling that for many countries, economic policies and governance still limit meaningful participation, reinforce the systematic deprivation of marginalized people, and undermine personal security.

Acknowledgments

We gratefully recognize the support each of us has received over the years from our home institutions: Claremont McKenna College, Yale University, the East-West Center, and Soka University of America. We are especially thankful to the Pacific Basin Research Center of Soka University of America, whose financial and staff support made this collaboration possible.

CHAPTER 1

Introduction

This book assesses the trends in development theories, doctrines, and practices for developing countries since World War II (WWII). Development thinking and practice span a wide range of areas. The broadest are development economics and economic management, and governance and development administration. No less important are the foreign assistance theories and approaches and the roles of key institutions within and beyond conventional government entities, particularly the armed forces and nongovernmental organizations (NGOs).

Our assessment will show that compelling theories and doctrines have been developed and applied at various times in various places. The agenda has expanded to address a much broader range of the dimensions of human development than what prevailed prior to the 1970s. More robust mechanisms for citizen participation and more effective and inclusive governance now characterize the development process. Foreign aid doctrines have shed some of their more problematic aspects. NGOs have refined their missions to more clearly distinguish themselves from governmental initiatives and, consequently, have more constructive interactions with governments. Yet the most troubling conclusions are that some compelling approaches, both old and new, have been abandoned, ignored, or distorted, and thorny issues remain unresolved. The balance between economic efficiency and the protection of vulnerable people, when not tilting in favor of the former, is often elusive, as are constructive roles of the armed forces. Foreign assistance is often caught between humanitarian and geopolitical motives and faces uncertainties regarding its effectiveness.

For this introduction, it is useful first to clarify our standpoint for assessing the progress and challenges of development. Success or failure, progress or regression, depends on the goals we seek. Unrealistic goals lead to unrealistic expectations, which almost always lead to poor results.

Next we outline key factors that drive the evolution of theories, doctrines, and practice. It is not enough simply to document changes and remaining challenges—further progress requires understanding of how theories, doctrines, and practices change and how constraints on them arise. To make sense of the complex changes and obstacles that have emerged over the past half century, we devote most of this introduction to the dynamics involved in the evolution of theories and doctrines and then to the translation of doctrines into practices, across all spheres covered in the book.

We define development theory as a body of causal propositions that in principle anticipate the consequences of specific policies, strategies, or other actions. Behind many development theories are analytic methods that are employed to generate the projected consequences. This is particularly relevant for the applications of development economics. A doctrine is a set of principles that can guide the selection of particular policies, strategies, or other actions. Practice consists of these policies, strategies, and other actions undertaken by government officials or other actors.

Our Standpoint

Recounting the patterns of development thought and action could be conducted as a straightforward history, to put the broader history of the developing world into context or to honor brilliant thinkers who revolutionized approaches to development. Many histories of development thinking have already been written (e.g., Chenery 1975; Dorfman 1991; Bevir, Rhodes, and Weller 2003; Meier 2004). However, few satisfy the requirements to be both problem-oriented and diagnostic—limitations we attempt to rectify.[1] Our concern is to recast and contextualize the obstacles to providing greater material and nonmaterial benefits to the broadest range of people and to identify approaches to overcome these obstacles. Undoubtedly, impressive progress has been made in improving the robustness of development theories, whether addressing economics, governance, or the roles of particular institutions. Yet there is also no doubt that social, political, and economic development has faltered badly in many countries. These facts raise crucial questions. If breakthroughs in theory and doctrine have been made, why have improvements in practice been so limited? Why has the identification of effective development approaches been so difficult? Why has so much slippage occurred between the creation of compelling approaches and the policies and programs that have been enacted? If the goal is greater human dignity, what further conceptual and intellectual breakthroughs, effectively enacted, would lead to its achievement?

Our intent is not to condemn; we mean to be as alert to development successes as to failures. Some Millennium goals have been exceeded, and some very poor countries as of 1950, such as South Korea or Norway, are now significant foreign assistance donors, although such countries are often overlooked in studies that excoriate development efforts

Uncertainty about how to evaluate the development record of particular countries extends to evaluating external institutions tasked with providing development guidance and operations. Bilateral aid agencies and international financial institutions come under heavy fire for every subpar development effort undertaken. One conspicuous failure may well overshadow dozens of successes. Accusations that the World Bank and the International Monetary Fund are pawns of Western governments in imposing unwanted conditions overlook the fact that both institutions (and others like them) usually work hand in glove with the economic ministries, planning agencies, and development institutions within developing country governments.

Our Goals

To understand and assess the results of this evolution of development thinking and practice, it is crucial to specify the goals of development that guide our evaluation. This is separate from determining the goals that guide the decisions of particular policymakers. The policy sciences approach offers useful guidance: *We seek and value anything that contributes to the broadest shaping and sharing of valued human outcomes.* This guidance is sometimes summarized as *working to do whatever is possible to increase human dignity* (Lasswell and McDougal 1992, 195).

Understanding the Evolution: Theories and Doctrines

The evolution of theory and doctrines emerges from many dynamics. At the base of this evolution is the formulation, acceptance, or rejection of theories about development and the roles of policies and institutions. These theories may be rooted in experience, in general theories developed for other contexts (such as developed nations), or in the interests that various actors are pursuing.

Theories and doctrines may originate domestically or internationally. External institutions—foreign governments, militaries, NGOs, and international organizations—play multiple roles in developing and disseminating theories and doctrines and in advising and pressuring governments to adopt particular doctrines and practices. Their impacts reflect both direct actions, such as providing resources, and their influence on perspectives within the country. For example, Latin American military officers trained in the US Inter-American Defense College are more likely to adopt the security doctrines presented in this institution. Economists from the same country trained in the same foreign university may carry back that university's distinctive economic orientation: the so-called Chicago Boys in Chile or the Berkeley Mafia in Indonesia. Based on varying combinations of prestige, funding, personal connections, political pressure, and threats of coercion, these external actors often have major impacts on the adoption of theories and doctrines. Yet their influence is frequently contested by both internal and external actors.

The Constraints of Limited Information and Analysis

Theory development depends on information that is available *and* reaches the focus of attention of those involved in thinking about development. In the early post-WWII period, fundamental information about crucial factors such as historical growth rates and income distribution was sparse even for developed nations. This is epitomized by the title of Wolfgang Stolper's 1966 book on Nigerian economic policies, *Planning without Facts*. There was only a dim understanding of Nigeria's income distribution, the productivity of various economic activities, or how the governance and economic institutions would perform postindependence. Therefore, theories and doctrines were based on highly stylized, often simplistic depictions of political and economic dynamics. Even today, with poorly monitored informal economies in many developing countries, it is difficult to know with any degree of precision how the economy is doing and hence whether the doctrines underlying economic policy or foreign assistance need to be rethought. The same is true for informal governance—the ad hoc decisions on how to distribute government benefits, the unreported transactions between government officials and others—that makes it so difficult to assess whether governance approaches need to be changed.

In some cases, the plight of certain peoples is beyond the ken of experts and policymakers. For decades the Chilean census lacked any questions that could gauge the portion of the population comprised of indigenous Mapuches whose language and geographic isolation warranted special attention.

Theory and doctrine development is often severely challenged by the difficulty of recognizing problems that either emerge in changed circumstances or had been ignored in the past. The analytic limits to determine where a group's interests lie, and to mount arguments in favor of particular doctrines or policies, often put low-income people at a serious disadvantage. The growing efforts to offset this deficit, by NGOs, government agencies, and international organizations, constitute an important part of the evolution of development practice.

The vacuum created by sparse knowledge regarding particular countries leads to dependence on importing theories and doctrines from other areas, frequently with perverse impacts. Without solid contextual knowledge, judging the appropriateness of imported theories and doctrines tends to be a futile exercise. A notorious example was the wholesale adoption of Latin American constitutions mimicking the US constitution, in countries with radically different political cultures and socioeconomic challenges. Today the East Asian Economic Miracle, with its latest Chinese incarnation, heavily influences many policymakers, just as the now-discredited Japanese model did in the 1970s and 1980s. Some prominent successes in radically decentralized governance such as Indonesia have spurred a large and growing number of decentralization initiatives, with mixed results.

Similarly, ignorance of local conditions can impart serious biases in theories developed at great distance. Governance and economic theories often focus on national-level patterns and embrace an implicit assumption that citizens have a more broad national identification than is the case in such countries as Afghanistan, Iraq, and Myanmar, and that the economy is more integrated than the on-the-ground reality warrants.

The Sway of Interests and Ideologies

Theories and doctrines reflect not only technical analysis but also the values and interests of the advocates. The generation of theories and decision routines is the province of development experts of many stripes who advance their careers, professional standing, and programmatic objectives in a competition for recognition, funding, and the standing of their institutions. Many suffer from the insularity of their academic disciplines and inability to undertake interdisciplinary analysis.

Within developing countries, the endorsement of theories and doctrines invariably involves advocacy efforts by stakeholders of different levels of influence, depending on the country's governance arrangements. The same holds for theories and doctrines formulated in developed nations. Thus, although the capital output ratio approach to estimate the volume of foreign assistance required for a given country to achieve self-sustaining growth rested on the simplistic premise that capital would straightforwardly generate output, ignoring institutional obstacles, it was decisive in convincing US officials to pour resources into big-push foreign aid.

Sometimes the link between interests and the support of theories and doctrines is obvious, as in the case of industrialists promoting a proindustry economic strategy. Yet sometimes the commitment to theories and doctrines reflects deep-seated aspects of cultural immersion or professional training that is perceived as technical but has underlying value implications. People of different cultural or professional backgrounds often find opposing views to be patently wrong, or even unethical, in terms of their own paradigms. The professional and institutional origins of the theorists can have major impacts on their theories and doctrines. Thus, the World Bank and International Monetary Fund largely recruit their economists from universities with neoclassical economics departments, while the UN agencies addressing economic development (UN regional economic commissions, the UN Development Programme, the Secretariat of the UN Conference on Trade and Development, etc.) are more likely to recruit from universities with heterodox economics, political science, and public administration departments. Governance theories reflect even more disparate origins and ideologies of the theorists, from Maoist enclaves that still exist in the People's Republic of China to the Heritage Foundation and other highly conservative think tanks. Programs on peace and conflict resolution studies produce very different mindsets than military academies.

Given the remarkable variety of outlooks, development theories—whether in general or in application—will be contested. Different conceptions of whether other approaches could have done better, different attributions of which policies are responsible for success or failure, and different goals preclude consensus on the appropriateness of a theory and related doctrine. At no point in the post-WWII period has universal acceptance been enjoyed by any theory of development economics, governance, civil society, national security, or foreign assistance. In the case of governance, shifts have taken place from traditional public administration to new public management, engagement of civil society and democratic (good) governance. Of course, given the varying contexts of developing countries and the complexity of the relevant processes, this openness is not necessarily bad. Yet interests at odds with maximizing human dignity may hide behind theories and their related doctrines even when inappropriate in a given situation. Evaluations are no less contested than the theories and doctrines.

The Filtering Dynamics

Whether particular theories or methods are translated into development doctrines depends crucially on yet other institutions and individuals who transmit this knowledge through journals, books, reports, teaching, etc. This transmission process diminishes the sway of some theories by filtering out knowledge that might support them and endorses others simply by transmitting them. Even the explicit labeling of a development initiative heightens the theories and doctrines associated with the labels: a *Basic Needs Program* endorses the theories behind the basic needs approach, though it may discredit the approach if the program fails or is perceived as failing.

The image of each institution developing or disseminating doctrines through advice or application may play a profound role in supporting or discrediting these doctrines. These images are often severely distorted, particularly through negative stereotyping by opponents. The doctrines of the International Monetary Fund and the World Bank are widely reviled because of their images as servants of the West-dominated global economic system. Some government officials reject NGOs' poverty alleviation strategies because they see NGOs as soft-hearted and soft-headed—or as intrinsically antigovernment. Many developing country governments prefer to work with UN agencies because of their perceived neutrality and partnership approaches, even though bilateral donors and international financial institutions often question their analysis. Nationalist Latin American military officers may look askance at doctrines of civil-military relations formulated at the Washington-based Inter-American Defense College. Many nationalist or leftist leaders may also reject a governance doctrine developed by the US Agency for International Development. The question is not whether these doctrines are appropriate but rather whether they would get a fair hearing. Whether they get a fair hearing or not, some special interest is always willing to advocate these doctrines.

Assumption Drag

After WWII, the developing world was remarkably unknown to development theorists, practitioners, and policymakers. The challenges these nations presented became somewhat clearer in the 1950s, though not quickly enough to avoid premature shortcuts and unfortunate misconceptions by development professionals. Consequently, certain questionable assumptions were established early, became anchored, and then persisted for years, misinforming and misdirecting development practice. Taking these anchored assumptions as givens inhibited sufficient examination and honest evaluation. A powerful and influential assumption in the early days of development theory held that industrialization was the obvious path to success. The premise was that agriculture was not promising except to finance industrial development. The possibility of a successful Green Revolution was neglected, and the bias against agriculture has persisted in many countries to this day.

Another persistent assumption was the reliance on modernization theory that presumed that economic growth would be self-sustaining and would bring poverty reduction and democratic governance in its wake. This view persisted in the face of mounting contrary evidence, and even today it may seem natural that industrialization is economic advancement and that a more advanced economy ought to bring a more advanced state of democracy. Another example is the tenacious premise that strong centralized governance is optimal to transform underdeveloped economies. Centralized natural resource management has been disastrous in many countries, as indicated by rampant deforestation. In Indonesia, for example, central government control over the nationalized forests led to excessive logging by private companies in violation of customary user rights of local communities that in many circumstances would have husbanded the forests more responsibly. In Vietnam, for example, the highly centralized funding for schools, health clinics, roads, and other inputs has often failed to meet the priority needs of communes and districts.

Bandwagon Dynamics

Yet old assumptions are sometimes replaced with new ones that may also be too broadly accepted to the neglect of specific conditions. In some circles the pendulum now has swung in favor of decentralization, which in many contexts is viewed as a panacea, without taking into account that whether a particular service ought to be provided by the national government or by subnational governments depends on a complex mix of considerations of efficiency, effectiveness, responsiveness, fairness, and conflict sensitivity. Agrawal and Ribot (1999, 2006) note the predominance of decentralization efforts and the paltry success in devolving meaningful authority to the local levels.

Dilemmas of Integrating Technical and Normative Considerations

No matter how technical the derivation of a development theory may seem, it inevitably has implications for value outcomes that development produces and the distribution of benefits. These implications may be only dimly recognized by analysts and others. Some theories and the doctrines derived from them seem particularly attractive because they are cast in noncontextual terms that seem to promise broad generalizability. Yet, rigid, context-ignoring theories in fishery or forestry (e.g., maximum sustainable yield management theories presuming that existing stocks ought to be maintained) neglect many factors, ranging from environmental implications to whether other species should replace the current ones. In other cases, theorists willfully disregard technical constraints and means-ends rationality as unworthy of ideological commitment.

From Doctrine to Practice

A development doctrine has to go through a tortuous path to become practice, and the resulting practice is often highly problematic in terms of the goals that motivated the doctrine. Some of the same dynamics that account for the patterns and problems of translating theories into doctrines also hold for the translation of doctrines into practice. Concrete development decision making is particularly demanding, due to the multiple layers and numerous specific parties involved in collaborative efforts. For large-scale and consequential problems, finding an acceptable solution may take inordinate amounts of time, attention, and resources as compromises are hammered out and modifications made to account for specific contextual details.

Impacts of Weak Analytic Capacity

Informational and analytic deficits are just as important to the translation of doctrines to practice as they are to the translation of theories to doctrines. For example, the feasibility of a combatant reintegration doctrine depends on the uncertain prediction of whether demobilized militias would be willing to take appropriate advantage of reintegration incentives. By the time a theory that is sound under one set of conditions finds its way from doctrine to practice, the original circumstances may have changed dramatically. For example, in Chile the theory that crucial groups would accept a constitutional land reform amendment may have been correct before the 1960s, but was a disaster after faith in government eroded.

Another problematic analytic issue arises in the translation from general policy to case-by-case application. Details matter—which requires a clear comprehension of context and may call for rethinking theories and doctrines. For example, ignorance of the incomes of families that may or may not meet the means test for eligibility for cash transfers can introduce arbitrariness in granting the transfers, creating resentment among those excluded, thereby undermining the cash transfer policies and the social safety net doctrines that are the bases of these policies (World Bank 2014a). By the same token, trying to prevent forest destruction in developing (usually tropical) lands demands clear knowledge about what motivates subsistence in or nearby the forests. Putting a payment for environmental services program into practice to save a forest, without provision for a fair and transparent sharing of the payments themselves, is bound to fail.

Some of these details pertain to the individual and organizational expectations and demands of stakeholders. Theories and doctrines developed for different locales or time periods are likely to be challenged in this respect. For example, economic policy liberalization doctrines may depend on high levels of labor mobility, but sometimes local attachments reduce the willingness of people to move to where jobs are available.

The transitions from doctrines to broad policies to specific applications are further complicated by the gaps between formulating policies and deciding *which policy* should be applied to specific cases. For example, in countries with race-based affirmative action, decisions, often ad hoc, must be made to determine whether a particular individual has enough of the racial characteristics to qualify for benefits. Similarly, if one budget rule has been formulated to benefit economically backward areas, and another is to allocate resources on strict population proportionality, a decision must be made as to which rule ought to be applied for each case.

Restricting the Full Force of Doctrines

One prominent dynamic is the widespread pattern of restriction by partial incorporation: the incomplete embrace of a doctrine despite an ostensible commitment to it. Two examples illustrate this pattern. First, as mentioned above, although many governments claim to embrace the democracy-enhancing governance doctrine of decentralization, often they have not provided the financial wherewithal for decentralized governance to be effective. The emphasis has been on the deconcentration of government officials into local areas and the delegation to semiautonomous development authorities, instead of devolution of political authority to elected local governments. For example, if decentralization requires subnational authorities to provide services for their communities, success is unlikely unless national authorities engage in fiscal decentralization as well, as has been done in Brazil, Indonesia, and China. In many instances, decentralization amounts to unfunded mandates—great fanfare about empowering local authorities.

Second (as discussed in chapter 9), the military in several countries seem to accede to the doctrine of civilian governance, yet the operational doctrines fall short. Some military leaders claim the role of safeguarding democracy by ousting a civilian government they deem threatening to democracy, as in Argentina in the 1950s through the 1980s (O'Donnell 1973).

The Technical Covering up the Political

Decisions with crucial normative and political implications often appear as simply technical. The valuations required for benefit-cost analysis, ostensibly the gold standard for selecting projects and programs, boil down to estimates of what people are willing to pay. Similarly, the World Trade Organization's highly detailed, seemingly technical trade rules reflect years of bargaining and politicking among national governments. It is common that apparently technical routines benefit those with enough power to establish these routines.

Promotion through the Manipulation of Expectations

Evaluating the applicability of theories and doctrines depends not only on experience but also on expectations, which are often subject to manipulation. For example, the need for military build-up may be promoted through exaggerated claims about security threats, sometimes reinforced by exaggerated assumptions of the effectiveness of the build-up. Best case assumptions regarding the effectiveness of foreign assistance, ignoring the weakness of institutions, supported the use of simplistic models of foreign aid magnitudes and economic growth.

This does not mean that all best and worst case inquiries should be avoided. Some problems or threats demand consideration of likely extreme events. Preparing for devastating, once-in-a-thousand-year natural occurrences is clearly more important in the Anthropocene Age. *Thinking about the Unthinkable* during the Cold War, as Herman Kahn (1962) and other nuclear strategists did, examined the worst possible case of nuclear war to reinforce the folly of mutual assured destruction.

The Dynamics of Termination

Unlike theories that last indefinitely whether embraced or rejected, doctrines and specific applications often warrant termination, as policies, programs, or organizations become dysfunctional, redundant, or outmoded. Terminations also encompass eliminating or reducing benefits for particular individuals and groups. Since existing commitments involve people whose lives are invested in past decisions and practices, termination often invites hostility, provoking passionate behaviors of those who stand to lose. Attempts to terminate may mobilize the masses, but often more importantly the educated elites. The developing world's educated individuals may be key resistors to loss of opportunities, especially when they are employed in bureaucracies. This explains the enormous difficulties that policymakers and planners face in enacting some interventions, even if they are based on rigorous analysis or are simply commonsense solutions. The costs and burdens of termination must be assessed according to many different human values, not simply wealth. Some terminations are unwarranted, often resulting from a change in government whose new leaders want to distance themselves from policies or programs associated with their predecessors. Certain terminations, such as disbanding environmental agencies or independent central banks that had disciplined macroeconomic policy, may set back sound development for many years. Many policymakers have notoriously short-term perspectives, often linking their policy choices with the electoral cycle. Long-term commitment to specific doctrines and policies is particularly weak in countries such as Myanmar, Indonesia, and Pakistan, which are going through political transitions from long periods of military-led government to civilian oversight of the military. Thus, some applications ought to be insulated from termination.

Preview of the Book

To explore how relevant actors can formulate and implement development strategies in the service of both growth and human dignity, we identify five major challenges, each highlighting interactions among government officials, civil society, and other actors. Effective government and equitable economic development are obvious, perennial challenges. The three remaining challenges involve the crucial roles of NGOs, the military, and international development assistance agencies. While these actors also play significant roles in shaping economic policies and governance, their interactions with government and the population in general are critical in themselves.

Economic Development

One challenge is to enact economic policies that promote economic growth and poverty alleviation while keeping environmental damage and income inequality within reasonable bounds. As chapters 2 and 3 chronicle, early doctrines for accomplishing these intertwined goals did not prove to be viable. The parallel political and administrative challenge is to improve governance and public administration so that citizens can participate meaningfully in shaping government policies and policies can be enacted with efficiency and fidelity. The early optimism that economic development would bring democracy, high levels of citizen participation, and more responsive policies has clearly given way to more skeptical understandings of political change. Thus the pursuit of economic growth and justice must be accompanied by the active pursuit of democracy and responsiveness.

The bewildering expansion of goals and considerations that development economics must incorporate has profoundly changed the challenges of balancing objectives and the complexity of the whole set of economic policy initiatives. Even if recently enacted policies are progressive, will the more privileged still be able to extract other assets and opportunities from the less privileged, offsetting the progressive policies? To what degree are legislative commitments to equity eroded by deliberate or inadvertent weaknesses of implementation?

A progressive government faces the dilemma that poverty alleviation in the long run requires economic growth, yet growth with equity requires a delicate suite of economic policies that require political forbearance by the many groups that face risks and uncertainty from such policies. No consensus on economic doctrines will prevail as long as shortcomings of any approach are regarded by some as perfectible defects but by others as insurmountable flaws.

Public leaders face a series of development challenges to promote growth with poverty reduction: vested interests that limit policy choices, rising expectations for both rapid democratization and economic dividends, and the need to accommodate diverse ideological, ethnic, and religious groups. This requires institutions that can mediate differences and promote outcomes considered by most to be sufficiently fair. Equally important are the personal integrity of political and administrative leaders and their commitment to political and economic reforms.

Governance

The governance challenge must take into account the particularly difficult path for progressive policies. Chapters 4 and 5 describe how the evolution of governance has entailed a greater role of civil society in public affairs, which increases the need for institutions that can channel representation into constructive discourse and decision making and constrain the advantages that wealthier groups typically have to influence the outcomes. It requires mechanisms to develop and implement policies and programs based on the principles of transparency, access, accountability, checks and balances, and appropriate balances of centralized and decentralized authority. Will policies approved legislatively be able to avoid diversions of attention and resources, corruption, or ineffective implementation? Can the legal system protect the poor? Does the court system have the integrity to enforce the rule of law; does the Supreme Court have enough power to check executive or legislative overreaching? Insofar as civil society can play a major role advocating just policies, how can civil society organizations constructively stand up to the demand for control pressed by many government leaders?

Nongovernmental Organizations

Following the euphoria over the initial rise of NGOs, development practitioners have become much more guarded about the potentials of both domestic and international NGOs. The roles and challenges of NGOs are critical to development: NGOs operating domestically constitute a major component of civil society engaged in efforts to enhance citizen participation and provide services. Chapter 6 examines whether NGO-provided services strangle the state's capacity to develop its capacity to provide services, whether the NGOs truly serve to maximize human dignity, and whether the NGOs' stance toward government—often highly critical—is constructive or destructive. International NGOs may also provide important services, bringing resources from developed to developing countries, guiding efforts to strengthen democracy, and protecting vulnerable domestic NGOs from repression. Yet international NGOs may not share the priorities of the population or the domestic NGOs and may also undercut the development of state capacity.

Foreign Assistance

Assistance doctrines have undergone considerable change, reflecting the end of the Cold War, the greater availability of private capital and nongovernmental donations, and the limited capacity of international actors to shape domestic economies. Yet external aid still provides the financial resources and expertise badly needed by many developing countries, and in some instances, foreign assistance has been an enabling factor in strengthening civil society and democracy. However, as chapters 7 through 8 document, the potential for divergent interests between foreign assistance agencies and recipient nations has been highlighted by critics of these agencies. The doctrines linking economic assistance and political-military commitments of recipient governments fluctuate in often problematic ways, according to the threat levels perceived by donor governments. Different approaches to economic policy have pitted domestic groups and often their governments against the assistance agencies requiring compliance with particular political positions or economic reforms. International security interests and the domestic economic interests of the nations of bilateral foreign assistance agencies have frequently colored the targets of foreign assistance. How have bilateral assistance agencies learned (or not learned) to navigate among these pressures? What roles should they play? Finally, how, in otherwise highly organized donor countries like the United States, has interagency and bureaucratic infighting colored practice, sometimes dooming even the best intentions?

Some multilateral financial institutions have had to cope with the successful defiance by recipient governments regarding the conditions that these institutions pose in exchange for their support and face possible tradeoffs between protecting the global economy and the interests of particular developing nations. Insofar as the doctrines espoused by international development agencies diverge, it is important to determine whether this provides a healthy opportunity to test different development approaches or simply adds to the confusion about best practices.

The Military

The optimal roles of the armed forces remain elusive. Chapter 9 demonstrates that this is not simply a matter of the varying contexts in different countries, though that must be taken into account for efforts to avoid overgeneralizing theories and doctrines. The more direct problems are the conflicts of objectives between the military and other institutions, such as civilian officials, civil society, and international actors, but also within the armed forces itself.

The military may be the institution mandated to protect democratic constitutions, but in many instances, it has intervened against democratically elected governments. That these risks remain despite the widespread decline in interstate wars brings into question whether the military forces should have any significant role in countries free of serious external threats. Perhaps a lightly armed police force would suffice for maintaining domestic order, yet few military leaders willingly consent to the dismantling of the armed forces, and without their acquiescence, the military will remain. If the military is a major drain on a nation's resources, should it be "put to work"? The question is whether the military should provide services outside of the conventional military roles: engage in training and socialization beyond military preparedness, or undertake nondefense infrastructure projects. In many countries the military is among the most effective institutions, sometimes *the* most effective institution. Yet serving these functions may drain the economy, threaten democracy, and retard the capacity building of civilian government and other service-providing institutions. Even if a nation's military can provide a global public good of serving as international peacekeepers in other countries, any nonconventional role ensconces the armed forces even if its size ought to be reduced. Eventually, civilian oversight of the military, willingly accepted by military leaders, is the most sustainable option to promote development and build democratic institutions. How should civil-military be managed to achieve the twin objectives of national security and reinforcing democratic institutions? How can the military's professional role be promoted without giving it undue role in influencing political decisions? What is the impact of the military engagement in economic activities?

On the international level, military concerns have played a highly significant role in the volume and shaping of foreign assistance. Chapter 10 demonstrates the intertwining of US national security and foreign assistance doctrines in three of the most crucial cases: Chile, Korea, and Vietnam. The more recent developments in these countries and parallel challenges reflect responses in foreign assistance and security doctrines in the dramatically changed global context.

While the assessments covered in this book reveal daunting challenges, progress made thus far and lessons that can improve better practices in the future are cause for cautious optimism. Our final chapter 11 takes stock and proposes recommendations for going forward.

CHAPTER 2

Evolution of Economic Development Theories and Doctrines since World War II

Introduction

Economic thinking and practice in developing countries have gone through striking changes in theories, doctrines, and practices over the post-WWII period. Yet many of the same issues and obstacles remain. To begin to assess this evolution, this chapter explores the linkages between theories and doctrines. These doctrines are only partially shaped by economic theories; they also reflect changing goals, political considerations, and administrative constraints.

The goals, evolving in the face of both gains and disappointments, have evolved from a predominant focus on aggregate economic growth to a vastly broader agenda. On the one hand, without economic growth, the opportunities to further human dignity, including alleviating poverty, are highly limited. There has been growing acceptance of the finding, most prominently reflected in the work of Dollar and Kraay (2001), that poverty alleviation largely proceeds in lockstep with aggregate economic growth. Thus, the United Nations, despite its enormously broad purview, emphasized growth in its several Development Decades, beginning with 1961–1970. Fukuda-Parr (2013, 8) notes that "the four decadal agendas up to the 1990s included social objectives but were dominated by economic growth priorities and focused particularly on issues of the international economic environment." Remarkably, the Development Decade economic growth targets were increasingly ambitious, despite mounting evidence of the daunting obstacles to rapid and sustained aggregate growth.[1] Roberts (2005, 114) writes that "The UN Development Decades of the 1960s, 1970s and 1980s proclaimed objectives that were over ambitious, and were known well before time to be unattainable." Perhaps the ambition was an exhortation to the developed countries to reduce barriers to developing country exports and to increase foreign assistance, investment, and technology transfer. Whether exhortative or simply naïve, the primacy of growth was apparent.

To understand the evolution requires identifying the relevant actors. Economic policymakers of developing countries are obviously relevant, but they must be divided into at least four sets: the highest political leaders of the government, the officials of the spending ministries and their agencies, the top executives of state enterprises, and the officials of the central economic, finance, and planning ministries who oversee the economy as a whole. The domestic actors outside of the government include business groups, labor unions, NGOs, and the general public.

The international actors include officials of bilateral foreign assistance agencies; the staffs of international financial institutions, principally the IMF, the World Bank, and the regional development banks; various agencies such as the UN economic commissions in each world region; and universities and other training institutions. Other international actors include global businesses and international NGOs.

While government officials at the highest levels of economic policymaking have to make the critical decisions on development strategies and the policies to enact them, the officials of external agencies that provide substantial economic support often penetrate the internal decision-making structures. Others influence policy by giving advice or exerting pressure to adopt particular economic measures. An underappreciated influence is imparted by the educational entities that shape the outlooks and analytic frameworks of their graduates, whether in technical or policymaking roles.

Economic theories and doctrines address two analytically separable issues. The macroeconomic questions address the optimal money supply, savings rates, government spending, government deficits, investment levels, interest rates, balance of trade, levels of inflation, and so on. While many of these are not directly controllable through government policy, they are certainly shaped by policy. Macroeconomic policies determine how much the economy is stimulated, at the risk of excessive inflation and decline in foreign exchange. The more specific microeconomic questions focus on which projects and programs, of which economic sectors, ought to be financed. Different methods, theories, debates, doctrines, and practices hold for each of these issues.

It is not surprising that many efforts have summarized the history of economic approaches in developing countries.[2] Our assessment is distinctive in two respects. First, we explore how theories are undergirded by analytic methods that orient and sometimes constrain the theories, doctrines, and decisions that policymakers take. We have discovered that the application of development thinking depends, to a surprising extent, on whether various considerations can be situated within the analytic and decision routines employed by policymakers and their advisors. The theories need to be operationalized within the decision routines that underlie policymaking. In addition, even if particular methods accurately reflect an embraced theory or doctrine, their use may be neglected because of the effort involved in employing it, or the vulnerability of those who do so. We shall see that some of the most compelling methods for incorporating income distribution considerations in the evaluation of development projects have been rejected or simply ignored because of these challenges. Thus, we examine how development economics theory and doctrines have evolved to address the two major challenges of economic development: overall economic growth and alleviating poverty. We also emphasize the continuity of the controversies between the mainstream neoclassical approach, which clearly dominates the training of development economists, and alternative perspectives.

Neoclassical Economics and Alternatives

It is important to clarify the meaning and implications of the neoclassical approach, which dominates in professional economics in general, and the structuralist alternative. The neoclassical approach posits that powerful mechanisms, based on the rational pursuit of profits by firms and greater welfare by individuals, will operate constructively if they are unconstrained as long as property rights are secure.[3] Through this pursuit of maximization, resources will be fully used through market signals that, if allowed to operate freely, would induce capital and labor to be fully supplied and their output fully purchased. The market will permit the economy to approach full employment, as long as the government does not distort wages or prices for goods and services. Investment will be optimally deployed, as long as impediments to savings and productive investment are not erected. According to neoclassical assumptions, if an economy does not behave in these ways, it is because market forces are not permitted to operate—the government's interventions break down the pursuit of maximum efficiency, maximum wealth generation, and full utilization of resources, including labor.

Structuralists see more need for government intervention. In seeing bottlenecks in production, the structuralist remedy may concentrate on instruments that stimulate supply as much as demand; these may include subsidizing promising opportunities for expanding production or engaging in state economic activity to fill the gap left by sluggish private sector supply. State economic activity, rather than crowding out private investment, may reinforce private investment by establishing reinforcing clusters to enhance productivity and by increasing confidence that an industry will grow. Rather than assuming that savings will be invested productively in the domestic economy, structuralists may assume that savings would be diverted overseas or locked up into unproductive investments that may provide social standing (such as owning a large estate) rather than the highest returns. Therefore, the government would have an important role in directing savings.

The Challenge of Economic Growth

In the early post-WWII period, three explanations—not necessarily mutually exclusive—seemed plausible to account for the low levels of economic development of the bulk of nations in Africa, Asia, and Latin America. Perhaps these economies lacked the resources—capital, technology, infrastructure, and so on—needed for growth. Perhaps the economies were locked into particular activities, such as raw material exports, that limited the potential for growth. Finally, perhaps the economic structures had rigidities that inhibited full and efficient use of resources. The economists who devoted their efforts to stimulating economic development applied the economic thinking that had emerged from the Great Depression to sort out the dilemmas of stubborn underdevelopment.

Macroeconomic Approaches to Growth in Developing Countries

This emphasis on the general stimulation of the economy directly engaged the differing perspectives on how to manage the macroeconomic aspects of an economy. Not surprisingly, the disagreements among economists in the studies of Keynesianism, monetarism, and so on in developed countries reverberated in the debates over development economics. Yet, for the sake of conciseness, it is necessary to confine this brief history to the most prominent shifts in dominant theories and doctrines.

One highly prevalent view was that the most important shortage was financial capital. The earliest preoccupation that drove development economists post-WWII was how to mobilize both international and domestic sources of capital. This situation was shared with war-ravaged Western Europe, for which the injection of aid through the US Marshall Plan was remarkably successful, as well as with the long-poor developing countries, some of which were just reaching independence. The belief was that poor countries lacked the savings to generate sufficient investment for economic growth. Although in some cases savings later proved to be greater than anticipated, it was often in forms that were not investible, such as the gold jewelry held by Indians. Outside transfers of capital would be the most direct way of propelling economic growth. Yet, for the transfer of capital to produce more than a short-lived effect, a theory of overall economic stimulus was desired. Several of the earliest post-WWII theorists provided reassuring models. Paul Rosenstein-Rodan (1944), later to become an early World Bank economist, argued that investors needed to see a substantial amount of investment in a particular sector before they would join in; a "big push" was needed to provide sufficient momentum. To determine how much external capital would be required to generate a sufficient degree of economic growth, the Harrod-Domar model[4] was developed to determine the financing gap—the amount of capital to be provided by foreign assistance to join the country's internal savings rate (assumed to be available for investment) to stimulate growth. Only the effectiveness of capital to generate growth was invoked, due to the assumption that labor shortages would not be a constraint. Given an estimate of the domestic savings rate, the Harrod-Domar model would estimate how much more capital would be needed to achieve a particular GDP growth target.

The priority given to dramatically increasing the level of investment dictated other macroeconomic doctrines. Some structuralists believed that the capital needed to finance sufficient expansion of key sectors would be provided neither by the private sector (which lacked either sufficient capital or the incentives to invest in the appropriate sectors) nor from tax revenues that could feasibly be extracted by governments (Felix 1961). Yet, if the sheer lack of savings and investable capital was a crucial (if not necessarily permanent) problem for developing countries, heavy international borrowing, deficit spending, easy money, and the resulting inflation may be necessary evils.

However, emphasizing the sociopolitical changes required for sustained economic growth, Walt W. Rostow, a key figure in US foreign assistance and security policy, posited his theory of economic takeoff: the economy could maintain its own momentum, culminating in sustained growth, but only after profound social, political, and institutional changes occur:

> Initial changes... require that some group in the society have the will and the authority to install and diffuse new production techniques; and a perpetuation of the growth process requires that such a leading group expand in authority and that

> the society as a whole respond to the impulses set up by the initial changes, including the potentialities for external economies...The take-off requires, therefore, a society prepared to respond actively to new possibilities for productive enterprise; and it is likely to require political, social and institutional changes which will both perpetuate an initial increase in the scale of investment and result in the regular acceptance and absorption of innovations. (1956, 25)

Rostow acknowledged that this process changes power relations and entails conflict over societal goals, yet he added a soothing caveat: "because nationalism can be a social solvent as well as a diversionary force—the victory can assume forms of mutual accommodation, rather than the destruction of the traditional groups by the more modern" (1960, 7). It is telling that Rostow saw nationalism as a social solvent without anticipating the highly contentious impact that nation building would have in the relations between majority and minority populations.

Killick notes that another institutional demand of the "big push" was the belief that government planning and direction are essential: "Under the influence of the 'big push' school of thought, planning was also seen as the only way to mobilize resources on the scale necessary for a successful development effort, and as the only practical means of binding the various strands of economic policy into a consistent whole" (1976, 3). This was reflected in the requirement for Latin American governments to submit medium-term (e.g., four to six years) comprehensive economic plans to qualify for aid under the 1960s' US Alliance for Progress.

The remarkable recovery of Western Europe under the Marshall Plan, which permitted Western European leaders to develop their own comprehensive economic recovery plans, seemed to promise that the will could be mobilized and a direction charted for developing nations to take advantage of available capital through institutional changes. Of course, the stable economic and political institutions that prevailed in most developed nations during their most successful economic growth periods were largely lacking for developing countries in the 1940s and 1950s. It was most insightfully recognized that preexisting economic, political, and social institutions had often been rigidly stable, but industrialization, wars, independence, and other political phenomena would undermine these institutions, resulting in economic and political instability.[5] Nevertheless, institutions in the nations defeated in WWII were working, despite the instability or complete lack of democracy prior to the war; it was not clear that the hopes for developing countries were worse. In many countries, these institutional changes entailed dramatic expansions of government involvement in the economy. According to the much acclaimed work on Western and Eastern Europe by economic historian Alexander Gerschenkron (1957), the more backward the economy, the more coercive the policies must be to reduce consumption in order to increase the savings needed to finance the new industries. In some nations, such as China, Cuba, and Tanzania, the government instituted an explicitly socialist structure. In many others, including most Latin American nations, state direction involved establishing or expanding state-owned enterprises, establishing development corporations to subsidize state and private firms, and dominating the supply of credit to favored firms. These efforts were frequently accompanied by government policies that controlled the prices of goods and services, regulated wages, and set interest rates.

Along with a strong role of the state in directing or undertaking economic activity was a structuralist rejection of the free trade doctrine and the classic premise that nations would prosper most if they pursued the economic activities that they currently could do most productively. This comparative advantage principle was challenged on several grounds. Harvard economist Hollis Chenery (1961), later World Bank Vice President for Development Policy, argued that static comparative advantage theory had to be converted into a dynamic conception, allowing for future changes in a nation's economic strengths. Gerschenkron also argued that economies had to remain open to new opportunities for comparative advantage; this was interpreted to mean that new comparative advantages should be created through state action.

The most direct attack on developing countries' existing specialization in raw material exporting was articulated through the so-called Singer-Prebisch thesis that raw material producers face deteriorating value of their exports in relation to their purchasing power for manufactured goods (Prebisch 1949; Singer 1950). Pessimism was based partially on theoretical grounds; unlike raw materials, technology adds to the value of manufactured products, with higher profits for the manufacturers and higher wages for the well-organized manufacturing workers. Others reinforced this pessimism by asserting the likelihood of more sluggish demand for raw than for finished products.[6] Moreover, agriculture, forestry, and other important aspects of raw materials face physical constraints to their future growth.

The concept of *potential* comparative advantage provided a rationale for "infant industry" subsidization or protection even if no measurable price signals were available. In terms of industrial strategy in the early post-WWII era, the most attractive industrial strategy was light manufacturing to replace the need to import these goods. This import substitution industrialization (ISI) typically protected industries through high import tariffs or outright import restrictions, as well as subsidies to the manufacturers. Similar advantages were extended to several nations adopting a heavy industry promotion strategy (see chapter 3). Self-sufficiency for national security was another consideration, as wasthe strategy of initially supporting domestic market-oriented manufacturing through protectionism and subsidies, widely prevalent at least through the 1970s. Although it has become far less explicit as a heralded strategy in the wake of severe criticism of the logic and record of ISI, its vestiges remain through a host of subsidies (e.g., in energy pricing and access to cheap capital).

Initially the critiques were few and far between. Peter Bauer (1957/1965; Bauer and Yamey 1957) epitomized the skepticism, asserting that state economic direction distorts prices and incentives, sapping efficiency and worsening poverty. The critique focused on both the intrinsic problems of undermining the efficiencies of market forces and the self-serving behavior of government and state officials.

The scarcity of criticism of policies violating mainstream neoclassical principles has several explanations. Invoking infant industries calls for patience before judging the strategy as a failure. Low economic growth could be attributed to the political turmoil that struck many developing countries. And the strategy seemed to have the legitimacy of expertise through the endorsement of several UN regional economic commissions. Raul Prebisch directed the UN Economic Commission for Latin America; Hans Singer directed the Economic Division of the UN Industrial Development Organization. The UN Economic Commission for Africa espoused the same doctrine

(Owusu 2003, 1655). Finally, even into the 1970s, hope persisted because of an abundance of capital originating from oil-exporting nations following increases in world oil prices. International borrowing fueled industrial expansion in many countries, even if the industries survived only on protection, subsidies, and borrowing.

As state-dominated economies faltered, criticism mounted. While some industries failed prior to the 1980s because of intrinsic inefficiencies or small markets, the drastic implosion occurred for the many heavily indebted countries in the 1980s, as real-world interest rates skyrocketed. Previously manageable debt service burdens became unmanageable; hard currency reserves evaporated; the threat of default deterred further lending. Table 2.1 demonstrates the 1980s' meltdown, but also illustrates the disappointing earlier growth rates of Africa and Latin America.

The 1980s, widely regarded as the "lost decade," brought enormous suffering as governments were compelled to cut back social services and employment opportunities contracted. The contrasting experiences among countries of vastly different exposures to indebtedness (e.g., India had eschewed international borrowing) carried a lesson of financial dependence as powerful as the lessons of trade dependence.

As lenders and borrowers wrestled with the risks of default, the retrenchments in heavily indebted countries were often overseen by the international financial institutions. The IMF and the World Bank have been the major international entities in establishing the mainstream doctrine for addressing the debt crisis and the problems of inflation, budget deficits, and bankruptcies. The importance of these institutions is magnified beyond their own funds, as their willingness to lend to indebted governments bolsters the confidence that other lenders have in the creditworthiness of these governments.

The "structural adjustment" advocated by these institutions was concisely summarized by Williamson (1990; 1999, 252) under the label "Washington Consensus": (1) fiscal discipline; (2) redirection of public expenditure priorities toward sectors with both high economic returns and the potential to improve income distribution; (3) tax reform; (4) interest rate liberalization; (5) competitive exchange rates; (6) trade liberalization; (7) liberalization of inflows of foreign direct investment;

Table 2.1 Growth rates of developing regions, GDP, and GDP per capita in constant (1990) dollars, 1950–1999

		1950s	*1960s*	*1970s*	*1980s*	*1990s*	*2000s*
Africa	GDP (%)	3.8	4.6	4.4	2.5	2.5	4.8
	GDP per capita (%)	2.1	2.9	2.6	0.8	0.7	3.1
Asia	GDP (%)	3.8	4.6	4.4	2.5	2.5	6.2
	GDP per capita (%)	3.3	4.1	3.8	3.1	2.9	4.4
Latin America	GDP (%)	4.9	5.2	5.7	1.8	2.9	3.6
	GDP per capita (%)	3.2	3.5	4.0	0.1	1.2	1.8
World	GDP (%)	4.3	5.0	4.1	3.1	2.8	4.3
	GDP per capita (%)	2.6	3.3	2.4	1.4	1.1	2.5

Source: The Maddison-Project, http://www.ggdc.net/maddison/maddison-project/home.htm.

(8) privatization of state-owned enterprises; (9) deregulation to abolish barriers to entry and exit; and (10) secure property rights.

Some of these goals run counter to structuralist doctrines. Inflationary government spending and heavy borrowing to spur economic activity challenge fiscal discipline. The requirement of high (current) returns on public spending runs counter to investments in the long-term pursuit of new comparative advantages. Tax reform,[7] interest and exchange rates free of government manipulation, and trade liberalization reduce the means to favor particular industries through tax exemptions, cheap capital, exchange rate manipulations, or import tariffs. State enterprise privatization reduces the capacity of the state to direct investment into areas of potential future comparative advantage. Foreign investment and eliminating barriers to entry would also reduce the protection of favored industries.

Nevertheless, Williamson (1999) forcefully asserted that these objectives do not entail market fundamentalism or rigid adherence to narrow neoliberalism; they reflect a convergence of economic policy views rather than simply or predominantly the views held and imposed by these international financial institutions.

Although only one of these objectives, fiscal discipline, is specific to financial crisis, the time of greatest leverage that the international financial institutions and other stakeholders have on borrowing country governments comes when the financial crisis puts the greatest pressure on the governments to comply with the conditionalities of the international financial institutions. And because the crisis almost always requires reducing the drains on foreign reserves, and often requires reducing domestic demand to counter inflation, the conditionalities entail not only structural adjustments but also austerity measures. This important distinction is often lost on critics of the conditionalities, as structural adjustment comes to be equated with the pain of economic contraction.

The criticisms of the structural adjustment conditionalities, often heavily colored by the conflation of structural adjustment with austerity, have spawned numerous counterdoctrines proposed to offset various perceived weaknesses. We later review efforts to protect the poor from the ravages of austerity; here we examine the two arguments regarding economic growth that have dominated. One argument is that conventional neoclassical analysis, especially as practiced by the IMF, underestimates the optimal levels of money supply, spending, debt ceilings, and so on. Thus, the pain of austerity is greater than necessary, economic recovery is delayed, and economic growth is weaker. Nobel Prize laureate Joseph Stiglitz (2000), former chief economist at the World Bank, launched a withering attack on the IMF staff who formulated the conditionality packages required of the Southeast Asian governments in the aftermath of the 1997 East Asian economic crisis.

Williamson's insistence that the Washington Consensus should not be taken as narrowly neoliberal has been challenged in many quarters. Mavroudeas and Papadatos (2007, 46–47) articulate the typical critique:

> Washington Consensus' macroeconomic discipline... [i]n almost all cases [has] led to austerity budgets and policies that favoured the wealthier and worsen the position of the lower strata. The same holds for the push towards a market economy and the opening of the economy. The first stems from a neoconservative

> conception of the economic role of the state and of its alleged inability to manage properly the economy. The second has the same origins complimented with the simplistic belief that it will lead to increased competition and thus consumers will in the end be better off.

The reference to the alleged inability of state economic management has a certain irony, coming from Greek economists, but the message is clear that the opponents of the Washington Consensus question both the free market premise and the impacts of the austerity measures. Again, structural adjustment and austerity measures are treated as one.

The second argument is that structural adjustment fails when the government fails to embrace it. The checkered compliance with conditionalities (reviewed in chapter 3) led to the "country ownership" doctrine. Boughton (2003, 3), of the IMF, reported:

> When the IMF embarked on a reexamination of it policies on conditionality in the millennium year 2000, a key objective was to promote national ownership of policy adjustments and structural reforms. It was clear from experience and from formal studies...that the main reason for failure of Fund-supported programs to achieve their objectives was that governments too often did not implement policies to which they had committed. Whatever could be done to deepen and strengthen commitment was likely to improve implementation and raise the success rate.

Boughton (2003, 3) acknowledges that "national ownership" is difficult to define "with sufficient precision to make it operational"; however, the IMF developed a working definition: "Ownership is a willing assumption of responsibility for an agreed program of policies, by officials in a borrowing country who have the responsibility to formulate and carry out those policies, based on an understanding that the program is achievable and is in the country's own interest."

However, ownership by the government is not an all-or-nothing. Typically, government officials mandated to maintain economic stability have already sided with the international financial institution pressing for the reforms. Yet the government's political leaders and spending ministry officials usually have different incentives than the financial managers. Political leaders may prefer the reforms if not for political consequences, but reneging will result if political costs are regarded as too high. Political leaders often state opposition to conditionalities, to reduce their accountability for the unpopular aspects of the reforms, even if they welcome being forced to accept them. Leaders of spending ministries typically pursue their mandates by spending more. Moreover, national ownership should involve consensus among key stakeholders, including the private sector and civil society. Officials of the ministries and departments may "own" these programs but lack the capacity to carry them out if political leaders cannot gain the support of political and economic forces in the society. Finally, the international financial institution may conclude that no technically acceptable set of conditionalities would produce sufficient commitment, resulting in the refusal of the institution to lend.

Macroeconomic Modeling

While the Washington Consensus and competing doctrines concern general macroeconomic management, the question of how to determine appropriate macroeconomic parameters is on a more technical level. Disputes over doctrines are reflected in disagreements about analytic tools. Initially, there was a general agreement that the development of a coherent national accounts framework for relating the various aggregate measures of economic activity was a great advance, followed by the input-output matrix approach,[8] which, based on tracking existing material flows, could link the flows of economic activity comprehensively (if somewhat statically) throughout the economy and even capture the flows of international trade. By the 1960s, these yielded the social accounting matrix, encompassing national accounts variables to represent all economic transactions at some level of aggregation. In principle, analysts could extend the intersectoral analysis to project the impacts of macroeconomic changes (e.g., increases in money supply) as well as expansions in particular subsectors.

The social accounting matrix is the basis for computable general equilibrium (CGE) models, intended to trace out economy-wide effects of policy changes. Yet, this methodological advance has not unified economic theory; CGE models reflect contrasting development theories. Some have a strong neoclassical pedigree, reflecting the conceptually simple equilibrium assumptions of mainstream economics.[9] In contrast, Mitra-Kahn (2008, 6) argues that true CGE models are not neoclassical, and although some CGE models are faithful to neoclassical assumptions, other development economists incorporate more complex relationships, beyond the simpler equilibrium assumptions of the neoclassical framework. In rejecting the validity of some neoclassical assumptions, these structuralist economists have tried to model the consequences of inefficient markets, immobile labor, the differences between effective demand and potential demand, and so on. These models are controversial in general (Acheson 2005). Nicholas Stern, who served as chief economist at both the European Bank for Reconstruction and Development and the World Bank, judged that "[s]uch models are no doubt useful tools, although due to their large appetite for assumptions and data and their lack of transparency, their usefulness should not be exaggerated" (Stern and Ferreira 1997, 556–557). Yet, the structuralist CGE models are controversial on distinctive grounds. The additional relationships they represent reflect the modeler's beliefs about how the particular economy under examination diverges from the neoclassical assumptions. Different economies, whether developing or developed, would be expected to have different rigidities violating the neoclassical equilibria. Minford (2011, 290), assessing the recent work by well-known structuralist Lance Taylor (2011), expressed the following reservations about the assumptions represented in the latter's structuralist model: "The trouble is that these ideas are frequently untestable as proposed because they have too many elements that can be fixed to suit the data…Where attempts have been made to set up such models in properly testable form and compare them with the data, they do not seem to do well." The ad hoc nature is likely to be even greater for developing economies, insofar as they have more—and more varied—deviations from the fluidity of the market economy.

This controversy reflects fundamental differences in the conception of the purposes of economic analysis. The development of neoclassical economics—theory and method—is dedicated to general applicability. Both neoclassical and structural CGE models require heavy data-gathering and estimation of constants. Yet, the neoclassical template provides rather clear guidance on how to proceed, at the risk of missing relevant complications, whereas structuralists must adapt as they see fit. Structuralists are far less concerned about the capacity to generalize than to identify and model the distinctive nature of each economy to project the impacts of alternative policy options.

Sectoral Strategies

The analytic challenge of determining how to invest in the most potentially dynamic sectors and subsectors is the bridge between macroeconomic and microeconomic strategies. Sectoral development consists of supporting or discouraging myriad ventures, but its goals often encompass enhancing or stabilizing macroeconomic balances.

The early post-WWII period witnessed heavily theory-dependent debates over which sectors had the greatest promise of dynamism and the sectors from which capital (and suitable labor) could be drawn. Some argued that economic growth (and therefore investments to stimulate growth) ought to be balanced among sectors.[10]

However, the dominant view was to pursue growth by stimulating industry resting on several beliefs. First, scarce capital and entrepreneurial expertise (Hirschman 1958) and capacity for technological innovation (Streeten 1959) implied that these scarce resources ought to be concentrated where they were believed to be most effective. Second, if "unbalanced growth" was necessary, structuralist views tilted toward industrialization: if free trade penalizes developing countries that follow the neoclassical doctrine of pursuing current comparative advantage through raw material export, industrial promotion was needed to surmount the expected deteriorating terms of trade of raw materials. The confidence in rapid industrial expansion was also reinforced by the belief that agriculture typically had underemployed labor that could be directed into industry and that agricultural growth was severely constrained by relatively fixed amounts of land and soil quality. Therefore, the industrial promotion doctrine called for reinforcing external capital with capital drained from the agricultural sector, directly (taxes on agro-exports, ranging from beef- and grain-exporting Argentina to cocoa-exporting Ghana) or indirectly (ceilings on food prices as in Brazil in 1950–1970; Adams 1971, 48). Agricultural taxation could underwrite cheap government credit to manufacturers; low food prices permitted manufacturers to lower industrial wages. Despite rhetoric about the importance of agriculture because of the huge proportion of labor devoted to this sector, the proindustry, antiagricultural bias was widespread.[11]

This perspective eroded sharply by the 1980s. Protected industrialization stalled, agricultural bottlenecks arose, rural poverty increased, and uncontrolled urbanization occurred in many countries. It became clear that agriculture had more potential than the conventional wisdom had assumed and that beggaring agriculture would lead to its deterioration as well as the general stagnation of exports (Valdés 1983,

97–98). Additional factors caused delays in recognizing and addressing this danger, attributing disappointing growth to the necessarily slow gestation of new industries or to external economic shocks. The precipitous decline of Sub-Saharan Africa's agriculture in the 1960s and 1970s reflected not only weather and civil war but also widespread biases against agriculture in pricing, investment, and capital-extracting institutions such as marketing boards.[12] Southeast Asian discrimination against agriculture was strong in the 1970s and 1980s, though declining since then (Anderson 1994, 21–22). Krueger, Schiff, and Valdes (1991) document similar patterns in Latin America. Today, the antiagricultural bias is largely dismissed. Bezemer and Headey (2008, 1345) conclude that

> there is overwhelming evidence from theory, history, and contemporary analysis that agricultural growth is a precondition to broader growth... agricultural growth is quintessentially pro-poor growth... agriculture is generally labor intensive and skill-extensive, so that agricultural growth creates additional employment with low entry barriers. Increased agricultural productivity also lowers food prices for both the rural and the urban poor, who typically spend most of their household budgets on food. Especially productivity growth on small family farms is very pro-poor.

Whereas the ISI strategy clearly emphasized light industry, several countries adopted a heavy industry promotion strategy. India, faced with grave security concerns, tried to pioneer a heavy industry strategy (the "Mahalanobis strategy") emphasizing such industries as steel production, machine tools, cement, and chemicals. This strategy dominated the First National Plan (with a largely symbolic emphasis on the micro-cottage industry in homage to Gandhi). South Korea's heavy industry strategy promoted steel production and ship-building. In both countries, security concerns seemed to justify mobilizing resources for industries needed for national defense.

Both industry promotion strategies precluded export promotion, as protectionism was met with retaliatory tariffs by potential importers. The thinking was that once the infant industries were nurtured and well established, protection could be removed and export possibilities would open at some point. This doctrine was convenient for industrialists and other business leaders. Their political influence, reinforced by the political costs to governments that allow once-favored businesses to fail, prolonged many industrial promotion policies that had outlived their relevance.

A proindustry sectoral strategy calls for investments in the physical infrastructure subsectors (transportation, energy, water, and communications) supporting manufacturing. Some infrastructure development might be devoted to support rural industrialization, but few countries emphasized rural industrialization in light of apparent economies of scale, and locational advantages had favored existing cities. Evidence (Duranton 2009) bears out the long tradition of viewing cities as engines of growth, because of economies of scale, industrial clustering, more effective matching of factors of production, greater opportunities for learning, efficiency service provision, and so on. Combined with the typically greater political power of urbanites over rural residents, urban-oriented infrastructure development, whether connecting cities or serving individual cities, generally strengthened the urban over rural areas. For example, in the early 1980s, Kenya and Liberia had urban electricity coverage of over 80 percent, but rural coverage of only 4 and 12 percent, respectively. In the mid-1990s, 99 percent of urban Ethiopians had access to telephone main lines, compared to only 1 percent of rural Ethiopians (Lanjouw and Feder 2001, 32).

International aid agencies found funding infrastructure projects to be particularly attractive. Large-scale projects facilitate industrialization generally, without having to bet on particular firms; large sums could be transferred efficiently; and engineering and project planning skills seemed eminently transferable to developing countries. Infrastructure development was also spurred by the preference of bilateral donors for infrastructure projects that required heavy equipment and engineering services from the donor country.[13] However, infrastructure was often compromised by subsidization through underpricing of infrastructure access and outputs, undermining cost recovery, and thereby weakening the capacity to maintain and expand infrastructure systems.

The widespread failures of inward-directed industrialization reflected neglect of a basic insight: protected and subsidized firms have little incentive to become efficient, as long as the government lacks the capacity or willingness to punish inefficiency or credibly threaten the withdrawal of privileged treatment. The exception was South Korea, where General Park provided both capital and protection to Korean industrialists but credibly threatened dire consequences for those who might take unproductive advantage of the privileges.

In the early post-WWII period, the social sectors (education, health, nutrition, family planning) received little direct attention, reflecting the presumption that abundant labor from rural areas—where education and health deficits were often the greatest—could easily be absorbed into industry where on-the-job training could upgrade worker skills. However, East Asian countries took another path, with South Korea, Taiwan, and Malaysia following the Japanese approach of human resource development. These Northeast Asian countries are rather homogeneous ethnically, and Malaysia's poorer, less-well-educated Malays are a politically dominant majority. This suggests that human resource development strategies may hinge politically on whether the poorer population segments are part of the majority ethnicity.

Project and Program Levels

The alternative to theorizing about sectoral priorities is through the methodologies of individual project or program selection. The logic underlying the methodologies of project-specific societal profitability (societal rate of return) would seem to challenge the broader theories of sectoral priorities. Rate-of-return analysis (or benefit-cost analysis) estimates the value of costs and benefits (not confined to narrow economic outcomes; e.g., a project may improve the population's health) for each year into the future. This income stream yields the return on investment, which typically has to exceed a minimum "hurdle rate."[14] Despite debates on how to value benefits and costs, and the technical details of calculating the rate of return, every organization assessing proposals for development assistance had worked out similar procedures.

One major contribution to valuing a project's benefits or costs was the introduction of "shadow prices" into the analysis. Price-controlled goods and services do not reflect their true values; estimating what the "shadow" prices would have been in a free market in principle offsets the distortions (Little and Mirrlees 1969, 1990; Squire and van der Tak 1975). This demonstrates that the mainstream economic approach does not have to pretend that markets are operating freely in order to assess the merit of specific projects.

It might seem that if the rate of return is the criterion for selecting projects or programs, broader sectoral theories would be unnecessary. Yet, a comprehensive assessment of benefits and costs requires understanding the interconnections of the economy. In principle, models could be devised to trace out both direct and secondary impacts. Rate-of-return analysis and theorizing about the myriad interactions within the economy can be linked—at least in theory.

However, the challenge became far more complex as the development agenda expanded. The narrowly defined economic rate of return is no longer a sufficient metric if poverty alleviation, health, environmental protection, respect for human rights, enhancement of women's roles, and so on, are additional goals of development, justified on the grounds of both their intrinsic value and their contribution to economic growth. Even if the analysis could estimate the contribution to growth from each of these improvements with reasonable accuracy, estimating the intrinsic value of each improvement is extremely daunting. For this reason, certain types of projects and programs are typically excluded from rate-of-return analysis. Even the World Bank's project appraisals typically do not apply economic rate-of-return analyses to education, environment, finance and private sector development, health, nutrition, population planning, public sector governance, and social protection. World Bank practice is to confine economic rate-of-return analysis largely to agriculture and rural development, energy and mining, transport, urban development, and water projects (World Bank Independent Evaluation Group 2010, 7).

However, even for projects where rate-of-return analysis is expected to be conducted, the challenge remains to account for impacts on goals other than the directly economic ones. For example, a road project may provide greater health service access for low-income people, but it is likely to have environmental impacts and may displace people from cherished sites or enable outsiders to enter the area, thus increasing the potential for intergroup conflict. Estimating these societal benefits and costs, typically characterized by highly uncertain probabilities and magnitudes of impacts, is often daunting. Therefore, the usual approach to evaluating projects with these possible impacts is a hybrid of benefit-cost analysis, qualitative invocations of the potential added benefits, and a separate assessment of whether the potential negative impacts are great enough to kill the project. To incorporate the potential governance impacts of programs and projects, as well as how they ought to be formulated, the UN Development Programme has developed a governance assessment framework to identify or anticipate the consequences for inclusive, participatory, and responsive governance (Hydén 2007), going far beyond traditional cost-benefit analysis.

Equity Issues and the Poverty Alleviation Imperative

While increasing overall prosperity is easily embraced as a consensual goal—all other things being equal—the distributional question of who gets what is inevitably contentious. Five distributional questions have preoccupied development economics since the 1970s. The first is how to define the general relationship between equity and growth—a concern that has pitted those who believe that a country prospering through unequal income and wealth distributions will ultimately find the means to lift up the poorest segments of the population against those who either reject the effectiveness of growth strategies based on highly unequal distributions or demand more equitable distribution even at the cost of aggregate economic growth. Second, how should the objectives of addressing equity issues be defined? Third, for those who are intent on addressing income disparities immediately, how can this be incorporated into technical methods of assessing policy alternatives (the "methodology of macroeconomic planning and policy formulation"; Ahluwalia and Chenery 1974, 39) and what doctrines would accomplish this effectively? Fourth, what ought to be the role of the state in shaping the income distribution? Fifth, what are the existing structural inequalities, whether economic, political, or social, that impede propoor, equity-oriented interventions? These last two questions raise the methodological question of whether the economist's toolkit underestimates the political economy constraints—such as local and national power structures, centrifugal forces in the society, and ethnic and cultural differences.

The Growth-Equity Relationship

As mentioned above, one reason why the early post-WWII development thinking paid relatively little attention to income distribution was the presumption that overall economic growth would lift the poor along with everyone else. One of the most influential concepts in this regard was the dual-sector model, positing two largely independent sectors. The traditional, low-technology, predominantly rural sector is characterized by low productivity and therefore low incomes. It was thought to be commonly plagued by severe underemployment. The modern sector, employing more advanced technology, generally urban-centered, and with more interaction with the international system, enjoys higher productivity and compensation. Workers in the modern sector, largely employed in medium- and large-scale private and state firms, are more likely to be organized vis-à-vis their firms.

Within this clearly simplified conception, positive development entails the expansion of the modern sector and consequently the movement of people from the traditional sector to the modern sector. As this process unfolds, the overall economy expands. The optimistic expectation in terms of income distribution is that competition to hire sufficiently skilled modern sector workers, or workers' organizational strength, will elevate their wages. Income would increase for the population segment transitioning from the traditional to the modern sector.

The potentially problematic assumptions underlying this scenario include the need for incentives for employers to add labor as firms expand, which may not prevail if high wages provoke the substitution of machinery for labor. This also hinges on whether the expansion of the modern sector labor demand can keep up with the expansion of the workforce to absorb sufficient labor such that employers must compete for workers. Rapidly expanding populations in developing countries have often exceeded workforce expansion. Finally, the existence of surplus labor does not guarantee that these workers have the skills or geographical access to take on modern sector jobs.

Moreover, even if this transition from the traditional to the modern sector does occur, the short- and medium-term implication is a deterioration in income distribution. This was foreshadowed by Simon Kuznets's (1955) demonstration that the poorest and the richest countries of the day had more equal income distributions than the countries of intermediate per capita income levels. While Kuznets did not have time series data to trace the income distribution patterns over time, his analysis is consistent with the dual-sector theory,[15] with the countries with intermediate per capital income levels reflecting the inequality resulting from the income differences between the traditional sector and the modern sector. In short, even if the healthy process of expanding the modern sector proceeds, income inequality will increase for at least a while. Relative inequality rises, even as the poverty of traditional sector workers is alleviated as they move into the modern sector. From a doctrinal perspective, this model also implies that the transition can be hastened by directing investment and policy advantages to the modern sector, which is largely the industrial sector.

In many countries, the transition did not go as hoped: industrialization faltered or other factors undermined economic growth, industrial firms were reluctant to hire, or industrial wages stagnated because of an ample unemployed urban labor. Therefore, even in some of the most dynamic developing countries, with the exception of South Korea and Taiwan, the overall income distribution became more skewed. This was particularly acute in Latin America, where it was not until the decade of the 2000s that the fastest growing nations (i.e., Brazil and Chile) finally experienced improvements (Solt 2010).

The overall implications of the distribution of income remain heavily disputed. One argument is that high-income individuals can be a boon to the overall growth of the economy, because they have a higher propensity to save and invest. It is ironic that this proposition—derogatorily labeled the trickle-down approach and widely attributed to neoclassical development economists—was originally a creation of the structuralists, such as Arthur Lewis (1954), who saw the top income-earning decile as the key to domestic investment. If the government does not throw impediments such as distorting taxes, expensive licensing, or administrative red tape, domestic investments would be embraced. Because developing countries typically have underemployed labor, labor inputs will be competitively attractive for investors, thus bringing the poor into more remunerative employment.

A sharply opposing argument is that the luxury goods and services demanded by the wealthy draw savings and investment away from production that would make goods more available and affordable for the poor. The wealthy are also more likely to seek imported goods, attractive in part because they are imported, sending earnings out of the domestic economy and undermining the potential of domestic production of these goods. For goods and services that the upper- and middle-income households do seek from the domestic economy, the capital tied up in producing them underfinances the production of "inferior goods" that only the poor would consume.

Framing the Objectives of Addressing Equity Issues

The framing of distributional issues has been a crucial aspect of how development economics has addressed these questions. The more provocative framing focuses directly on the relative distribution of income, wealth, and other aspects of well-being within a society, thereby emphasizing the differences. Even if complete equality is not a widely held goal, the call for greater equity is widely shared. The question is how drastic the efforts to achieve equity ought to be. The most extreme efforts at direct redistribution, ranging from land reform without full and immediate compensation to confiscation of factories and savings, have often led to long-term economic stagnation (China, Cuba) or even civil war. Even less extreme efforts, if they strongly deter investment and production, will not provide the productivity needed to sustain the gains of the poor. Therefore, the doctrine that emerged from the development agencies in the 1970s, led by the World Bank, was to promote strategies that would not wrench away income already earned by wealthier people, but rather would favor the economic growth of lower income groups. World Bank Vice President Hollis Chenery asserted that "distribution is not simply a concern with income shares but rather with the level and growth of income in lower-income groups. Distributional objectives therefore cannot be viewed independently of growth objectives. Instead they should be expressed dynamically in terms of desired rates of growth of income of different groups" (Ahluwalia and Chenery 1974, 38).

The challenge, then, is to devise macroeconomic policies that favor productivity gains by lower income people and to target them for programs and projects that increase their incomes. The structuralist approach endorsed by Chenery and others emphasized asset distribution as key to income distribution. Such policies would include: directing government spending to improve the human capital (essentially health and education) of the poor, benefiting them directly while improving the overall capacity of the economy; land reform and other policies to enhance the productivity of small-scale agriculture; and promotion of the urban informal sector rather than capital-intensive industries (Chenery 1974, vii, xvii). This is clearly a more gradualist approach than direct redistribution, reflecting caution in light of the risk of political backlash.

Political resistance to asset redistribution makes this approach unlikely to succeed on any large scale in most countries. However, in areas such as land ownership and security of tenure, some degree of asset redistribution is essential for any program to make the rural poor more productive (see chapter 3). Beyond this essential minimum, a vigorous policy of investment reallocation in a rapidly growing economy may be more effective in increasing the productive capacity of the poor than redistribution of existing assets, which would have a high cost in social and political disruption (Ahluwalia and Chenery 1974, 49). Yet there is even milder framing of the challenge as poverty alleviation. Whether or not the incomes of the rich are growing faster than those of the poor, the focus can be simply to lift the poor out of poverty. This absolute rather than relative conception has a far more consensual normative appeal. Many policies hold the promise of reducing poverty without undermining the incomes of the nonpoor even in the short run and may enhance the income potential of all in the long run.

Macroeconomics of Distribution and Poverty Alleviation

On the macroeconomic level, most of the attention has been focused on the impacts of the combined austerity and structural adjustment programs, because of their short-term impacts on the poor. This is a misleading focus, because the whole range of macroeconomic policies shapes the income distribution. For example, macroeconomic policies that trigger inflation differentially affect the savings of different groups, with lower income families generally less able to buffer their incomes from inflation. Tax policies that provide special privileges for the wealthy (and typically well-connected) reduce the tax revenues that the government could use to finance social services for the poor. Trade policies increase or decrease the protection of particular subsectors. Yet the structural adjustment programs that have largely been introduced along with austerity measures during economic crisis have borne the brunt of attention and criticism.

During the 1980s, the Washington Consensus approach to resolving macroeconomic imbalances through the combination of austerity packages as well as structural adjustment measures reinforced the perception that mainstream, largely neoclassical economics was associated with income regressivity. In the long run, structural adjustment is progressive, insofar as it dismantles the distortions offering special advantage to the well-connected (Mikesell 1983). However, austerity programs reduce two very visible indicators often regarded as addressing the poor: social spending and wage policy. In fact, for many countries, both affect not the poorest segments of the population but rather the higher (if not highest) income recipients of government benefits and of wages directly influenced by government policy. Their objections to austerity programs make such programs seem highly regressive.

By the mid-1980s, the realization that austerity measures were having highly adverse impacts on the poor, especially poor children, gave rise to an explicit doctrine of "adjustment with a human face" (Cornia, Jolly, and Stewart 1987). The liberalization initiatives and general cutbacks would be accompanied by social services or cash transfers targeted to the poor. The most current variation is the conditional cash transfer, entailing payments to low-income families as long as the families' children are enrolled in school and receive regular health care. Conditional cash transfer programs were employed in conjunction with Argentina's drastic austerity program in 2002, Indonesia's fuel price liberalization in 2005, and the broader Brazilian liberalization undertaken by President Lula da Silva throughout 2000–2009 (Widianto 2007; Grosh et al. 2008). Additional benefits are the improvements in human capital and the potential reduction in unrest by the poor.

Microeconomic Approaches to Poverty Alleviation

Quite diverse approaches have been proposed to select and enact poverty alleviation programs and projects: emphasizing sectors that redound more to the poor, integrating the provision of goods and services through "integrated rural development," the "basic needs" approach of direct targeting of goods and services to the poor, adjusting the calculated rates of return to favor propoor projects, and exempting poverty alleviation projects from meeting minimal rate-of-return standards. Intergovernmental organizations, such as the international financial institutions, have played a leading role in advancing, debating, and implementing these microeconomic approaches to poverty alleviation.

Sectoral strategy as poverty alleviation strategy. The World Bank, explicitly embracing poverty alleviation as its first priority following President Robert McNamara's highly publicized Nairobi speech in 1973, increased its emphasis on rural projects, particularly agriculture, explicitly to enhance the productivity of the rural poor. A distinction was made between rural development projects, requiring at least half of the benefits targeted to households below the poverty line, and other projects (including agriculture projects). Projects designated as "rural development" were favored over "agricultural" projects.

Integrated rural development. Recognizing that the rural poor lack a broad but varying range of assets, bilateral and multilateral aid agencies experimented with a new doctrine of "integrated rural development" to provide the full set of interventions in rural areas, from retail outlets for farm inputs and production credit to extension education and family planning programs (Mosher 1976, 54). It typically entailed the coordinative mechanism of a dedicated project administration, often specifically designated and recruited, to secure inputs from higher level mainline administrative units, such as the national or subnational ministries. In some cases, the initiative included efforts to help organize the poor so that they could exert pressure from below to ensure that government programs and facilities are accessible to them.

Because urbanization in many countries has involved the movement of rural people with few assets, governments have been compelled to formulate doctrines to address the urbanization of poverty. In keeping with the structuralist doctrine of strengthening the assets of the poor in order to generate more income, one prominent initiative was to develop microenterprise programs—with direct government funding, or commitments of state banks to earmark funds for micro-borrowers, or requirements that private banks do so.[16] However, it was—and remains—controversial as to whether credit should be subsidized and whether NGOs are more appropriate for organizing and supporting microenterprise programs.

Another painful dilemma that arises from the urbanization of poverty is how to address the tendency of the poor to occupy fragile peri-urban land that the government has declared off-limits for residence. The common initial reaction, out of concern for health, safety, and environmental concerns, was to ban unauthorized settlements, forcibly move migrants back to rural areas, and bulldoze the illegal settlements. The ineffectiveness of this reaction, as well as the conflicts that it created, typically led to more "accommodationist" approaches of contributing to upgrading of the settlement sites, developing peri-urban satellite settlements, and other policies recognizing the futility of forcibly limiting urbanization (Laquian 1981).

Basic needs. In the mid-1970s, several international organizations (the International Labour Office, the World Bank, the UN Development Programme, the World Health Organization, and the UN Food and Agriculture Organization) and the major bilateral donors formally embraced the "basic needs" doctrine, which Hoadley (1981, 150) defined as "the acceptance of the goal of providing all persons with a certain minimum standard of these basic needs as a central priority of development. A basic needs strategy is the deliberate adoption of a set of policies designed to provide, or help people provide for themselves, these basic needs." Meeting basic needs could enhance the productivity of the poor through improvements in human capital, but basic needs advocates were adamant that this is secondary to fulfilling consumption needs.[17]

The two obvious questions regarding basic needs are who should be targeted and how much expenditure is needed to meet their needs. No technical definition exists to address either question. While it is possible to determine some of the minimal standards of nutrition and (more arguably) housing, the other elements are unavoidably subjective. Paul Streeten and Shahid Burki (1978, 413), prominent basic needs proponents, acknowledged that

> there are no objective criteria for defining the contents of a basic needs bundle. While certain minimum physiological conditions are necessary to sustain life, basic needs vary between geographical regions, climates, cultures and periods. Even such a basic requirement as nutrition for the same sex, the same age and the same activity varies between different people. Housing requirements also show wide variations and so do all other basic needs.

One approach to addressing this dilemma is to leave the judgment to whoever is regarded as reflecting the preference of the society as a whole.[18] In practice, an external foreign assistance agency would have to rely on either the government as the voice of the society or on domestic NGOs through which the agency is channeling its assistance. For a government, a calculation of the required expenditure to meet some level of basic needs would require identifying the total resources (domestic and external), but also to examine tradeoffs with other objectives (Streeten and Burki 1978, 415). One example that illustrates the logic is a calculation of the resources required to meet the basic needs of the poor in Bangladesh—an enormous challenge given the abject poverty in that country. Following an analysis of the resources needed to overcome the lack of adequate nutrition, clothing, sanitation, housing, and power, Streeten and Burki (1978, 471) concluded that "[if] the basic needs target were to be met in the next ten years, within the present structures, this would imply an 8% rate of real growth in the *average* incomes of the absolutely poor. Meeting the target over a period of 25 years implies a real growth in personal incomes of the absolute poor at the rate of 3.1% per annum."[19] However, from this analysis, they concluded that "such high rates of growth in incomes do not seem possible without a fundamental change in development policies" and require a "global compact" among wealthy and poor countries that requires not only the mobilization of sufficient resources but also changes in the "domestic policies of recipient nations" (418–419). Although domestic policies have changed in many developing countries, the global compact has certainly not emerged.

These macro-implications of the basic needs doctrine focused attention on consumption deficits rather than on aggregate economic growth targets. Although in theory it was to apply "wherever people were found who fell below certain standards of basic needs" (Hoadley 1981, 152), in practice the resource limitations meant that it was largely applied on the project level. Working back from consumption targets rather than from growth objectives, "the consumption targets are translated into specific programme goals[, for example] a life expectancy of 65 years or more; a literacy rate of at least 75%; an infant mortality rate of 50 or less per 1000 births; and a birth rate of 25 or less per 1000 population" (Ruttan 1984, 397). Thus, the analytic approach is to specify targets for the project area, not necessarily to bring all of the poor in the area out of the throes of basic needs deficits.

Conditional cash transfers. Although the basic needs approach as initially conceived soon fell out of favor, the normative commitment to the lowest income families resurfaced in the doctrine of cash transfers, calling for regular payments to eligible low-income families. Rather than having to coordinate inputs from multiple providers for myriad local projects, cash transfer programs can be mounted on a broad basis, and the recipients have greater choice over the benefits they receive. Behind unconditional cash transfer programs is a consumer sovereignty theory that the family decision makers are in the best position to meet the family's needs. When cash transfer programs are conditional, in requiring families to engage in training or workfare programs, keep their children in school, or provide them health care, they offer explicit promise of greater productivity for the society as a whole. This is also a means of avoiding the problem of unconditional transfers that families may end up spending the cash on nonproductive activities (e.g., dowries, drinking, or gambling).

In transferring the cash to the female head of household, the typical cash transfer program also engages one corner of the complex challenge of recognizing and enhancing the role of women in development, which emerged as a focus around 1980. Recognizing the previously undervalued role of women in economic development as well as in family well-being was straightforward enough, but recognizing—and addressing—the negative outcomes of *intrafamily* politics between male and female family members (Agarwal 1997) created demands for policies to strength the power of women within families, a sphere commonly regarded as off-limits to government policy. Yet targeting cash transfers exclusively to mothers, despite criticism that this is cultural interference, reflects the recognition of these dynamics.

Deciding the amount of cash transfer also engages the issue of how much resources ought to be devoted to poverty alleviation. Rather than establishing a target such as increasing average life expectancy, the magnitude of conditional cash transfer could, in principle, be based on how much each eligible family would require to obtain proper nutrition and health care. In practice, however, the government must also consider the magnitude of available budgetary resources, often less than a given basic needs target.

Amartya Sen's capability approach has a clear linkage to the basic needs doctrine; it is cast in terms of fulfilling what is needed for individuals to have meaningful choices regarding both physical well-being and social interactions (including political participation). The doctrine emphasizes the capacity of the individual more than on the inputs (according to Sen [1985, 7], "a functioning is an achievement of a person: what she or he manages to do or be, such as being nourished or participating in the political process"). However, there is significant overlap—Streeten and others emphasized both the capacity to engage in the full range of social activities and the individual's centrality in determining what these capabilities are. Sen also agrees that the ranking of capabilities depends on the particular society and, therefore, must be left to public discourse. In essence, the capability approach embraces the same imperative of bringing the poorest segments of the society up to a satisfactory level.

The technical manifestation of the capability approach has been a series of composite indicators, developed by Sen, Mahbub ul Haq (former World Bank official and Pakistani finance minister), and others for the UN Development Programme's annual *Human Development Reports*. The most publicized is the Human Development Index, which ranks nations according to a composite measure of indicators (aggregate levels of health, education, and standard of living), but neglects the far richer set of needs. Far more enlightening is the Poverty Index, a composite measure of deficits in education, nutrition, child survival, electricity, potable water, sanitation, nondirt flooring, clean cooking fuel, transport vehicles, and appliances (UN Development Programme 2011, 172–173).

Second, the Millennium Development Goals (MDGs), endorsed by most national governments and the major international development organizations in 2000, established time-specific targets for specific countries and for the international development community: eradication of extreme poverty and hunger; universal primary education; gender equality; reduction of child mortality; improvement of maternal health; combating HIV/AIDS, malaria, and other diseases; ensuring environmental sustainability; and developing a global partnership for development.

Both the capabilities approach through the development indices and the MDGs strongly convey a poverty alleviation doctrine. Although many countries have fallen short of the targets, the consumption focus rather than aggregate economic growth focus has endured.

Social benefit-cost analysis. The rate-of-return analysis for assessing development projects or programs can also be directed to equity and poverty alleviation considerations. Beginning in the mid-1970s, the World Bank's management tried to implement poverty-alleviating approach to calculating the rates of return of projects.[20] The conventional approach is to regard every unit of benefit the same regardless of the beneficiary. Yet the poverty alleviation emphasis that arose in the 1970s moved development agencies toward benefit-cost analysis that favored projects targeting the poor. Several prominent development economists associated with the World Bank tried to introduce "social benefit-cost analysis": "the impact of a Project on income distribution is estimated by giving weights to the incremental consumption accruing to each income group benefitting from the Project, including labor" (Cleaver 1980, 14).

The rationale for attributing different weights to incomes received by people of different income levels is based on the welfare economics principle of maximizing utility rather than income, and a given unit of income typically has greater utility to a poor person. Therefore, a project that directs income to the poor contributes more utility than one that directs the same income to the wealthy. Squire and van der Tak (1975) developed an approach based on the work by Little and Mirrlees (1969) to explicitly adjust the rate of return of projects targeting more of their benefits to the poor.[21] To make the adjustment, the analyst would have to posit the income level at which the actual and adjusted benefit are equal and how steeply the adjusted benefit increases or decreases according to the beneficiaries' income levels.

However, several challenges arose. First, any effort by the World Bank or other international organization to set the adjustment factors independently would be condemned as far too heavy-handed. The only viable alternative was to infer the distributional preferences within the target country, possibly from examining the nation's expenditures on welfare, the distribution of tax burdens, or other indications of redistributive preferences. Cleaver (1980, 14–15) implemented the World Bank's premise that the weights ought to "reflect Government's income distribution objectives." Yet these "revealed preferences" hinge on often volatile governmental policies that certainly do not necessarily reflect the society's ideal distributions of benefits or burdens, even assuming the meaningfulness of the concept of societal preference.

Second, the analysis of who would receive how much of the project benefits, and what their incomes are, is typically prone to high levels of uncertainty. Third, the time-consuming analysis and the professional risk of making the heroic assumptions required for the analysis were regarded as unwarranted burdens if the results do not determine project approval (Leff 1985b). The approach was basically abandoned.

The Ongoing Evolution of Goals

The substantial success in meeting the poverty reduction objectives of the UN-led initiative MDGs has generated another round of goal-setting beyond the MDG target year of 2015. The UN Secretary General's "High-Level Panel of Eminent Persons on the Post-2015 Development Agenda," after lauding the MDGs for reducing absolute poverty, raised the same question of how to address the expanding development agenda that has challenged theory and practice since the 1980s:

> But to fulfil our vision of promoting sustainable development, we must go beyond the MDGs. They did not focus enough on reaching the poorest and most excluded people. They were silent on the devastating effects of conflict and violence on development. The importance to development of good governance and institutions that guarantee the rule of law, free speech and open and accountable government was not included, nor the need for inclusive growth to provide jobs. Most seriously, the MDGs fell short by not integrating the economic, social, and environmental aspects of sustainable development as envisaged in the Millennium Declaration, and by not addressing the need to promote sustainable patterns of consumption and production. The result was that environment and development were never properly brought together. (United Nations 2013, i)

The Post-2015 Development Agenda, recently approved by the UN member states, proposes eradicating absolute poverty completely by 2030, but also shifting international efforts from short-term, sectoral approaches to systemic treatment of inclusion, participation, environmental quality, and human rights. This obviously requires stronger institutions, to which we now turn.

Institutions and the Role of the State

The "discovery" of institutions. Aside from external factors, positive or negative economic performance rests on the quality of economic policies and their implementation, in turn depending on the quality of policymaking and administrative institutions, which in turn depends on the quality of governance and overall societal institutions.[22] Daniel Kaufmann, a World Bank researcher in the 1990s, demonstrated that sound macroeconomic policies are essential for productive projects and programs; by the 2000s, as the director of the World Bank Institute's Global Programs, Kaufmann's focus shifted upstream to the importance of sound governance as essential for sound macroeconomic policies as well as sound projects and programs (Kaufmann 1992, 2009; Kaufmann and Wang 1995; Isham and Kaufmann 1999). Indeed, all economic development approaches have come to recognize the importance of "good governance," which is considered in more detail in chapters 4 and 5. Nevertheless, alternative approaches define and try to address institutional inadequacies differently.

Within the neoclassical paradigm, many of the failures of economies to behave according to neoclassical assumptions are attributed to weak institutions that undermine efficiency. Poor institutions may channel factors of production into efforts of lower societal returns; the interactions of hiring, production decisions, financial exchange, trade, and so on, may be unnecessarily costly. The question is whether the inefficiencies are caused by the interventions of the state or whether these interventions can, at least in principle, resolve inefficiencies.

The neoclassical position is that unwise government policies and inefficient state structures interfere with market forces. From this perspective, the problem is not that neoclassical equilibrating mechanisms of supply, demand, price, wages, etc., would fail to operate if left to their own devices, but rather that these mechanisms are blocked by policy failures.[23] Therefore, the doctrine consistent with neoclassical economics calls for the reduction of the role of government and state institutions in obstructing the operation of the market or even dismantling some of these institutions.

In contrast, structuralists emphasize that the rigidities attributable to the socioeconomic characteristics—such as immobile labor, poor information, or unwillingness to put savings into productive investment—unresolvable without governmental action. They also emphasize how national and local power structures constrain the ability of the poorest segments of the society to adequately benefit from market mechanisms, calling for interventions to level the playing field.

The most prominent challenge to the neoclassical perspective is the new institutional economics, enjoying the prestige of Nobel Prizes awarded to theorists who inspired the approach.[24] This perspective emphasizes interactions among economic actors, particularly the costs of arranging contracts broadly conceived (transaction costs[25]) and the importance of institutions that determine these costs. Oliver Williamson (2002, 178) controversially claims that "[t]he contract/private ordering/governance (hereafter governance) approach maintains that structure [of firms] arises mainly in the service of economizing on transaction costs." If arranging transaction cost-minimizing structures were simply at private actors' discretion, state intervention would not only be unhelpful but also stifling of the private sector's flexibility in moving to these structures. For example, some regulations on firms, such as minimum wage requirements or greater auditing burdens, may discourage the formation of more efficient firm structures. On the other hand, some potentially beneficial transactions may not be feasible because of high transaction costs imposed by the overall structure. Institutions beyond the control of the firm may be suboptimal because of poor information, unclear property rights, or other impediments; therefore, revamping the relevant institutions becomes compelling. Government (which Williamson [2002, 175–76] calls hierarchy) could be instrumental in restructuring so that firms can reduce their transaction costs: "The problem of economic organization is properly posed not as markets or hierarchies, but rather as markets and hierarchies. A predictive theory of economic organization will recognize how and why transactions differ in their adaptive needs, whence the use of the market to supply some transactions and recourse to hierarchy for others."

The new institutional critique of neoclassical economics is that the neoclassical framework cannot identify the institutional arrangements that would raise or lower transaction costs. Eggertsson (1990, 14–15) asserts that the narrowness of the "frictionless" neoclassical approach limits the capacity to identify optimal arrangements. The broader perspective requires multiple demanding tasks: securing information about prices and quality of materials and labor; searching for potential trading partners and information about them; bargaining among potential partners; making, monitoring, and enforcing contracts; and protecting property rights, whether from private actors or the government itself.

Yet the debate on the roles of state institutions in developing countries has not changed in practical terms. Government and the state institutions (e.g., state-owned enterprises and quasi-governmental corporations) have the potential to reduce transactions costs. Yet, they often set prices at variance with market prices, make unprofitable investments with less constraint than profit-maximizing private firms, and unjustifiably favor some economic activities over others through subsidies or tax policies. Government and state officials, from high to low, may extract payments from private actors seeking excessive profits through restrictions to competition that these officials enact; this rent-seeking both squanders firms' resources and distorts markets. While some theorists of rent-seeking are within the new institutional economics tradition,[26] the prevalence of corruption based on rent-seeking in developing countries supplements the neoclassical doctrine of freeing markets from government intervention.

Conclusion

Policymakers are left with ambiguous guidance in terms of both theory and doctrine. Economic patterns are far better known than at the beginning of the post-WWII period, when national accounts were in their infancy, but the means for selecting optimal poverty alleviation strategies lack consensus. Technical ambiguity leaves open the choice of economic strategy and practice to whatever ideological stance and political calculations top political leaders may have. In the next chapter, we examine the record of practice.

CHAPTER 3

Economic Policy and Program Practice

Introduction

We have seen that economic policy doctrines often do not correspond with the latest theories of development economics. Yet, to an even greater extent, economic policy *practice* has departed from theories and doctrines. Examining how and why these departures have occurred will build our understanding of the challenges that policy-makers continue to face. We shall see that the decision-aiding methods derived from economic theory do not provide the guidance required to select optimal economic policies, programs, or projects. We shall also see the faltering progress of the major policy reform designed to further both equity and efficiency. These include liberalizing the economy to reduce rent-seeking, reforming or privatizing state-owned enterprises, enhancing tax collection, creating propoor social safety nets, extending social services, stimulating propoor regional development without provoking violence, managing natural resources soundly, redressing the bias against agriculture, developing sound physical infrastructure, and decentralizing economic decision making. While poverty alleviation has made progress in most developing countries, the full potential to redress poverty has been hampered by the failure to translate these propoor doctrines into practice. Finally, the inability to specify technically which sectors to promote for greatest societal gain leaves the field open to organized interest groups, often at the expense of the most vulnerable families.

The Uneven Path of Liberalization

Although liberalization has been the most important global trend in economic doctrines in the modern era, in many instances liberalization reforms throughout the developing world have been stalled or even partially reversed. The movement to greater market openness has been limited despite the long-standing consensus among professionally trained economists that trade openness, market pricing, macroeconomic stability, tax reform, privatization (or at least reform of inefficient state enterprises), and a competitive but well-regulated banking system are all necessary for economic growth.

The adoption of liberalized policies such as deregulation and trade openness varies widely from country to country. Yet the Economic Freedom of the World assessment (Gwartney, Hall, and Lawson 2014) reveals that developing countries in general have undertaken far less liberalization than developed nations or transitional European nations. Of the 98 developing countries in the data set, excluding Hong Kong and Singapore and the oil-rich Gulf states and Brunei, only 18 were scored at or above the median for developed countries in terms of deregulation; only ten were at or above the developed country median for trade openness.

Over the post-WWII period, the greatest liberalization was among Northeast Asian nations. Their dismantling of impediments to trade and business operations in general and the promotion of an export orientation have clearly been a key to the rapid economic growth and poverty alleviation, even if state direction and (in the case of Japan) protectionism remain prominent.

Substantial—though less dramatic—liberalization occurred in Latin America in the 1990s and 2000s. Lora (2007) notes that orthodox approaches prominent in the 1980s gave way to more flexible policy reform, with broader stakeholder participation, targeting specific market and government failures. These second-generation reforms began with fiscal reforms of the early 1990s, followed by mid-1990s' pension reforms, and fiscal responsibility reforms in the 2000s.[1] The fact that more reform occurred in the 1990s than the 1980s, despite the high visibility of structural adjustment and austerity agreements in the 1980s, reflects the hostility created by a combination of structural adjustment and austerity in the 1980s. Yet with a few notable exceptions,[2] liberalization in Latin America never reached developed world levels. Only five of the 23 Latin American nations in the Economic Freedom rankings are at or above the developed country median rank on deregulation, and only six are at or above the developed country median in trade openness. Moreover, several Latin American nations have seen dramatic reversals, particularly the populist regimes of Argentina, Bolivia, Brazil, and Venezuela.[3]

African nations have liberalized the least. The Economic Freedom assessment ranks 11 African countries[4] in the 20 lowest ranked countries among the 152 nations in the study, despite the rhetorical commitment to the New Partnership for Africa's Development (NEPAD). The North African nations do no better than the Sub-Saharan countries.[5] The World Bank Group's Doing Business survey, with broader coverage (189 nations) to include smaller African nations and focusing on liberalized economic activity in practice, shows similar results. Only seven of 54 African nations are above the overall median in overall ease of doing business; nearly two-thirds are in the bottom quartile (World Bank Group 2014, 4). Nevertheless, the assessment notes that in 2013/2014 alone, 39 Sub-Saharan African countries conducted regulatory reforms to reduce "the complexity and cost of regulatory processes" (5) and 36 countries reformed business-related legal institutions.

Africa's lack of trade openness is particularly striking. Only four of the 40 African countries are above the Economic Freedom median, and 11 are in the bottom 20.[6] Of the 54 African nations in the World Bank assessment, only five are at or above the median of all 184 countries, with 12 African countries among the bottom

20 countries.[7] All North African nations were below the median in their formal policies as gauged by the Economic Freedom rankings, yet the World Bank Group assesses the North African nations as relatively more open in practice.[8] This contrast implies that the formal bureaucratic and regulatory trade rules are less constraining in practice.

The major Asian oil-exporting countries (Saudi Arabia, the United Arab Emirates, Qatar, Bahrain, and Oman), with the exception of Kuwait, have high liberalization rankings and trade openness. Oil exporters can count on hard currency flows regardless of their degree of protectionism and have relatively little domestic manufacturing to protect. Yet of other Arab Asian nations, the most telling ranking was Syria's in the 2010 assessment (Syria was excluded in the most recent assessment). Its extremely low rankings in both openness and general liberalization may reflect the protection of Assad supporters' businesses.

The poorest Asian nations have made the least movement toward deregulation. The persistence of heavy regulation in South Asia and the poorest nations of Southeast Asia is counterintuitive—why are their leaders the least eager to try new approaches or more compelled to accept liberalization directives from the international financial institutions? However, it is plausible that in very poor nations the limited number of stakeholders with the opportunity to capture the limited economic surplus would cling tenaciously to the privileges that permit this capture.

The discrepancies between assessments of formal regulations and the survey-based Doing Business evaluations reveal how red tape can be used to favor insiders. The European Bank for Reconstruction and Development (2010), in evaluating liberalization in Mongolia, Turkey, and the former Soviet republics, gave Armenia, Azerbaijan, Georgia, Kyrgyzstan, Mongolia, and Turkey the maximum score for movement toward liberalized trade policies. Yet the Doing Business rankings reveal Azerbaijan, Kyrgyzstan, and Mongolia still face severe trade obstacles, reflecting regulations and bureaucratic processes that can be used to facilitate rent-seeking or favoritism.

Finally, liberalization timing in Asia compared to Latin America illuminates the importance of global events in shaping orientations toward liberalization. The 1980s' Latin American debt crisis produced more sound and fury than real change. Yet in Southeast Asia, liberalization during the 1980s and into the 1990s was rapid, but the 1997 Asian financial crisis stalled it. Sally and Sen (2011, 574) observe that "Southeast Asia fits the developing-world pattern of fast trade-and-FDI [foreign direct investment] liberalisation through the 1980s and first half of the 1990s, followed by a slowdown of momentum after the Asian crisis," but they note that "with the exception of Singapore, government enthusiasm for further liberalisation declined markedly."

Impact on the Poor

The most controversial aspect of liberalization is its impact on poverty and income distribution. It is crucial to understand that poverty trends and income distribution trends can run in opposite directions. *In some* contexts, poor people get richer even while wealthier people gain more than the poor do. *If* liberalization succeeds in freeing markets from competition-inhibiting policies, it can unleash economic growth that alleviates poverty through employment, as well as increase resources for propoor social programs. Even the limited liberalization has spurred economic growth across the developing world. From 2000 to 2014, Latin America's real GDP per capita in purchasing power parity terms increased at an annual rate of 4 percent, Sub-Saharan Africa by over 5 percent, the Middle East and North Africa by 4.5 percent, and developing Asia by an astonishing 9 percent.[9]

Even more dramatic than economic growth has been the decline in severe poverty in some regions. As table 3.1 illustrates, from 1990 to 2005 and projected to 2015, severe poverty (the proportion of population living on less than $1.25 per day) has shrunk in every developing region. The most impressive decline by far has been in East Asia and the Pacific: 55.2 percent in 1990 to a projected 5.5 percent in 2015—largely because of Chinese growth. Likewise, those living on less than $2 per day declined to 21.6 percent from 81.0 percent. For Latin America and the Caribbean, with far less severe poverty to begin with, the proportions are on track to decline to less than half from 1990 to 2015 and nearly half for the population living on less than $2 per day. Similar declines are projected for the Middle East and North Africa. For South Asia and Sub-Saharan Africa, both with over half their populations living in severe poverty in 1990, this poverty level has been projected to be less than a quarter in 2015 for South Asia, though the proportion in Sub-Saharan Africa will remain above 42 percent. For populations living on less than $2 per day, the proportions for both regions will still exceed half. This combination of economic growth and poverty alleviation is accompanied by varying patterns of income distribution trends, complicated by conflicting results that reflect the use of different measures of inequality (e.g., shares of deciles vs. the overall Gini index), how regions are defined, and which countries are excluded. Nevertheless, the patterns vary within each category; Alvarado and Gasparini (2013, 25) caution that "even in this [2000] decade of widespread social improvement, the country performances in terms of inequality reduction were quite heterogeneous." A global UNICEF study (Ortiz and Cummins 2011) concludes that

Table 3.1 Proportions of population living below $1.25 and $2 per day in developing regions

	Percentage of population at poverty levels (in 2005 purchasing power parity)	*1990*	*2005*	*2015 projected*
East Asia and Pacific	<$1.25/day	55.2	17.1	5.5
	<$2/day	81.0	39.0	21.6
Latin America & Caribbean	<$1.25/day	12.2	8.7	4.9
	<$2/day	22.4	16.7	11.8
Middle East and North Africa	<$1.25/day	5.8	3.5	2.6
	<$2/day	23.5	17.4	9.3
South Asia	<$1.25/day	53.8	39.4	23.2
	<$2/day	83.6	74.1	56.6
Sub-Saharan Africa	<$1.25/day	56.5	52.3	42.3
	<$2/day	76.2	74.1	60.8

Sources: World Bank (2012a, 2014b); World Bank PovcalNet databank.

the 1990–2007 period was marked by growing income inequality in low-income countries but declining inequality in middle-income countries.[10]

Regional trends largely confirm this pattern. For relatively low-income regions, Alvarado and Gasparini found generally rising inequality from 1990 to 2010 for South Asia, the second poorest region, but almost no change overall in Sub-Saharan Africa, the poorest region. For middle-income regions, Latin America experienced a substantial increase in inequality until the new millennium, when inequality began to decline substantially. They report that 95 percent of Latin American countries had declining inequality in the 2000–2010 decade, and inequality in the Middle East and North Africa region declined substantially over the 1990–2010 period (Alvarado and Gasparini 2013, 29).

This may mean that the low-income countries are entering the same classic growth pattern that first separates the income levels of workers with higher productivity from those still in traditional economic roles; later the growing proportion of modern sector jobs restores some degree of equality. Whatever the explanation, it is clear that despite the fears that liberalization would worsen income distribution, the more vigorous pursuit of liberalization in the middle-income countries has not coincided with either greater poverty or greater distributional disadvantage for the poor.

While these trends alone do not demonstrate that liberalization contributes to poverty alleviation, Graham (2002, 3) notes that liberalization reforms did translate into important material improvements for the poor in Latin America:

> In the 1980's and 1990's, market reformers focused on getting the macroeconomic fundamentals right, and on narrowing the scope and increasing the efficiency of public expenditures. This included a major effort to target public social expenditure to the poorest groups, an effort which was very effective at protecting the poor during crisis and adjustment in many countries... At a time when governments were stabilizing high and hyper levels of inflation and reversing years of negative economic growth, the focus on the poorest made economic and political sense, and in many cases resulted in the poor benefiting substantially from public social expenditures for the first time.

However, Graham also notes that these reforms also increased the economic insecurity of low-income Latin Americans, insofar as reduced regulation increases the exposure to competition.

Explaining Limited Liberalization

The Economic Freedom assessment, going back to 1970 for many countries, shows an overall liberalization trend, but also oscillation between liberalizing efforts and illiberal policies. We can conclude that liberalization has been highly contested, as its impacts differentially affect different groups. The churning itself often triggers fierce confrontations among presumed winners and losers, reflecting several factors.

First, as mentioned in chapter 2, governments when most in need of international loans and grants are most likely to consent to strong liberalization via structural adjustment programs. This need is typically greatest when the economy is in disarray, requiring strong austerity measures. Therefore, the conditionalities required by the international financial institutions are often a combination of structural adjustment and painful austerity. The structural adjustment may be seen as punitive along with the austerity measures. The pain and political unpopularity of austerity provide government leaders with strong incentives to renege on the commitments to conditionalities, including some structural adjustments.

Second, no economic policy approach can escape unscathed from the external shocks that undermine economic progress—global recessions, natural disasters, sharp increases in crucial import prices, technological advances that undermine the nation's exports, etc. The openness of the liberalized economy exposes it to the fluctuations of the world economy, even if overall economic growth is superior to that of the closed economy. During the downturn periods, the pressures for government to do something can be irresistible. This has given rise to a permanent tension between the liberalization commitment and the perceived need to recapture control. Obviously the periodic global economic crises provoke these efforts to recapture control by the government by reversing some of the liberalization reforms, whether or not the real problem resides in liberalized policies. These efforts provide opportunities for economic actors who benefit from protectionism and subsidies.

Third, the economic policy regime following liberalization is typically subject to piecemeal erosion as a progression of concessions to interest groups is introduced. These special incentives are typically rationalized on the grounds that the favored industries are promising for the long term. Tax exemptions and credits are frequently reintroduced after a liberalizing tax reform; fundamental tax reforms often have a surprisingly short half-life.

In addition, doctrinal opposition to liberalization persists. Even as Japan's credibility as a thought leader on development strategy has declined along with the dynamism of the Japanese economy, the rise of the state-directed economies of China and India has received enormous attention. And some multilateral international organizations have not abandoned important aspects of the statist positions they held in the 1960s. For example, the Secretariat of the UN Conference on Trade and Development (2007, 10–11) recently made the following appeal regarding the role of the state in Sub-Saharan Africa:

> African countries need a "strong State" to carry out the continent's development agenda. States should re-engage in the development business from which they have been marginalized…Strategic intervention combines subsidies, protection and free trade in proportions that are determined in accordance with the specific national situation. All industrialized and industrializing economies implemented various forms of protection of their infant industry in early stages of development. However, there should be time limits to protection so that, once an industry becomes reasonably competitive, it should be allowed to face world competition.

The promise that infant industries would become reasonably competitive was the trap that led to so much wasted investment and economic distortions since the 1950s. The fact that it is still invoked in this era is quite remarkable.

It is also important to note that the rethinking by international financial organizations regarding the pace and maintenance of liberalization has reduced the friction between these organizations and governments, but has also reduced the pressure to pursue structural adjustment reforms. The IMF has embraced a doctrine of country ownership—that the adjustment package must be owned by the government of the country. This implies that the government officials involved in the negotiations and higher officials who are concerned with the ultimate agreement have to signal explicitly that they are responsible for the reform package. Boughton (2003, 9) notes that "[t]he new guidelines affirm the principle that the country authorities are responsible for the design of their own policies, subject to the understanding that those policies much be acceptable to the Fund if the country is to qualify for financial support." In theory, the IMF could deny the support unless the government is willing to accept the stringent conditionalities that the IMF would have required under the earlier doctrine. In practice, however, the pressure on the IMF to reach agreements tends to lessen the stringency of the conditionalities.

Another major factor is the vulnerability of those hurt most by the special privileges of an illiberal economic policy regime. The sad irony of the persistence of subsidies is that the benefits typically go largely to higher income beneficiaries, and yet the elimination or reduction of subsidies is often resisted by lower income people, through riots over hikes in bus fare, food prices, electricity rates, and so on. These reactions have often led to the restoration of subsidies to quell the disruptions. This reversal has been common, even though freed-up budget resources could be devoted to propoor social services. The explanations typically lie in the lack of awareness that the resources could be used to provide benefits to the poor, the skepticism that the resources would be used for them, or the vulnerability of the poor to higher immediate costs that price liberalization would bring.

Finally, the potential of political disruption unleashed by strong liberalization efforts often deters government leaders from enacting these efforts. Many countries have lacked so-called adjustment-with-a-human-face measures that would protect lower income groups from serious losses. In these cases, the threat of disruption can be a powerful deterrent to strong structural adjustment efforts. Later in this chapter, we review the social safety net measures that in some countries have provided credibility that the potentially disruptive low-income group will be shielded from heavy losses due to structural adjustment programs.

Still Losing the Rhetorical Battle

In the twenty-first century, somewhat liberalized China and India have grown dramatically. Sub-Saharan Africa, finally partially embracing the free market doctrine of the NEPAD, beginning in 2001, has grown at an unprecedented rate. Yet liberalization arouses deep suspicion and hostility, among low-income groups as well as leftist intellectuals. Undoubtedly, skepticism arises from privatizations that benefit wealthy cronies of government leaders and from structural adjustment programs that provide little protection for the poor. However, the continued rejection of even the principle of liberalization reflects a failure by its advocates to convey the long-run benefits to low-income populations. Although government manipulation of the economy has often been defended as the most direct way for government leaders to channel benefits to lower income populations, this premise ignores the unfortunate fact in most developing countries that the wealthy have more resources to influence policy. Many if not all state interventions—preferential tax treatments, protective tariffs, price subsidies, subsidized credit, state enterprise actions diverted from profit maximization, minimum wages for limited segments of the working population, etc.—privilege the well-off at the expense of the rest of society *and* distort the economy at the expense of long-term growth. These points have long been recognized in the mainstream economics literature.[11]

Surprising Reversals of Privatization

The sale, elimination, or partial ownership reform of state-owned enterprises has often been associated with broad liberalization initiatives, but privatization has its own logic and dynamics. John Nellis, who served as one of the World Bank's foremost experts and policymakers on privatization, asserted in 2001 that "[p]rivatization has swept the field and won the day."[12] Yet in 2012 he concluded that "[f]rom a triumphant high in the late 20th century, esteem for privatization has significantly declined, post-2000" (Nellis 2012, i). This retrenchment varies across sectors. State enterprises in manufacturing (unless controlled by powerful armed forces) have been privatized in many countries. Perkins (2001, 268), though pointing selectively to "a number of highly efficient state enterprises in Asia" (citing South Korea, Singapore, and Taiwan), also points to the difficulties of achieving the essential needs of state enterprise autonomy and profit orientation:

> Stringent conditions were needed to achieve success with these state firms…All enjoyed a high degree of autonomy. Management's performance was judged mainly or even solely on its ability to generate long-term profits for the company. The multiple objectives—so often imposed on state enterprises elsewhere in the world—were mostly absent. Autonomy and profit orientation were difficult to achieve. POSCO [of South Korea] was run by an individual politically more powerful than most government ministers at the time. Singapore was able to isolate these enterprises completely from local politics.

The difficulty of making state enterprises profitable and the allure of revenues from selling off the enterprises can explain why manufacturing privatization proceeded and has largely not been reversed. For physical infrastructure, often entailing enormous costs, many governments have adopted public-private partnerships that place financing, building, and operating into private hands. The eventual transfer of control back to the government at a predetermined date and the attractiveness of not having to raise taxes to finance infrastructure projects have minimized opposition to this form of privatization.

However, privatization of natural monopoly services, such as electricity and potable water, has often been rolled back. Private providers are typically sought initially because subsidized state provision of these services, undermining cost recovery, leaves systems degraded and of limited coverage. Private companies are presumed to provide greater efficiency and relieve the government of blame for poor service. The possibility that privatized services would improve reliability and coverage, without greatly increasing the vulnerability of the poor, rests on the premises that (a) the government would provide vouchers or cash transfers for low-income users or permit service providers to use differential pricing, generally on the basis of consumption levels, to make the service affordable for low-demand users; and (b) the revenues of the provider would be enough to expand the system to previously uncovered areas and maintain overall quality. However, when governments permit higher rates but not steeply differentiated pricing to protect the poor and do not compensate the poor, the private providers are castigated as exploitative. When governments squeeze the rates that private providers are permitted to charge, coverage expansions and service improvements cannot occur. In short, the reversals lay in the absence of compensation mechanisms for the most vulnerable to higher prices and restricted coverage required by the private company to cover its costs plus a reasonable profit.

Similarly, many governments have reversed privatizations of extractive industries. Privatizations intended to increase investment and technical expertise in oil or mining sectors typically also reflected the logic that developing countries should not devote scarce capital to risky resource extraction investments that diversified international firms can more easily tolerate, nor should they place major revenue-earning activities in the hands of less transparent, easily corruptible parastatals. These advantages are sacrificed by renationalization, which often also deters foreign investment overall.

The reversal typically has several origins. Private companies may lose interest in further exploration and extraction in a given country, if world prices decline or other countries' resources are regarded as more promising, whereas state enterprises can often be directed to continue exploration and extraction. Second, governments often use state enterprises to serve distributional functions, such as providing services or choosing locations for operations where the government can gain political support. Finally, "resource nationalism" often has a potent appeal when populist leaders argue that the nation's patrimony will be stolen by foreigners, even though the government can capture the value of the extracted resources by charging appropriate royalties.

This costly oscillation may be avoided through mixed ownership. Brazil's Vale mining company, the world's second largest, was partially privatized in 1997, but with de facto control by the national government. Private mining companies also operate in Brazil's mining sector, while Vale expanded rapidly internationally. Even Brazil's leftist leaders have resisted demands to renationalize Vale, in part because the government maintains substantial influence over the company. The government, directly and through the shares of the state development bank and the state pension funds, controls a majority of shares, and has a golden share that gives it the power to demand or veto changes in Vale's administration. Although technically the government is not to interfere in strategic decisions, it has pressured Vale to reduce layoffs and increase investment in Brazil. The Brazilian government has a majority stake in the huge Brazilian oil company Petrobras, also partially privatized in 1997. In Chile, the state copper company Codelco shares the mining sector with private multinational companies. These hybrids of state and private participation convey the important lesson that when multiple objectives animate policies, pragmatic compromises may hold the key to stable and effective arrangements.

Strengthening Tax Effort

To combine efficiency and equity, governments must be able to finance benefits for people who do not have the assets to compete for higher incomes against those who are better endowed. Unless huge revenues from natural resource royalties come to the central treasury—which is not guaranteed even in resource-rich countries—equity requires a strong tax effort, relative to the country's tax capacity.[13] Yet this is a chronic weakness in most developing countries. Low tax effort precludes the capacity to enhance equity through services or income support for the poor. Bird, Martinez Vazquez and Torgler (2008, 55) note that "[m]any developing countries need to spend more on public infrastructure, education, health services and so on, and hence they need to increase their tax effort—tax revenue as a percentage of gross domestic product (GDP)—if they want to grow and to be less poor."

Of the 96—54 developing and 42 developed—countries included in the tax effort study by Pessino and Fenochietto (2010), all but one in the lowest third of the rankings are developing countries (the exception, Singapore, is the second lowest of all, reflecting the city-state's reliance on revenues from foreign corporations with regional headquarters there). Forty-two of the 48 countries below the median are developing countries, and only 12 of the 48 above the median are developing countries.[14]

Even if taxing capacity is constrained by the feasibility of collecting more revenues—for example, because of widespread poverty or a largely agricultural economy—the weakness of tax effort can typically be traced to tax policies as well. The advice of tax policy experts—broaden the tax base, reduce marginal rates, eliminate exemptions, reduce reliance on evadable or avoidable income taxes, and simplify the tax system—often falls on deaf ears, particularly when government leaders try to convey an impression of progressivity by emphasizing income taxes even if they are ineffective and are not progressive due to legal avoidance or illegal evasion.

It is striking that countries more heavily dependent on natural resource exports tend to have weak domestic tax effort (Henry and Springborg 2001; Devarajan, Le, and Raballand 2010). Although Pessino and Fenochietto's (2010) rankings do not cover the major Middle Eastern oil-exporting countries, they do reveal low domestic tax effort in the hydrocarbon-exporting nations of Egypt, Indonesia, Malaysia, Mexico, and Peru. Only Syria, with rather modest net oil exports, was above the median, ranked at 52.

Many analysts interpret predominant reliance on resource export revenues as a governance problem as well: governments may be less accountable to their citizenry if government revenues do not come from the citizen's pockets (Bates and Lien 1985; Moore 2004, 2007; Bornhorst, Gupta, and Thornton 2009; Fjeldstad and Moore 2008). However, this premise presumes that citizens are less aware, or less caring, about what happens to revenues coming from raw material exports. Yet in countries that are heavily dependent on these exports, the focus of attention may be strongly centered on how export revenues are used. For example, few Venezuelans are unaware that the disposition of oil revenues is crucial to their prosperity.

Second, governments beholden to taxpayer may be more restrained in providing services to the poor. Ravallion (2002), examining the Argentine case, points out that when budgets are tight, budget allocations for propoor programs may be hit harder than other programs. In contrast, in Indonesia, a huge expansion in propoor social service and income support programs occurred in the mid-2000s when the government could rely more heavily on hydrocarbon exports following the reduction of regressive energy subsidies. Put simply, while accountability to citizens is desirable, all other things being equal, accountability to the tax-paying subset of citizens does not always dovetail with progressive policies, particularly when their attitudes toward the poor are more critical than sympathetic (Graham 2002).

Social Safety Nets

The goal of poverty alleviation depends heavily on providing social safety nets for the poor, most constructively going beyond providing education and health care by supplementing income directly. This is especially true when austerity programs and structural adjustment are underway, because the poor often have less maneuverability to protect their incomes in times of change and have less economic cushion against state service cutbacks.

Social safety nets require government spending, which would seem to entail progressive redistribution. However, safety nets—whether direct income support, vouchers for goods or services, designated food assistance, pensions, etc.—are often regressive. Without modern sector jobs, many of the poor are ineligible for contributory pension plans or minimum wages. They may benefit to a certain degree from subsidized energy, food, or other goods and services, but wealthier people generally consume more energy and often more of the other subsidized items. Efforts to target subsidies to the poor often fail due to corruption, difficulty in identifying eligible people, or resale to wealthier people. If contributory programs such as pensions are included, government spending on social protection often favors higher income beneficiaries insofar as retirement funds are subsidized. Lindert, Skoufias, and Shapiro (2006) report that the transfers to cover the deficits run by Latin American pension plans exceed the government funds dedicated to social assistance for the poor. Government spending on social protection is regressive for all eight countries they examined (Argentina, Brazil, Chile, Colombia, the Dominican Republic, Guatemala, Mexico, and Peru). They note that for Brazil, despite the vaunted Bolsa Familia Program of conditional cash transfers for low-income families, government spending to cover the deficits in the federal pension plan in 2006 dwarfed the spending on Bolsa Familia—3.7 percent of GDP vs. 0.4 percent—and more than half of these benefits go to the wealthiest 20 percent of the population (Lindert, Skoufias, and Shapiro 2006, 7–8). Grosh et al. (2008, 32) conclude more generally:

> The idea that governments cannot afford to redistribute income to the poor must be contrasted with the evidence that they regularly redistribute income to the non-poor. Energy subsidies are highly regressive and often more costly than safety nets. The Arab Republic of Egypt spent 8 percent of its GDP on several energy subsidies in 2004…and Indonesia spent up to 4 percent of GDP between 2001

> and 2005 on fuel subsidies... Similarly, countries dedicate resources to bailouts of insolvent contributory pension funds by transferring general revenues to support them.

Therefore, in many countries the doctrine of poverty alleviation targeting has hardly been followed.

Conditional Cash Transfers

However, an emerging initiative may redress this neglect through a combination of multiple objectives served by conditional cash transfer programs. These programs, providing cash transfers to families that comply with conditions typically requiring school enrollment and healthcare visits for the children, are an increasingly prominent type of social safety net. They directly alleviate poverty through the cash transfer, but they also enhance human capital. Even without conditions, cash transfers targeted to vulnerable populations have special significance in their potential to link social spending with the political feasibility of macroeconomic reform. Government leaders—otherwise reluctant to initiate sound policy reforms out of fear of disruption by economically threatened people—may be emboldened if well-publicized cash transfers reduce the potential for disruption.

Yet another generally salutary effect of conditional cash transfer programs from a poverty alleviation perspective is that they often put greater pressure on governments to expand educational and healthcare capacities to meet the increased demand. In Indonesia, for example, the conditional cash transfer program coincided with significant increases in education and healthcare spending (both financed in large part by severely reducing energy subsidies). The overall social service improvements for the poor will occur as long as the government does not simply permit the deterioration of service quality by capping education and healthcare spending.

In light of these many virtues of conditional cash transfers and the concerted efforts of the international development banks to promote these programs,[15] the number of programs has expanded dramatically. From 28 programs as of 2008, by 2013 there were 52 (World Bank 2014a, xiii). However, given the dependence of these programs on government budgets, their sustainability and the magnitude of transfers are often uncertain.

Extending Social Services

Despite the grave difficulties (discussed later in this chapter) in demonstrating the value of the intrinsic benefits and greater productivity induced by providing more education, health care, and income support for the poor, there is compelling evidence that the social sectors have been neglected in many developing countries.

Education

In light of the phenomenal success of investment in education in Northeast Asia, the neglect of education investment in other regions is striking. Although the rates of return on general education spending is greater for developing countries than for developed countries (Psacharopoulos and Patrinos 2004), government spending on education in developing countries is generally anemic. According to World Bank regional statistics, the public spending on education as a percentage of GDP for Arab countries, Sub-Saharan Africa, and Latin America is only three-quarters that of high-income countries. The percentage for the developing countries of East Asia is less than two-thirds that of the rich nations; South Asia's education spending is only slightly under half that of high-income countries. Only the Middle East and North African region comes close to the proportional investment as the wealthiest countries (World Bank 2012b). Sub-Saharan Africa, with gross secondary school enrollment at 40 percent, has the lowest educational attainment of any world region; South Asia, the next lowest region, has a proportion of 58 percent; North Africa and the Middle East are at 77 percent (World Bank 2012c).

A 2008 IMF simulation estimated that if average developing country education spending had been increased by 1 percent of GDP beginning in 2000, by 2015 it would have increased the net enrollment rate from 90 percent to 99 percent and reduced the child mortality rate from 76 to 65 per thousand. It would also increase GDP per capita by 0.5 percentage points annually[16] (Baldacci et al. 2008, 1335). This is a very attractive return on capital.

There is also a troubling imbalance in how education spending is allocated, in terms of the subsectors of primary, secondary, and tertiary education. Colclough (1980, i) enumerated the multiplicity of primary education payoffs: primary education "increases productivity in all sectors of the economy…[with] economic returns…in many countries considerably greater than those arising from other levels of schooling…it reduces fertility, improves health and nutrition, and promotes significant behavioral and attitudinal changes at the level of both the individual and the community."

Yet while Sub-Saharan Africa's low spending on education is not surprising in light of the region's low per capita income, it is striking that the governments of the region privilege tertiary education far beyond reasonable levels in light of the overall educational profile. At least 16 African governments spend more than a fifth of the public education budget to the tertiary level; half devote more than a quarter of public education funding to this level, with Botswana at a remarkable 43 percent.[17] To put this into perspective, all of the nations of South Asia, the next poorest region in per capita income terms, devote less than a fifth of public education spending to higher education. In nations with high levels of functional illiteracy and low levels of secondary school enrollments, heavy proportions of education spending going to universities entail a questionable neglect of more basic education. The skew favoring higher education also holds in North Africa, despite suffering from high levels of unemployment among college-educated people (World Bank 2011a, 3). Marino (2011, 24) points out that Tunisian graduates with diplomas had a three to four times greater chance of being unemployed than people without diplomas.

When tertiary education is heavily promoted through government subsidies, it becomes regressive in terms of income distribution by virtue of the fact that higher income individuals are far more likely to enroll. For Latin America, Lindert, Skoufias, and Shapiro (2006, 31) note that "public spending on tertiary education is highly regressive in all [Latin American and Caribbean] countries in terms of its direct absolute incidence…The redistributive issue for higher education is whether or not public tax resources should be used to finance it, given the high degree of regressivity (as well as the inefficiencies surrounding publicly-financed and protected higher education systems)."

Health Care

The brief for greater spending on health care is based on the premise that higher health-care spending yields better health, and better health not only has an intrinsic value but also generates greater educational attainment, higher worker productivity, lower job absentee rates, and cumulatively greater economic growth. Bloom, Canning, and Sevilla (2004) trace out these impacts, concluding that the relatively high levels of economic gain demonstrate that health care suffers from underspending.

It is certainly true that many countries have failed to meet the health-care spending targets that their governments embraced as explicit goals. For Sub-Saharan Africa, the 2001 Abuja Declaration formally committed 46 African region member states of the World Health Organization to allocate at least 15 percent of the national budget to health care; other, equally strong statements of commitment followed (Harmonization for Health in Africa 2011, 9–10). According to the World Health Organization's latest pan-African data (2014, 142–148), only six of the 48 Sub-Saharan countries reached the 15 percent level by 2011; the average proportion of government budgets was 8.3 percent, only a fraction above the 8.2 percent in 2000. Moreover, 14 countries had lower levels than in 2000.[18]

Regional Development

Regional development strategies, typically consisting of favoring a particular region through government investment and infrastructure, have proven to be highly complex and problematic when economically backward areas are targeted. They often entail resettlement programs or at least large-scale migration of people trying to take advantage of the expected opportunities. Natural resource exploitation is frequently a component, as is rural industrialization, billed as a multiobjective vehicle for rural poverty alleviation, industrial development, and deconcentration of industrial production away from overcrowded metropolitan areas, reducing environmental stress and stemming urbanization. The argument is that with apparently abundant land, natural resources, less costly labor, and new capital, less developed regions can outstrip more developed regions.

The problems encountered by these programs have centered on the risk of economic failure whenever investment is at variance with market signals and, perhaps surprisingly, the resistance of local residents within the favored areas. While in some cases an economic argument can be made that economically backward regions have the potential for much greater growth that thus far has gone untapped because of market failures, poor economic policies or lack of physical infrastructure, whatever obstacles to productivity exist, can be quite daunting. For example, in Thailand, despite efforts to deconcentrate economic activity away from the much wealthier central region, Pansuwan and Routray (2011, 40) note that "[t]he very high level of industrial development pattern is still scattered around the [Bangkok Metropolitan Region] because of its comparative advantages. Interestingly, almost all provinces falling within the very low industrial development level are located in the northeastern and northern regions of the country mainly due to lack of physical and human as well as capital resources." Moreover, the areas selected for regional development promotion may be chosen for reasons that depart from, or even run counter to, economic growth potential: they may be poor areas in need of poverty alleviation, border regions targeted for geopolitical reasons, areas with secessionist movements, or jurisdictions to which government leaders wish to attract newcomers who support the government.

Rural industrialization—in principle a promising centerpiece of regional development strategies targeting economically backward areas—has had surprisingly little traction outside of China. The most intensive efforts have been in Asia, in such countries as India, Turkey, Indonesia, and Thailand. Shifting population and pollution to economically backward areas puts greater pressure on the infrastructure and the environment; the question is whether the area's growth will generate sufficient resources and willingness to address these stresses. An additional source of contention arises from the need to provide more energy and transport infrastructure to the backward area, when this displaces local people to make way of hydroelectric dams, highways, or other infrastructure (Fernholz 2010).

An often unanticipated obstacle to regional development programs is the intergroup conflict that arises in many countries as outsiders flock to the targeted areas, provoking resentment and defensive reactions from existing residents threatened by encroachments on their property rights, environmental degradation, greater competition over public goods, and so on. Rigg (2003, 229) argues that rural industrialization also frequently displaces local people if they sell their land; he suggests conceiving of rural industrialization as "industrial extension and rural displacement." The antagonism is likely to be even stronger when the original residents suspect that bringing in outsiders has the political motive of diluting the power of the original residents.

Natural Resources

Government investment and other means of promoting natural resource exploitation are complicated by the fact that typically the resources already exist. The question for both renewable and nonrenewable resources is whether the government should invest in (or otherwise encourage) the development of the capacity to extract the resources. For renewable resources, the additional question is whether the government ought to discourage their extraction to some degree in order to maintain a more sustainable supply of the resource and the environmental services that it provides.

The guidance that policymakers have received regarding government promotion of natural resource exploitation has been severely muddied by the neglect of crucial distinctions between resource abundance and resource dependence, by confusion between whether resource wealth is an advantage and whether government should encourage the exploitation of resources to be had. Resource dependence can signify resource wealth, as in the case of major oil exporters such as Saudi Arabia or Kuwait, but it can also signify the relative weakness of other economic activities, as in the case of the Democratic Republic of the Congo (DRC), which heavily relies on mineral exports even though its per capita subsoil *assets* (as distinct from annual revenues) have been estimated at less than US$80 in 2005 dollars, ranking the DRC in 89th out of 141 nations for which the World Bank developed subsoil asset estimates. The DRC's rank for total natural assets is 138th out of 149.[19]

If governments invest directly in natural resource extraction, it could take the form of supporting the discovery of resources (e.g., by financing geological surveys), providing capital to state-owned resource extractive enterprises, or devoting funds to purchase resource output to make exploitation more attractive. As mentioned earlier, financing state-owned exploration and production may prove risky to a developing country, when international corporations are prepared to provide capital, expertise, and marketing chains. Yet from 2000 through 2013, 18 developing countries expropriated international hydrocarbon holdings, mining holdings, or both (JLT Specialty Limited 2014), and this figure does not include the instances of governments simply buying out the international holdings without resorting to expropriation or unilateral cancellation of contracts. This vacillation between state and private ownership erodes efficiency and investor confidence.

Whereas the economic theory pertaining to state ownership of natural resources is straightforward in calling for payments to the state equivalent to the intrinsic value of the resource before extraction and processing (the resource rent), rent capture often falls far short, becoming a major trigger of nationalizations. Manzano, Monaldi, and Sturznegger (2008, 96), in reviewing "some of the main factors that help explain the recent wave of nationalizations and tax hikes in the Latin American hydrocarbon sector," note the following:

> A key force behind these trends is the distributive conflicts that arise between the governments and the producing firms. These conflicts occur, to a large extent, because the tax systems used in the region have not taken into account fundamental contingencies—in particular, price changes. As a result, the producers retain an increasing share of oil rents when oil prices rise significantly. This generates powerful incentives for governments to renegotiate, renege on contracts, or nationalize the sector. The optimal contract properly should include price contingencies. The policymaker may consider tax and royalty rates that vary according to the price, but implementing such a scheme is not easy.

The key point is that the failure to build in contractual contingencies for surprisingly high prices or unexpectedly high production has undermined the stability of extractive governance vis-à-vis the private sector, which has the resources to keep hydrocarbon extraction from being undercapitalized.

When oil and gas extraction is conducted by state enterprises, additional problems arise from manipulations that are not among the professed rationales. The temptation to borrow beyond a nation's sovereign debt ceiling through the state enterprise has resulted in massive borrowing in such oil-exporting countries as Indonesia, Mexico, Nigeria, and Venezuela. When the government taxes the state enterprise to gain control over the additional funds or directs the enterprise to finance operations beyond the extractive industry, the enterprises become undercapitalized in their primary activities. This is one of the key reasons why state extractive enterprises are often inefficient and why they are frequently compelled to bring in international companies to do much of the operational work.

These problems are not confined to subsoil assets; they also hold for forestry, with additional complications. While in the bulk of developing countries the subsoil assets are constitutionally reserved to the state, the control over forests is usually heavily contested. When the state succeeds in controlling commercially exploitable forests, the risk is that it will award concession rights without fully capturing the value of the timber. The failure of the state to achieve full rent capture not only denies the nation's riches to the people, it also prompts overexploitation. Inadequate rent capture is problematic for forestry for an additional reason: low rent capture from state forests provokes overexploitation that often degrades the ecosystem. In many cases, the concession fees are less than half of the value of the extracted timber. The causes of low concession rates range from successful rent-seeking by timber companies, whether domestic or foreign, to the government's desperation to bring in some revenues when the other nations have more attractive forestry prospects.

A final problem with forestry policies frequently occurs when governments return control over forests to the local communities, sometimes out of genuine recognition of the virtues of local control, sometimes out of the cynical effort to shed responsibility for degraded lands. If the community is not well defined (after all, in many circumstances, multiple decades have passed since government took control over the forests), conflicts may well ensue. If the resources for sound forest management are not transferred along with the responsibility, the forests may degrade even further. This risk is one manifestation of the risks of decentralization that amounts to underfunded mandates.

Industry vs. Agriculture

In the classic tradeoff between investing in industry vs. agriculture, public investment in industry is still widely favored even when the productivity of additional investment is considerably less favorable. For Vietnam, Fesselmeyer and Le (2010, 167) note that the latest estimate of the agricultural capital-output ratio was four times more favorable than that for industry, and yet the government investment in industry was twice that of agriculture. Murty and Soumya (2006, 10) estimate India's capital-output ratio for agriculture as more than eight times as favorable as that of industry. Anderson, Cockburn, and Martin (2010) summarize the long-standing antiagricultural bias due to price distortions, though the biases have been reduced to a degree in many countries. More generally, Eswaran and Kotwal (2006, 117), noting that the agricultural "sector has been much neglected in developing countries," enumerate the multiple mechanisms for favoring industry at the expense of agriculture. The neglect of agriculture is also indicated by the fact that many small-scale agricultural promotion programs remain unfunded even though the programs that have been funded often yield strikingly high rates of return. Alston et al. (2000, x, 59) found that the returns on agricultural research and on extension—two of the most important drivers of agricultural productivity—had a mean of 60 percent for developing countries, with no evidence that the rates were declining over time.[20] In short, assuming that the existence of project possibilities with such high rates of return means that other projects with high rates of return are still to be had, attractive agricultural investments have been neglected, as capital is still directed toward industry. In addition, huge investments in agriculture over the next decades are needed to wean Sub-Saharan Africa from the labor-intensive agriculture that has been holding back their productivity and incomes (Schmidhuber, Bruinsma, and Boedeker 2009).

What is perplexing is that the neglect of agricultural opportunities was recognized decades ago as an enormous impediment to the success of broad development initiatives. Ruttan (1984, 398) concluded that:

> By the early 1980s the new "basic needs" and "integrated" approaches to development were coming under severe questioning. The decline of integrated rural development and basic needs programmes did not reflect a retreat on equity goals as much as growing recognition that the programmes, particularly in Africa, were not solving one of the most fundamental rural problems—"achieving a reliable food surplus."

In short, for many countries the initiatives to provide basic necessities, the full range of productivity-enhancing input, and greater capacity for community cooperation were inadequate in the face of adverse pricing for agricultural outputs, neglect of agricultural infrastructure, and other manifestations of the proindustry bias. Farm incomes clearly suffer under such circumstances, which has been a major consideration in the promotion of cash transfers.

In Latin America, small-scale agriculture has suffered the most due to the promotion of industry. By the 1980s, export promotion focused not only on export-oriented manufacturing but also on agro-exports through cheap credit and lower agricultural tariffs. Lastarria-Cornhiel (2006, 17) notes,

> These policies, particularly trade liberalization policies together with agricultural policies that favor export products, stoked the growth of agri-business, particularly in the production of high-value horticultural crops, and agricultural exports, provided a growing demand for wage labor. At the same time, liberalization policies have resulted in higher input costs, lower farmgate prices, and significant cuts in access to credit and extension services for the smallholder sector that produces mostly food for local and regional markets.

Physical Infrastructure

Despite the emerging opportunities for the private provision of physical infrastructure through build-operate-transfer arrangements with governments, the overall provision of physical infrastructure—predominantly the responsibility of government—has fallen disappointingly short in the bulk of developing countries. Latin American nations spend, on average, less than 2 percent of GDP on physical infrastructure, although the economic returns on improved infrastructure would be enormous. Kohli and Basil (2010, 68) estimate that Latin American infrastructure investment would have to be 3.8 percent of GDP to maintain business-as-usual growth through 2040, and developing Asia would require 6.3 percent of GDP to maintain its business-as-usual growth. World Bank estimates indicate that if Latin American nations could improve infrastructure to the level of South Korea, GDP growth would increase by nearly 4 percent annually and reduce income inequality by 10–20 percent (Fay and Morrison 2007, 4). Rioja (2003) estimated that Latin America's productivity losses due to poor infrastructure *quality* are equivalent to 40 percent of real per capita income. Sub-Saharan Africa's underinvestment in power and telecommunications *alone* during the 1980s and 1990s had cost the region an estimated 1.3 percent of GDP (Briceño-Garmendia, Estache, and Shafik 2004).

The fact that gains can be made by investing more in physical infrastructure—which is largely in the government's domain—does not, in and of itself, mean that all governments have been remiss by neglecting physical infrastructure. Some countries, most notably China (building before demand),[21] have been dramatically expanding physical infrastructure. The question is whether highly productive infrastructure projects are being neglected as governments invest in other things. This indeed seems to be the case, but only in some countries. Canning and Bennathan (2007) have calculated the rates of return of two major aspects of physical infrastructure—electricity generation and paved roads—with models that capture some, though not all, of the contributions that these aspects make to productivity.[22] For the 49 developing countries for which sufficient information was available, the mean rate of return on investment in electricity was 42 percent, with a median of 36 percent. However, the average overall rate of return on capital was 40 percent, and in fact for 29 countries, the overall return on capital was greater than the return on electricity expansion. For the 25 developing countries with enough information on the returns on road building, the mean was 213 percent and the median was 104 percent; there was an even split between the number of countries with road building having a higher or lower return than the overall return on capital.[23]

One quite disturbing finding on physical infrastructure development is that public investment is often highly inefficient even in the ratio of the investment and the value of the infrastructure created. Gupta et al. (2011), examining 52 developing countries, estimate that the infrastructure value is roughly half the invested capital; they cite problems throughout the stages of infrastructure development.

This means that government promotion of power generation and road building makes sense in some countries, but not in others. The mixed results reflect that physical infrastructure can be done poorly in some developing countries, with remarkably low rates of return. The most compelling diagnoses of these failures entail the implementation of major infrastructure projects that are unsound from the start, on the one hand, and the neglect of maintenance to retain the usefulness of the project, on the other. Flyvbjerg, Bruzelius, and Rothengatter (2003) and Flyvbjerg, Garbuio, and Lovallo (2009) point to the general problems (in both developed and developing countries) of delusion—the belief that past problems can be cleverly overcome—and deception—the tactics of infrastructure entrepreneurs, whether or not in government, to make infrastructure projects appear more promising than they really are.

The neglect of maintenance is particularly acute in the poorer developing countries, where spending on infrastructure building often swamps spending on maintenance, as Benmaamar (2006) demonstrates for Sub-Saharan Africa. Here again, the limitations on liberalization reforms calling for cost recovery of the services provided by major infrastructure (e.g., electricity, irrigation water, potable water) also erode the capacity to maintain the usefulness of physical infrastructure.

Uncertainty about Economic Priorities

If we dig deeper into why some sectors have been severely neglected, we can identify an analytic limitation blocking the capacity to discredit special pleadings by interest groups to subsidize their subsectors. In contrast to the clear persistence of the unwarranted bias favoring industry over agriculture, the general challenge of identifying which sectors deserve more or less promotion has been beyond the capabilities of explicit economic analysis. In theory, government policies, which inevitably encourage investments in some sectors over others,[24] ought to direct investment to sectors with the greatest positive impact on economic growth and other national objectives. In strictly economic terms, public investment and other forms of promotion ought to go where the greatest output gains can be achieved per unit of capital, taking into account the synergies of indirect impacts (e.g., education promoting greater awareness of healthy behavior). Thus the diagnostic question is whether some sectors are neglected in terms of overall investment, such that more government promotion is warranted—or less for sectors promising lower social returns.

In principle, several measures can indicate whether a particular sector is deserving of more promotion. Assuming that a sector deserves more promotion if the returns (output) on investment have recently been considerably higher than those of others, a favorable incremental capital-output ratio serves to identify where promotion is more efficient. For directly productive sectors like industry and agriculture, the incremental capital-output ratio is a standard (though not widely estimated) measure; lower rates indicate more favorable investment prospects in that a given sectoral level of investment (capital) would produce more output. For physical infrastructure, which largely makes an indirect contribution to productivity, the ratio of overall investment to economic growth is a parallel measure. For education and health, the ratio of overall spending to economic growth is also relevant, in addition to the direct benefits to the individuals with improved education or health.

Although for some sectors (e.g., industry vis-à-vis agriculture) it is relatively straightforward to determine whether government promotion is misdirected, the methods for determining whether other sectors ought to be receiving more or less promotion are found wanting. Calculating the societal returns on education[25] or health is complicated by the indirectness of the potential impacts. It is obvious that more education has the potential to spur innovation, improve the transmission of agricultural techniques, improve overall manufacturing productivity, and induce more knowledgeable behavior regarding health and nutrition; healthier people are likely to be more productive, and for that matter, healthier children are likely to have higher educational attainment. Yet, valuing these impacts has been daunting for individual countries, and the cross-national correlations between education and economic growth, or health and economic growth, have too many potentially confounding factors.

Thus, some education advocates point to underinvestment in government education spending in developing countries by trying to capture the host of potential benefits of higher educational attainment. For example, Appiah and McMahon (2002) make rather heroic assumptions to include benefits ranging from longevity and lower fertility rates to democratic participation and crime reduction; they estimated that for Sub-Saharan Africa, "[t]he direct costs of two per cent of GNP translates into a per capita increase of two per cent of per capita income." Yet Psacharopoulos and Patrinos (2004, 118) note that "externalities or spillover benefits... are often hard to identify and even harder to measure... [E]vidence is not unambiguous. In fact, some estimates give negative values, while others give very high estimates."

Similar constraints are relevant to the micro-level decisions of project and program choices. Education projects and programs with higher rates of return than other investments, whether in education or other sectors, ought to be given priority, but the benefit-cost analysis necessary to determine rates of return is often absent. Jimenez and Patrinos (2008, 2–3) judge that

> at the micro level, [cost-benefit analysis] has also not been used extensively in justifying specific education projects. Why not? We argue that some key methodological shortcomings have been responsible: specifically, the difficulty of estimating social, as opposed to private, benefits; the complexity of measuring the costs and benefits of other dimensions of education other than access to a year of attending an educational institution; and attribution of outcomes to actual interventions.

This is a limitation to project selection undertaken not only by governments but also by international development agencies. The World Bank's Independent Evaluation Group (2010, 7) found that only 1 percent of the education projects funded by the World Bank from 1970 to 2008 was subject to an economic return analysis.

Very similar uncertainty pertains to determining the advisability of greater promotion of the health sector in general or individual health programs and projects. Some economists (e.g., Bloom and Canning [2003, 47]) assert that governments of developing countries typically underspend on health care. One approach using standard regression analysis is to demonstrate, across a broad set of countries, that those with higher health spending have better health indicators, or that countries with either higher health spending or better health indicators have had more rapid economic growth. The alternative is to simulate the impact of health improvements on greater educational attainment, higher productivity due to greater physical capacity, lower job absentee rates, and so on. Bloom, Canning, and Sevilla (2004) employ such a production function model to argue that underspending on health has sacrificed impressive opportunities to increase labor productivity and hence economic growth.

Yet Jack and Lewis (2009, 3), pointing to the inconsistency and methodological difficulties of both approaches,

> caution the reader against expecting to find consensus in the empirical literature on the links from health to growth or even from health policies to health. A number

> of papers present unambiguous results but contradict one another... [T]he literature is a mix of rigorous scientific investigation and well-motivated advocacy on both sides... Further, when attempting to untangle the link from health to growth, or vice versa, econometric issues of endogeneity and measurement error are particularly problematic, and the validity of even the most innovative approaches continues to be debated.

Thus, as with education, the adequacy of spending in health in the context of scarce financial resources across all sectors remains undeterminable unless the direct, clearly reliable measures yield rates of return that are much higher than those related to investments in other sectors. This does not appear to be the case (Weil 2007).

An additional reason why the findings on social returns on education and health care are inconsistent is that nations with poor governance and failing economic policies leave little opportunity for economic improvement even if the workforce is better educated and healthier.[26] As mentioned above, several World Bank assessments conclude that poor macroeconomic policies can easily undermine otherwise sound projects and programs. For cross-national estimates of the returns on efforts to promote sectors, programs, or projects, the pooling of cases of nations with sound and unsound macroeconomic policies will obscure the potential value of social service investment—and indeed of the potential returns across all sectors.

The lack of definitive, explicit results is not the fault of development economists; rather it reflects the intrinsic complexity of the linkages and uncertainties as to whether even sound sectoral investment can withstand external shocks, changes in government priorities, and internal turmoil. However, without definitive determinations of which sectors would use government investment more productively, the input from experts of different sectors is often regarded as special pleadings, no different from the standard budgetary discourse in which officials of the various ministries vie for higher allocations.

The most ambitious claim that economists can make to comprehensively tracing out the impact of policies on all levels is the computable general equilibrium model.[27] The objective is to project impacts through "completely-specified models of an economy or a region, including all production activities, factors and institutions, including the modeling of all markets and macroeconomic components, such as investment and savings, balance of payments, and government budget. These models incorporate many economic linkages and can be used to try to explain medium- to long-term trends and structural responses to changes in development policy" (World Bank 2013). There have been claims that these models are widely used; Mitra-Kahn (2008, 2) asserts that "Computable General Equilibrium (CGE) models are probably the most utilized tool globally for development planning and macro policy analysis."

The phrase completely specified models might seem to imply that the use of these models would provide unambiguous, consensually accepted results as to the optimal macroeconomic policies and sectoral promotion strategies. Yet, as mentioned in chapter 2, Gibson and Van Seventer (2000) demonstrate in an analysis of policy alternatives for South Africa that different versions of computable general equilibrium models will yield very different policy recommendations. In fact, the practical uses of computable general equilibrium models are far more limited than Mitra-Kahn presumes. For the reasons outlined in chapter 2, the simpler computable general equilibrium models may reduce the need for data, yet they rest on assumptions that many practitioners would regard as implausibly simplistic. In contrast, the structuralist models require more data and more specifications that may be seen as ad hoc.

However, one might still hold out hope for country-specific determinations of at least the rates of return on specific projects. And the sum total of the projects and programs would constitute a major part of the sectoral promotion strategy; if these are the optimal projects (i.e., those with the highest rates of return of all potential projects), the nation's portfolio would be the optimal sectoral policy. Yet, as early as the mid-1980s, when the use of social benefit-cost analysis was supposed to be the method of choice to evaluate potential projects or programs, Leff (1985b, 68) noted that "[t]he methodology [of social benefit-cost analysis] was designed for choosing the investments with the highest returns, not those that were preselected on other criteria but pass minimum standards." But he then pointed out "that policy-makers usually determine the main lines of the investment budget on the basis of *sectoral* priorities. Thus the de facto approach to investment choice usually involves a two-stage process: first designation of high-priority sectors, and then selection of projects within those sectors. The major allocation decisions, however, are taken on the basis of choices between sectors rather than between projects" (1985a, 337).

In short, the benefit-cost analysis, if done at all, is applied to justify projects selected through a sectoral balance that was not based on explicit methodology. In addition, it cannot be determined whether the projects selected to address sectoral needs are those with the most favorable social benefit-cost estimates, because project identification cannot proceed from calculations of rates of return of all plausible projects; it proceeds through the choice of candidate projects selected on the basis of essentially implicit economic criteria, political considerations, bureaucratic politics, and so on.

The neglect of formal project evaluation by governments is, ironically, increasingly reinforced by the international organizations that governments look to for setting the standards for economic analysis. Thus, the World Bank's own Independent Evaluation Group (2010, ix), based on a sweeping examination of the Bank's project evaluation process, very recently came to the following conclusion:

> Cost-benefit analysis used to be one of the World Bank's signature issues. It helped establish the World Bank's reputation as a knowledge bank and served to demonstrate its commitment to measuring results and ensuring accountability to taxpayers . . . The percentage of Bank projects that are justified by cost-benefit analysis has been declining for several decades, owing to a decline in adherence to standards and to difficulty in applying cost-benefit analysis. Where cost-benefit analysis is applied to justify projects, the analysis is excellent in some cases, but in many cases there is a lack of attention to fundamental analytical issues such as the public sector rationale and comparison of the chosen project against alternatives.

The Independent Evaluation Group's (2010, ix) assessment also provides a crucial insight into why the benefit-cost analysis is frequently neglected or sloppy: "The Bank's use of cost-benefit analysis for decisions is limited because the analysis is usually prepared after the decision to proceed with the project has been made." This limited use of the analysis in decision making would be less serious if the role of the World Bank were simply to approve or reject funding requests that meet the minimum rate of return.[28] The World Bank, like the regional development banks, works closely with governments in identifying and shaping projects and programs, often eschewing explicit benefit-cost analysis in selecting projects or programs.[29]

Economic Aspects of Decentralization

The issue of decentralization is obviously a concern of governance, as discussed in chapters 4 and 5, but the choices of which services ought to be provided at which levels, and how the services would be funded, are important economic decisions as well. Economic policies and programs can be scaled from the smallest units, such as funding particular initiatives within a neighborhood, to the national or even international levels, with implications for the breadth of policy application and, often more importantly, whether the fiscal resources transferred to the subnational level are adequate.

One reaction to the demand to reduce the central government's budget (a key component of many austerity packages) has been to devolve responsibility for services to subnational governments, but without the commensurate transfer of fiscal resources to undertake these services at the same level of scope or quality. We have already seen this with respect to natural resource management, but it is often severe for social services as well. Three adverse consequences for the poor have been noted. First, the services may simply deteriorate and the poor, who are least capable of finding supplementary sources for these services, typically suffer the most. Second, in order to keep up with the service requirements (generally scope rather than quality requirements, such as number of years of schooling), the subnational governments may impose additional taxes and fees. Bossert and Beauvais (2002) report that the decentralization of health care in the Philippines and Uganda shifted health care away from the most vulnerable populations. Bernstein and Lü (2008, 90) report that poor peasants in China were subjected to illegal taxes in order for rural schools to meet the unfunded mandate of lengthening the number of years of mandatory schooling. A hybrid of these two patterns was seen in rural Vietnam, where low central budgetary support for schools provoked the imposition of fees and the de facto partial privatization of primary education as teachers truncated the formal school day in order to provide "private tutoring" in the afternoon.[30] In short, decentralization without adequate financial support often imposes greater burdens on families that had received social benefits from the central government. In short, the economic parallel to the governance aspects of decentralization may beggar the poor insofar as the local authorities in poorer areas may lack both the resources and the willingness to channel benefits to the most disadvantaged.

Conclusions

Where definitive technical analysis is lacking, other considerations hold sway. Although the theories and doctrines developed during the post-WWII period were molded to pursue poverty alleviation while still recognizing the importance of economic efficiency, the deviations from these theories and doctrines, reinforced by the lack of definitive methods to identify optimal policies, limited how the quite impressive economic growth over the past decade contributed to elevating the most vulnerable populations.

The failure to do more for these most vulnerable populations lies, in large part, on the persistent weaknesses in governance. We now turn to these governance issues to identify the theories and doctrines that have shaped governance practice.

CHAPTER 4

Evolution of Governance and Development Administration Theory

This chapter examines the evolution of theories and doctrines of governance and development administration, especially over the post-WWII period. Because sound governance is the bedrock of responsive and effective policy, no other aspect of development, whether economic, political, or social, can escape the ravages of poor governance. As mentioned in chapter 2, economists have joined other social scientists and all shrewd practitioners—that institutions matter. This is certainly the province of governance.

Our review of governance and development administration theories captures the evolution of thinking on public administration reform, development administration, public sector management, the New Public Management (NPM), and democratic governance. This evolution has not occurred in a vacuum; it reflects the influence of academic fields (public administration, political science, international relations, and economics), but more directly, we demonstrate the linkages between theories and doctrines of socioeconomic development and the theories and doctrines of governance and development administration

For the past three decades, the term governance has been a significant part of the national and international development discourse among academics, policymakers, and development practitioners (United Nations 2011). Yet, the term and its multifarious practices across the globe are ridden with theoretical ambiguities, elusive meanings, and policy incongruities.

It continues to be used by different actors to mean different things ranging from a set of tools and mechanisms to improve public management, to political reforms and the engagement of diverse stakeholders with conflicting political objectives. Over the years, the concept has been redefined by scholars and development practitioners along with the paradigm shift in our thinking about development. Furthermore, the new definitions of governance have often reflected a narrow disciplinary focus of the academics or national and international development partners. The term is, therefore, undertheorized.

Rethinking Development and the Linkage with Governance

Theories of development governance have focused on a broad range of topics, including the structure of formal institutions involved in economic and social policymaking, public administration, and citizen participation. Because of the colonial heritage of emerging African and Asian nations, and the pressures for economic growth in Latin America, governance theory of the early post-WWII period presumed that strong development initiatives required a strong state. The corollaries were limited public participation, centralized governance, and the administrative challenge of strengthening and rationalizing the central governmental apparatus, relying on the standard, preexisting ministerial structures. Elsewhere, corporatist theories—giving formal standing to economic segments such as industry confederations and labor unions but designating the government as the ultimate authority[1]—were developed to justify authoritarian governance, in many instances through military dominance based on a convergence of developmental and national security concerns. Regarding development administration, the norms of professionalism, policy neutrality, impartiality, etc., were adopted straightforwardly from the Wilsonian model of public administration that simply implements directives from higher authorities.[2] One of the assumptions was that professional training would provide both the expertise and socialization to create competent and honest administrators.

The evolution of governance theories is linked with the shift in the theories and practice of development over the past 50 years. Different dimensions of development have been emphasized by scholars and practitioners over the years reflecting national and international priorities. In light of the vast network of UN-affiliated development institutions, it is useful to adopt the UN emphases as a lens for understanding the context in which governance and administrative doctrines were developed.

While each of the UN Development Decades has included main dimensions of development—economic, social, and political—their foci shifted from the trickle-down theories of economic growth to growth with equity and participatory and sustainable development. This has been accompanied by four phases in the evolution of governance: (1) traditional public administration to maintain law and order and promote national integration after independence; (2) development administration, public sector management, and NPM to improve effectiveness of government-initiated programs and projects and allow greater role of the market and competitiveness among the government entities; (3) from government to governance—to facilitate the engagement of the actors from the government, civil society, and private sector in managing public affairs; and (4) democratic governance to infuse the principles and values of democracy in the governance systems and processes.

Economic Growth Focus

By the 1960s, the importance of economic development was rising on the UN agenda, greatly stimulated by the many newly independent countries that had become or were about to become UN members. The first UN Development Decade was launched by the General Assembly in December 1961 (United Nations 1962, 10–11). It called on all member states to intensify their efforts to mobilize support for measures required to accelerate progress toward self-sustaining economic growth and social advancement in the developing countries. With each developing country setting its own target, the objective would be a minimum annual growth rate of 5 percent in aggregate national income by the end of the decade.

A notable feature of development policy, as it emerged during the 1960s and 1970s, under the influence of both structuralism (see chapter 2) and socialist-oriented models of development, was the need for the state to play an active role in the development process by orchestrating wide-ranging interventions in the economic sphere aimed at accelerating the pace of development. Indeed, this had become an important article of faith not only of the developing countries, but the concept of a developmental state was also embraced by international development institutions, such as the World Bank, as well as the regional development banks (Streeten et al. 1981). The emphasis on the role of government in the development process was also reflected in the formulation of five- and ten-year development plans that were premised on government intervention aimed at stimulating and regulating investment, particularly in the industrial sector, designed to achieve higher levels of growth.

Pairing Social and Economic Development

Despite these efforts, the UN Secretary General acknowledged at the midpoint assessment of the First Development Decade that only limited progress had been made, with the ultimate goals still distant. In light of this shortcoming and in recognition of the growing needs of the developing countries, the objectives of the Second Development Decade were to promote sustained economic growth, ensure a higher standard of living, improve social services, and facilitate the process of narrowing the gap between developed and developing countries.

Social development was conceived as best pursued if governments actively promoted empowerment and participation in a democratic and pluralistic system respectful of all human rights and fundamental freedoms. Efforts to sustain broad-based economic growth have to reinforce the promotion of social development. It was also declared that processes to promote increased and equal economic opportunities, to avoid exclusion and overcome socially divisive disparities while respecting diversity, are also part of an enabling environment for social development (Ishikawa 1967).

International Economic Order and Human Development

In the new international development strategy adopted by the General Assembly for the Third Development Decade, beginning on January 1, 1981, governments pledged to fulfill their commitment to the concept of a new international economic order based on justice and equity. They agreed (though compliance was checkered at best) to subscribe to the goals of the strategy and to translate them into a coherent set of interrelated policy measures. The concreteness of the strategy was indicated by such targets as reaching, by the year 2000, full employment, universal primary school enrollment, and life expectancy of 60 years as a minimum, with infant mortality rates no higher than 50 per 1,000 live births.

Sustainable Development

The concept of sustainable development was a major milestone in rethinking development. The 1987 Brundtland Commission defined sustainable development as "development that meets the needs of the present without compromising the ability of future generations to meet their own needs" (United Nations 1987, 154). The objectives of sustainable development have been summarized as a virtuous cycle with five components: (1) promoting equality, (2) improving the quality of life and well-being, (3) sustaining natural resources—along with sustainable jobs, communities, and industries, (4) protecting human and ecosystem health, and (5) meeting international obligations (Khator 1998).

Compared with the strategies of the first four Development Decades, which clearly emphasized the importance of economic growth in the overall development equation, the Strategy for the Fourth Decade (1991–2000) placed much greater emphasis on poverty eradication and social development. Quite significantly, it stated that while growth is desirable, developing countries need not wait for high per capita income to do away with poverty (Labini 2000). Moreover, notwithstanding the fact that during the 1990s the United Nations sponsored a series of global conferences on issues such as human rights, population, gender, and social development, it has failed to recapture its previous holistic conception of development. Indeed, the MDGs summarized in chapter 2 fall within a social development universe rather than within the narrower development conception that the UN system had previously embraced. Qualitative concerns replaced older quantitative ones as development projects moved from absolute to relative measures of poverty.

Human Development Reports

One UN initiative deserving special comment is the publication of the annual UN Human Development Report (HDR) (UNDP 1990). Published since 1990, the HDR seeks to advance the concept of human development as an overarching philosophy. It posits a substitute perspective on development, compared to the traditional growth-oriented development models based on per capita GNP.

At the heart of the original HDR is the Human Development Index (HDI), a composite index comprising three elements: life expectancy, educational attainment, and an income element, measured in terms of purchasing power parity to control for the costs of living across countries. Over the years, a number of other indices have been added to the original HDI, including a human poverty index, a gender-related development index, a gender empowerment measurement, and a technology achievement index, intended to measure technological progress based on four criteria (indigenous creation of technology, diffusion of old technology, diffusion of new technology, and human resources development).

The HDI, as articulated by Mahbub ul Haq and Amartya Sen, is underpinned by a new conception of development as a process of enlarging people's choices and ensuring greater freedom. Successive HDRs over the 1990s broadened further the development agenda by exploring what a human development approach would mean for a number of priority areas: the concept and measurement of development, development financing, human security, women's equality and gender, economic growth, poverty, consumption, globalization, deepening democracy, and human rights. Although various international organizations had focused on each of these issues previously (e.g., UN agencies, funds, and programs dating back to the 1960s or even earlier; the World Bank with its explicit commitment to poverty alleviation and its office on women in development in the 1970s), each of these was regarded not as an add-on to orthodox economic development but as part of a more integrated conception of human development, resting on fundamentally different foundations (Sen 2000). This concept of human development emphasized specific principles of people-centered governance, that is, accountability and transparency, access, social justice, human rights, and inclusion in the political and economic processes. The *National Human Development Reports* are published annually in over 100 countries. The World Bank programs have built on Robert McNamara's redirection of the Bank to poverty alleviation in the 1970s; they have incorporated participatory mechanisms for developing country assistance documents, governance indicators for estimating the magnitude of loans and grants that can be constructively used, and conditional cash transfer programs that address children's education and health in addition to straightforward increases in income (see chapters 2 and 3).

Toward the end of the 1990s, developing countries had taken steps to liberalize their economies and integrate them into the world economy. Against that backdrop, international attention was focused on the benefits of globalization and the growing interdependence in the world economy. The focus had shifted to a number of institutional preconditions for development, including good governance, transparency and accountability, decentralization and participation, and social security. When the UN General Assembly adopted the Millennium Declaration in 2000, the goals and targets it set in the section on development ultimately became known as the MDGs. However, the MDGs were not part of a new agenda, but an attempt to refocus years of debate, efforts, and struggle to advance the economic and social development of the world's poorest nations. The MDGs represented a more concentrated attempt to bring together all those activities, undertakings, and initiatives in a common focus, underlining their interrelationships and the need to make progress on all of them in order to succeed in any one of them.

Governance-Poverty Linkage

The HDRs promoted the idea that governance is the necessary condition for people-centered development. However, the contents of what is regarded as governance and development administration have been changing with shifts in development thinking. During the early post-WWII period, it was widely recognized that many of the problems that developing countries were facing at the time resulted from deficits in state capacity. This refers to poorly managed public institutions; inadequate public sector human capacities in terms of knowledge, skills, motivation, and commitment; inability to collect and manage public financial resources; and lack of knowledge, innovation, and technology strategies. With the evolution of development theories, scholars and development practitioners began to recognize the need for the capacity of the state to create an enabling environment for private sector development and for the full participation of civil society in policymaking processes.

The focus on human development and poverty alleviation required a holistic and multisector approach, because poverty is related not only to employment and income but also to access to basic public services. Access to safe drinking water and sanitation facilities is vital and requires efficient water management. Access to and the promotion of universal education, especially for women, is crucial in giving people the means to emerge from poverty and requires that the state administer these services efficiently. Access to health services is another fundamental ingredient in the fight against poverty in which the state has a pivotal role to play. In brief, how the public sector is structured, administered, and operated; what policies and programs are adopted; and the extent to which civil society is engaged in development processes have a great impact on people's well-being (UN Development Programme 2002).

Human development and poverty alleviation are facilitated by effective and transparent legislative bodies that can represent the demands of the citizens and check the powers of the executives through oversight and participatory local government systems that can facilitate active engagement of community-based groups. Effective local administration and community-level mechanisms are necessary conditions for the implementation of universal primary education.

To sum up, as the contents of the theories of development changed over the years, so did the concept of governance and development administration—from a technocratic, public administration focus to a broader concept encompassing the participation of all actors to achieve economic and political development. Hydén (2011) identified six shifts in the global discourse about governance: from technical/managerial to political aspects, global to country level, numerical indicators to narrative trajectories, quantitative to qualitative methods, top-down to bottom-up approaches, and representative to monitory aspects of democracy.

From Public Administration to Democratic Governance

With the paradigm shift of the theory and practice of development, the concept of public administration has gone through four interrelated phases: traditional public administration, development administration including public sector management and NPM, governance, and democratic governance. This section presents the evolution of theoretical perspectives.

Traditional Public Administration

After WWII, policymakers and scholars of development focused on the maintenance of law and order, the provision of basic infrastructure and services, and the promotion of national integration. This led to a strong role for the state and government bureaucracies. It also entailed centralized, legal-rational, and hierarchical bureaucratic systems. The separation of the political from the administrative was the key. Because of the prominence of Wilson's advocacy and implementation of these ideas, and the more general appeal to scientific management, public administration was considered the most efficient mode of organizing the public and the private sectors during the industrialization era in the developed world. This approach argued for insulating administration from politics, the need to improve efficiency, and the importance of improving public services through training civil servants and merit-based assessments (Wilson 1887, 74).

Thus, core elements of traditional public administration were deemed as suitable to achieve economic growth, with social and economic development as the core development objective. Furthermore, the provision of basic social services and the promotion of national integration by curtailing centrifugal political forces required apolitical public administration structures and processes. Neutral public administration structures and processes for day-to-day interactions with citizens were regarded as essential to promote trust in government.

Development Administration

Faced with the mammoth tasks of nation building, newly independent states in Asia and Africa needed a civil service with the skills to undertake core development functions, including the generation of policy and program alternatives in different sectors, program planning, implementation monitoring, and service delivery and access. Public administration institutes were established in most of these countries, with technical and financial support of the donors, including the United States and the United Kingdom, to introduce management skills and tools that had been successfully applied in these countries. The Development Administration Group of the American Society of Public Administration and the UN Division for Public Administration initiated technical assistance programs to strengthen the capacity of these new institutions.

It also became apparent, both in developed and developing countries, that the Wilsonian conception was seriously flawed in assuming that policymaking and its implementation are separable. Administrators have to determine which laws and regulations are appropriately invoked for each case; they make policy in each instance and establish precedents for future cases. Countries adopting the British practice passing minimalist skeleton laws leave even more discretion in administrator's hands, and unstable governments leave the public administration with even more responsibility to govern in the midst of instability. Thus, if public administrators are policymakers as well as just implementers, their training ought to include policy analysis to be able to anticipate the consequences of their decisions. It is striking that the premier US public administration programs that traditionally have played a major role in training civil servants of developing countries, such as Syracuse University and the University of Southern California, have been converted into public policy schools.[3]

Closely related to development administration, the public sector management approaches emphasized people-centered development processes. It dealt with such aspects as leadership training and development, team building, and motivation. It also addressed macro-organizational issues such as agency structures and interagency arrangements. Theories and practices of public sector management could be divided into two sets of issues: (1) intraorganizational issues, such as civil service reform, accountability, institution building, management training, and strategic management; and (2) interorganizational issues, including decentralization and the engagement of civil society and the private sector in government-initiated programs (UNDP 1995; Blunt and Jones 1992).

The post-1980 NPM paradigm addressed the limitations of the top-down, line ministry approach. In light of interagency rivalries resulting in mutual undermining rather than healthy competition, and the absence of incentives for hard work or creativity, the NPM paradigm called for smaller, more autonomous mission-oriented units, greater flexibility for innovation, deliberately structured competition among agencies, and use of incentives for public administration personnel.

The NPM professed to respond to the challenges posed by the top-down forms of governing, in both political and economic fields. It involved reorganizing public sector bodies to bring their management, reporting, and accounting approaches closer to business methods and making them more competitive, efficient, and effective (Dunleavy and Hood 1994, 9). Proactive managers as opposed to reactive bureaucrats were favored. Accountability as compliance to preset rules was substituted with accountability based on performance management toward benchmarks of success (Rondinelli 1990, 43–59). The boundaries between the administrative and the political, and the public and the private, became blurry as governments undertook public-private joint ownership and management of certain public services. Centralized bureaucracies were foregone to give preference to smaller, decentralized units with direct links to citizens, who were now perceived as consumers of public services (Falconer 1997). The six core characteristics of NPM, based on the business model of governance, were enumerated as: (1) productivity, (2) marketization, (3) service orientation, (4) decentralization, (5) policy orientation, and (6) accountability for results (Kettle 2000, 59).

From Government to Governance

Since the early 1990s, the concept of governance has been predominant in the discourse on local, national, and international development. The shift from development administration to governance is based on the premise that three sets of actors are engaged in the public policy process—state actors are responsible for creating enabling legal and political environments; private sector actors create jobs, income, goods, and services; and civil society actors facilitate political and social interactions. Thus, the governance concept comprises complex processes, relationships, and institutions through which citizens and groups articulate their interests, exercise their rights and obligations, and mediate their differences. The key elements of governance include public administration inclusive of the civil service, systems and processes of partnerships between different levels of government and administration, legal and judicial systems, parliamentary processes, and the mechanisms for the engagement of private sector and civil society to achieve national objectives. It should be added that the process of governance is not always orderly but one of accommodation, resistance, conflict, and assimilation.

Theories of bureaucratic corruption influenced the doctrines that emerged in the conceptual transition from government to governance. These theories converge on one central point: corruption increases where there is administrative discretion without accountability and control. There are, however, different theories about determinants of corruption (Halim 2008, 236–237). The "informal control" theory posits that institutions within the bureaucracy, including the presence or absence of merit-based recruitment, explain the degree of corruption. The "formal control" theory emphasizes the role of political institutions, including electoral and parliamentary bodies and checks and balances of power, in combating corruption and ensuring the accountability of politicians. The cultural control theory focuses on the role of civic engagement in reducing corruption through citizens' responsibility to press for honest government and bureaucrats' awareness of their obligations. The intertwining of factors makes it impossible to determine which theory is correct even for a single case, and the cross-national comparisons deny the possibility of knowing how much corruption undermines meritocratic recruitment, civic engagement, and political institutions. For example, one cross-national statistical study concluded that formal control "via electoral accountability and judicial efficacy" constrains bureaucratic corruption, whereas informal control and cultural control do not. Yet, uncertainty about causal directions renders the results only "suggestive" (Halim 2008, 253–254, 244).

With the recognition of the concept of governance as necessary to achieve development objectives, the advocates of decentralization argued that decentralization could help accelerate economic development, increase political accountability, and enhance public participation in local economic development. Furthermore, it is argued, decentralization could help break bottlenecks in hierarchical bureaucracies and assist local officials and the private sector to cut through complex procedures and get decisions made and implemented more quickly. Under certain conditions, decentralization could increase the financial resources of local governments and provide flexibility to respond more effectively to local needs and demands (Cheema and Rondinelli 2007). In the broader context of governance, those who promote decentralization see it as increasing the capacity not only of local governments but also of the private sector and civil society organizations to extend services to larger numbers of people. It was also seen as giving greater political representation to diverse political, ethnic, religious, and cultural groups without destabilizing the state. In addition, many proponents now see decentralization as an instrument for building institutional capacity within local governments and NGOs to achieve the MDGs and improve the chances of successfully implementing propoor policies that depend on local communities taking ownership of poverty alleviation programs.

However, governance experts recognize that several factors can undermine the gains from decentralization (Hydén 2007; Rondinelli 2007; Wekwete 2007). The subnational jurisdictions may lack the technical capacity to provide certain goods or services efficiently. Some goods and services may be provided more efficiently by taking advantage of higher level governments' economies of scale. Resource control by wealthier jurisdictions may reduce the overall equity of receiving state benefits across the country. The central government may put the burden of provision of goods or undertaking certain services onto lower levels of government without transferring the budget resources or the capacity to collect the resources through taxation or other means; underfunding can undermine the standing of subnational governments held accountable for providing the goods and services. Subnational authorities may be more corrupt than national officials. Placing responsibility for regulation in the hands of subnational governments may worsen the negative impacts that one jurisdiction imposes upon another, such as downstream water pollution. The current thinking is that different goods or services can be provided best by different levels of government, best in terms of efficiency, equity, accountability, responsiveness, and minimizing negative externalities (Cheema and Rondinelli 2007).

Democratic Governance

The current focus on democratic governance embraced by Western nations, some developing nations, and enshrined in the bulk of constitutions worldwide brings together two dimensions of managing public affairs: the interaction among actors (i.e., governance) and universally recognized core values (democracy). Governance is a means to an end. Democracy is both a means to an end and an end in itself. Democracy entails broad participation, rule of law, transparency and accountability, representation, and a system of checks and balances (Cheema 2005). Democratic governance entails the translation of these principles into tangible practices and formal institutions—such as free, fair, and frequent elections; a representative and accountable legislature to make laws and provide oversight; and an independent judiciary to interpret and enforce laws. They also translate into prosecutions for violations of human rights and the rule of law. Good democratic governance also decentralizes *some* authority and resources to local governments to give citizens a more direct role in governance. Finally, good governance ensures that civil society plays an active role in setting priorities and making known the needs of the most vulnerable. Thus, governance is good if it supports a society in which people can expand their choices in the way they live; promotes freedom from poverty, deprivation, fear, and violence; and sustains the environment and women's advancement.

The evolution of democratic governance doctrines has been influenced by liberal democratic theory. A liberal democratic political system is defined as one where regular elections determine the legislative and executive power of the government; there is the protection of civil liberties by law and constitutional safeguards that are impartially enforced by the judicial system; and the powers of and relationships among branches of government and public offices are specified (Wintrop 1983, 83–132; Dryzek and Dunleavy 2009, 18–32). The shift from governance as simply effective administration to democratic governance reflects the changing dynamics within developing countries that are presently electoral democracies. Some pressures for the shift to democratic governance have emanated from internal factors in developing countries, such as higher education and income levels and a more active civil society. These are reinforced by external factors, including the need for the rule of law in the context of globalization, and external pressures from donor community, including foreign aid conditionalities and democracy assistance programs (figure 4.1).

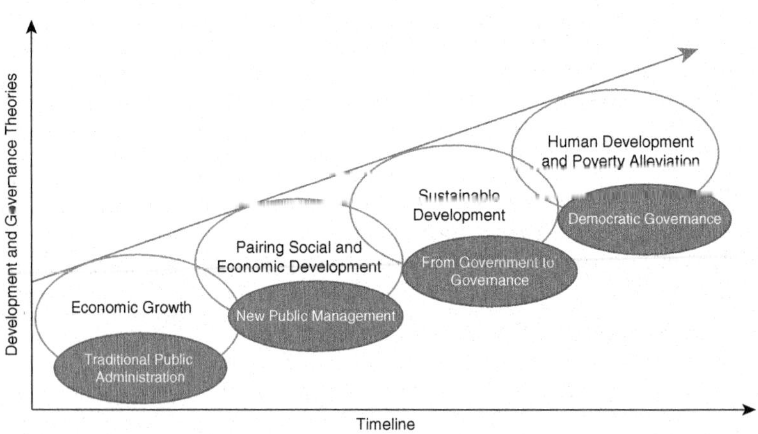

Figure 4.1 Shifts in the paradigms of development and government.

Concept-Driven Analysis: Definitions across Disciplines

This section describes the striking convergence of thinking about governance across multiple academic disciplines, demonstrating that there is now far less disagreement about what should be done to improve governance, though accomplishing the improvements is, of course, another challenge. Currently, the scope of what is considered governance can be as broad as all dimensions that shape who gets what, when, and how—as an umbrella term covering different schemes of collaborative efforts, policy entrepreneurship, and participatory initiatives (Duit and Galaz 2008). A more substantive definition is the sum of ways that individuals and institutions in both public and private spheres manage their affairs (Commission on Global Governance 1995; Knio 2010). Some focus on an instrumental conception of governance such as the method of public sector management, including financial accountability and the creation of internal and external control mechanisms (Thirkell-White 2003; Hout 2010). Many scholars have approached the concept from the perspective of a decision-making process characterized by interorganizational relationships and dependencies requiring multilevel and multiorganizational responses to societal problems (Lasswell 1971; Kooiman 1993).

Discipline-oriented definitions can be grouped along the following dimensions:

- Public administration sees governance as efficient, effective, and citizen-centric public sector management and service delivery based on capacity development and toward transparent, accountable, participatory, and inclusive state-society relations.
- Public policy views governance as the process, institutions, and distribution of authority that shape policy outcomes. As the most comprehensive public policy framework, the policy sciences (Lasswell 1971) note the political and technical issues involved in each of the functions involved in policymaking and define the ultimate humanistic objective of human dignity as the broadest shaping and sharing of valued outcomes (Lasswell and McDougal 1992, 720). This framework, dating back to the 1950s, anticipated both the equity and the participatory goals.
- Political science views governance from the perspectives of regime type and quality, particularly vis-à-vis democratization, and the issues of sovereignty of the nation-state and global governance when it comes to international relations.
- Economics emphasizes the relevance of transactions costs in how governance is structured (more details and implications are provided in chapter 2).
- Sociology applies governance to social movements of newly emerging power groups, such as NGOs, civil and noncivil society activists, and grassroots organizations.
- Anthropology and philosophy offer perspectives on governance via their focus on power discourses, subjectivity, and governmentality.[4]
- Business administration uses governance to understand and improve the running of corporate entities, organizational departments, and sectors based on open and visionary management.

Public Administration and Public Policy

The public administration literature views governance as the art of directing, guiding, and regulating individuals, organizations, and nations. When applied to public administration and public policymaking, it encompasses ordered rules of the government, its exercise of authority, patterns of decision making, and collective action undertaken along with its partners. In this understanding, governance is a statecraft with government as its principal actor (Robichau 2011, 116). The nature and scope of action assigned to government as opposed to nongovernment actors in the exercise of this art have been discussed extensively. As the central theory of public administration moved from old public administration to NPM, and from there to public governance or new public governance, these debates around the nature and scope of government's role within governance have shifted with significant consequences for practitioners (Osborne 2010, 9).

The traditional (Weberian or classic) public administration advocacy of centralized, legal-rational, hierarchical bureaucracy sees bureaucratic routine and impartiality as crucial advances of modernization beyond the ad hoc favoritism exercised by many governments (Gerth and Mills 1946). Accordingly, rule-following bureaucrats execute the orders received by their superiors to whom they are accountable, based on a set of clear rules and regulations flowing from public law where the separation of the administrative from the political is the key aspect (Pfiffner 2004).

In this conception, governance denotes legal frameworks through which contracts are negotiated, monitored, adapted, enforced, and terminated. Its major characteristics are (1) an apolitical civil service, (2) hierarchy and rules, (3) permanence and stability, (4) institutionalized civil service, (5) internal regulation, and (5) equality (Peters 2001). As early as 1925, scholars studied the best methods to make comprehensive empire-running policies in Great Britain, pointing to a series of resolutions as governance steps (Guest 1925). This legalistic approach gradually yielded to a more organizational perspective on governance. The variables of agency judgment, planning personnel's commitment to the implementation of the project, and more general social conditions such as crime and health hazards pointed to the growing importance of positivism and behavioralism in the 1960s.

From 1948 until approximately 1970, public administration equated effective administration with the state, which was seen as the most important engine of socioeconomic development. By the 1970s, given the fiscal and financial bottlenecks at both national and international spheres, drawbacks of this state-dominated and administrative approach became apparent (UN Department of Economic and Social Affairs 2009). Then the focus shifted from administration to management, and reducing the size of the state became the goal of almost all policymakers. The new term governance was explicitly used by Cleveland (1972) to denote the horizontal bonds of consultative forms of governing.

Following Cleveland and the incipient NPM wave, governance was increasingly used synonymously with this new paradigm of public administration. Progovernment scholars favor the state-centric approach of the traditional public administration paradigm, maintaining that despite fiscal belt-tightening, government still occupied the central role. They regard the diatribe against the primary role of government as misleading, because governance cannot take place without government. The progovernance scholars are more society-centric in emphasizing the roles of nonstate actors in the provision and delivery of services. For them, governance is a new form of government. Only a few scholars position themselves between the two camps, creating the state-centric relational path, maintaining that the state was not governing less but differently (Robichau 2011, 120). These middle-ground scholars prefer more in-depth analysis of new tools and modalities of state administration, as opposed to joining the debate over the primacy of administration versus the markets, or government versus governance.

The "4P approach" to governance (public-private-people partnerships) and the increasing disenchantment with the Washington Consensus led scholars and policymakers to look for new and more eclectic understandings of governance. Asian policymakers talked about the need to maintain a healthy balance between the invisible hand of markets and the visible hand of good governance. The solution was not a return to old public administration and its sole reliance on state, but a more balanced approach between the heavy touch of central planning and the light touch of laissez-faire markets (MacAuslan and Addison 2010). These evolving debates and the continual demands of civil society groups for the broadest possible humanistic agenda and the broadest participation by civil society[5] exemplify challenges to the conventional public administration approach. The basic premise of 4P or network-based governance is collaborative networks and partnerships formed around policy issues (Stoker 2006). In this system, governance is no longer dominated by a central government or free markets, but is co-produced and co-shared along with NGOs, businesses, and citizens (Castells 2006, 16) for the creation and implementation of policies using negotiation, persuasion, and collaboration.

Using the concept of regime, governance can be viewed as resting on the development of shared expectations about acceptable behavior: the convergence to "sets of principles, norms, roles, and decision making procedures" (Frederickson 2004, 18). The regime concept focuses attention on how governance must create cooperation to overcome the collective action problems of reluctance to provide public goods when other actors are not trusted to cooperate. Achieving the common good is feasible if government agencies fulfill the expectations of responsiveness, interest groups behave responsibly within reasonable legal limits, and NGOs act according to the norms of constructive engagement rather than obstruction. This depends on instituting and sustaining regimes aimed at establishing shared expectations about desired behavior. However, this broader range of goals accounts for the tensions that arise in the purposes and design of governance institutions. Hydén (2011, 8–10) points out the tensions between the concerns over effectiveness—the traditional preoccupation of public administration—and the concerns over participation and legitimacy: the distinction between getting things done, to be judged by *accomplishments*, vs. the *process* dimensions that encompass the politics of the policymaking. Yet a crucial intersection of these concerns has been embraced by international development agencies: inclusive socioeconomic development, sound public finance management, building democratic institutions, promoting social justice, and strengthening human rights.

Comparative Politics and International Relations: Dimensions of Democratic Governance

In political science, governance has been used to refer to democratization, emphasizing the principles of participation, transparency, accountability, rule of law, effectiveness, and equity (Cheema 2005). For Hydén (2011), democratic governance goes beyond the implementation of specific policies to encompass the normative framework—the regime—within which policies are formed and executed. To illustrate the integration of the political and technical aspects of democratic governance, one only has to recognize that transparency and accountability to the public require strong technical capabilities to monitor, assess, and communicate.

However, other elements of good governance go beyond democracy and efficiency, to preclude the risks of majoritarian democracy dominating minorities or imposing controversial moralistic strictures that impinge upon individual rights. Because both representative and direct majority pose these risks, inclusive democratization must strike a difficult balance between protecting the rights of the privileged—after all, secure property rights are key to both economic growth and personal security—and concern for the less privileged (Hydén 2012).

In contrast, Bang and Esmark (2009, 20–23), addressing the increasingly complex and contentious issues facing the European Union, point to the danger that if good governance is taken to mean technically effective policies and their implementation, democracy may be rejected out of the frustration that politics is getting in the way of sound decisions. This leaves open what constitutes a sound decision unless interests are articulated in an open, democratic process. Good democratic governance accepts "policy-politics," but it is best when based on negotiation, communication, trust, and solidarity (Castells 2006, 17). The attack, then, is not on politics per se but rather on exclusionary, poorly informed, excessively confrontational politics, often leveled at the formal party systems and politics. The assumed shift from democratic to good governance is one from decisions to action, politics to policy, input to output, and from legitimacy to the more specific concept of legitimacy as policy acceptance by citizens.

On the other side, international relations theory has examined governance from the perspective of state sovereignty and the changing role of the nation-state in the international system. Global governance has emerged to denote dissatisfaction with the conventional realist and liberal institutionalist theories dominating international relations theory from the 1970s to the end of the 1980s. Accordingly, global governance has been defined as the fragmentation and integration of societal interactions and authority patterns (Weiss 2000) or as the "burgeoning information, communication, market, finance, networking, business activities" (Finkelstein 1995, 368). Initially, it was equated with the increasing impact of nonstate actors, leading to the realization that the international system is no longer dominated by nation-states.[6] Later, more elaborate definitions of global governance evolved as a system of rule at all levels of human activity with transnational repercussions, or as a multilateral system of policy domains from the subnational to the transnational converge via the multiscalar relational organization of networks of governance (Weiss 2000; Campbell, Kumar, and Slagle 2010). In international relations, governance is essentially considered as the shared understanding of regime behavior, which entails managing principles, norms, and decision-making procedures that facilitate the maintenance of an international order.

Sociology: Civil Society and Social Movements

With the end of state-centric conceptions of power in the 1970s and the growing distrust of citizens in their government institutions and representatives, scholars and practitioners increasingly turned to civil society as a new actor needed for democratic and effective governance. Of course, this reflected the growing role and activism by nongovernmental actors, including civil society and citizen groups as well as diverse social and protest movements. The shift in focus from management, efficiency and performance to everyday societal problems faced by citizens as the nucleus of governance issues has gone in tandem with sociological studies in governance. In sociology, governance was defined as the newly emerging models of joint and continuous action by social actors from diverse milieus acting to influence policies in line with their shared values (Swyngedouw 2005).

In this sociological perspective, ideal governance arrangements are based on: (1) horizontal interaction among equal participants without distinction between their private and public status; (2) citizen-led initiatives to empower the poor and the disadvantaged; and (3) learning by doing and by discourse through regular and iterative exchanges among interdependent actors. In this process, civil society and social movements have been the two leading actors in forming networks of governance among themselves and with the state and the private sector.

As is apparent in the long-standing dichotomy of the liberal tradition of individual autonomy and the republican tradition calling for the state to lead the effort to establish civic and ethical responsibility, the controversy revolves around whether a vibrant and healthy civil society can transpire without a strong and facilitative state structure in place (Skocpol 1996). Others are zealous advocates for an independent and autonomous civil society (Hydén and Samuel 2011). The state-centric civil society scholars are also structuralists in maintaining that organized civil society cannot flourish apart from an active government genuinely interested in the strengthening of civil society through inclusive democratic politics. The structuralists find supporting evidence for their argument in US political history, where the openness of the Congress and state legislatures in organizing petition drives, the spread of public schooling, and the establishment of US post offices in towns and cities were essential in building and strengthening civil society organizations. In the opposite camp, some political scientists and sociologists cite examples from the Latin American and Eastern European experiences to flesh out the pernicious impact of authoritarian governments on civil society, and how the latter was formed and grew independent of, and sometimes despite, the state. The roles and activities of mass organizations in transitions to democracy, such as the trade union movement against the Uruguayan dictatorship and Poland's Solidarity movement, are examples of this second new social movement that scholars draw on (Blind 2009).

Scholars who have emphasized the increasing partnership between civil society and markets have proposed a third approach of reinventing the state. Reinventing the state meant conceiving government as a catalyst and facilitator between all sorts of actors and in finding solutions outside the public sector. In this approach, government should not necessarily privatize or work to reduce the state at all costs. Nor should it simply follow a recipe of redistribution of wealth toward the mitigation or elimination of inequity. Government should do what it does best: raising resources and setting societal priorities through a democratic political process while creating the conditions auspicious for the private sector to organize the production of goods and services. In that process, government should be both customer-driven and community-owned. Customer-driven government means an open, transparent, and accountable government. Community-owned government signifies an inclusive, participatory, and engaging government. A government that is catalytic, customer-driven, and community-owned is also holistic. It values entrepreneurship, markets, and the third sector composed of voluntary associations, civic communities and non-profit organizations (Osborne and Gaebler 1993).

Social movements and mass-based nonprofit organizations have also contributed to civic engagement and democratic governance. Globalization and the emergence of new social movements asking for greater political and social rights and freedom have gone hand in hand (Tarrow 1994). In their new role as channeling agencies of international aid, many NGOs are now representing interests, building participation and checking the state all at the same time. In other words, they are doing what political parties do to seek power. Special interest groups gradually replace civil society organizations, raising the dilemma of whether civil society should be nonpartisan or not.

Social movements like Brazil's Rural Landless Workers Movement, the Unemployed Workers' Movement of Argentina, the Zapatista National Liberation Army, and Mothers' movements, although sporadic and at times chaotic, have also contributed to both the building of social capital among the clusters of their members and the building and consolidation of democracy, as have labor unions in France and burial societies in South Africa. The impact has considerably increased as these movements have been interlinked with each other and with other societal actors, such as NGOs, women's organizations, and academic research institutions. Through the UN World Summit for Social Development in 1995, Social Watch brought together these movements and organizations through a network architecture of social watchers who mobilize societal action in the field.

Michel Foucault, a philosopher and historian of ideas, has coined a new term to describe the changes noted in governance. Governmentality, a broad term denoting the expansion of government and sources of governance, along with the set of beliefs about government and governance, entails a particular rationality of governing combined with new technologies, instruments, and tactics of conducting processes of collective rule-setting, implementing, evaluating, and monitoring (Foucault et al. 1991). The governmentality approach stresses the implicit irony that the socially innovative forms of governance were encouraged and supported by the very same agencies that have pursued a neoliberal—market-dominated—agenda and that they are still operating within a broadly liberal political economic order (Swyngedouw 2005).

A parallel vocabulary within this same field is humane governance, defined as a type of human rights-oriented governance that supports the creation of participatory, responsive, accountable polities embedded in competitive, nondiscriminatory, yet equitable economies (Ul Haq 1999). Components of humane governance were enumerated as broad ownership, decency, accountability; its objectives are freedom, genuine participation, and sustainable human development.

Conclusion

From the 1960s until the 1980s, central state administration based on legal-rational bureaucracies was believed to be the engine of growth and prosperity. The decade from the mid-1980s to the mid-1990s, dismantling centralized administrations, top-down management, and state enterprises, was the governance reform parallel to the economic liberalization reforms. Competition among agencies was encouraged, in keeping with the strong market orientation of the period. This was largely focused on efficiency in accomplishing goals set in the conventional representative policy process. Starting with the mid-1990s, a gradual shift occurred in foreign assistance approaches, away from the priority of efficiency to a greater sensitivity to governance concerns like human rights, legislative support, judicial reform, and corruption. The concept of good governance, at both turning points, acted as the tipping point. It reversed the state-dominated socioeconomic development patterns in the 1980s. It then reversed the hegemony of the markets in the mid-1990s. Then, it gradually coupled its socioeconomic agenda with political leadership and democracy followed by foci on human rights, human dignity in general, and accountability. For foreign assistance approaches, governance has thus taken on humanistic qualities, increasingly abandoning its more mechanistic face as aid conditionality.

The common debates and concerns of policymaking and administration across disciplines in academia show that governance emerges as the interdisciplinary bridge linking the different variables of interest embraced by each discipline. It does so through a genuine focus on the "how to govern" question, but also with a parallel concern for broad sharing of benefits. In this sense, governance can be defined as the willingness and ability of governments, along with its national and local partners, to respond effectively and legitimately to citizen needs, wants, wishes, and concerns.

As we have seen, over the past few decades, academics have been making additional contributions to our understanding of the governance concept. However, the practice of governance has moved ahead of the theoretical analyses. Faced with daunting social and economic problems—including inadequate access to services, rising income disparities, and exclusion of the poor from the political processes—policymakers in developing countries, international development partners, and civil society organizations have been experimenting with new doctrines and tools to promote greater citizen engagement for an inclusive and responsive state. This has led to a series of proposals of governance best practices and the development of innovations at the national, subnational, and local levels. Governance today involves deliberative efforts and multilevel solutions from myriad actors. It includes boundary-spanning policy regimes with a multitude of actors collaborating around their needs- and rights-based issues. Governance, as such, can best be described as an interdisciplinary bridge, while its myriad context-dependent practices can safely be likened to a panoply of statecraft. The actual practice of governance in the developing world over the past few decades is the focus of the next chapter.

CHAPTER 5

Evolution of Governance and Development Administration Practice

The feasibility of applying abstract theories and doctrines of governance and administration to the complex realities, limited resources, and political polarization of many developing countries has generated various approaches and strategies that are frequently at variance with the classic conceptions and aspirations of the theories. Democratic rhetoric has often been accompanied by autocratic rule. Principles based on idealized conceptions of development management have been replaced by more pragmatic approaches to cope with the messiness of bureaucratic politics, local power structures, and administrative shortcomings.

The experience in developing countries suggests that practices of governance and development administration, driven by economic and political pressures, often were privileged over the theory and doctrines. There is great diversity in the nature of development institutions in developing countries. Yet, important trends are discernable. In particular, the structures and practices of foreign assistance agencies have undergone significant changes. This led to variable results, including uneven forms of development, inefficient pooling of resources, and chronic indebtedness.

Governance strategies and practices in developing countries have been influenced by two sets of factors: (1) internal factors, including greater access to information by citizens and demands for greater accountability and participation, and (2) external factors such as donor conditionalities and changes emanating from rapid globalization, including information and communication technologies and greater movement of people, goods, and services.

This chapter examines shifts in the strategies and practices of governance pursued by three sets of stakeholders: governments in developing countries, civil society organizations (CSOs), and donors and international organizations. It discusses reforms of public administration structures and processes and reviews how practice has dealt with the evolving concepts of public administration, NPM, governance, and democratic governance, including civil service reform, decentralization, electoral and parliamentary reform, and the engagement of civil society. The chapter also discusses emerging trends and issues in the practice of governance and the ways in which wide gaps between theory and practice are being bridged.

Reform Strategies by Governments

The evolution of governance doctrine and practices since the 1950s has reflected the changing nature of development problems and nation building after independence of most of the developing countries. Traditional public administration, modernization of the civil service, and militarized modernization strategies reflected the need to reconcile centripetal and centrifugal forces within the political system in order to promote national integration. Centralized economic planning, administration, and political decision making seemed like logical tools to achieve these objectives. There were instances where groups representing varying mixes of economic, regional, political, and social divisions failed to reconcile the differences among diverse interest groups, leading in many instances to military interventions that blurred the line between the civilian government control and the military as a professional force (see chapter 9). The early apparent success of centralized planning led to a greater role of the state, even in countries such as India that were otherwise characterized by democratic processes and free enterprise. The Soviet model of centralized planning was found to be an attractive approach to the leaders of many countries in Asia, Africa, and Latin America to mobilize and allocate resources, promote industrialization, and provide access to basic services (see chapter 2). The establishment of independent central banks and regulatory commissions was the first necessary step in developing countries to promote economic growth and development, although progress in establishing these institutions as truly independent was stifled in many cases.

During this period, the dominant role of the state in identifying development policies and designing and implementing development programs was rarely challenged, and the role of the market in promoting jobs, income, and providing services was largely ignored. Concentration of powers and development resources led to increased political corruption by politicians and rent-seeking behavior directed at the bureaucracy, due to its excessive powers to allocate economic privileges (see chapter 3). The resulting top-down approach frequently led to the erosion of political support among citizens and an inability of government decision-making structures to address the interests of different segments of the society to promote balanced development. There were insufficient mechanisms for the people to articulate their interests through interest groups, unions, civil society, etc.

Critical Issues in Public Sector Management Capacity

Due to the dominant role of the state, governments in developing countries recognized the significance of strengthening public sector capacity, including civil service reforms, to effectively design and implement development projects. Since their independence, developing countries have traditionally faced a set of critical issues in public sector management. Permanent public administration reform entities (such as public service commissions and public services departments) within the central government apparatus were created to cope with these issues. Civil service reforms to streamline bureaucracy were identified as a priority concern. Initially these programs focused on issues of remuneration, number of employees, performance appraisal, personnel recruitment, selection, placement, promotion, and related issues (UNDP 1995, 55). Pressures on governments to provide employment often resulted in overstaffing of government departments and semiautonomous organizations, especially in unskilled and semiskilled grades—in some cases, resulting in ghost workers. Other concerns were favoritism in recruitment and promotion processes; inability of the central civil service agencies to take action against those performing at subpar levels due to political and ethnic pressures; and low salaries that led to low morale, fewer incentives to work, and moonlighting (UNDP 1995, 55–57).

Reformers also recognized other critical issues for public sector management:

- Good leadership and vision was often lacking for the essential task of mobilizing government organizations to produce quality services by identifying clear ways to overcome environmental constraints, inspire staff to achieve organizational objectives, encourage innovation, and adopt a mission-driven rather than rule-bound operation (Osborne and Gaebler 1993).
- Strategic management capacity was frequently critically weak, in terms of planning how to achieve the organizational mission and how to direct agency activities to serve the interests of stakeholders (UNDP 1995, 68).
- For virtually all developing countries, human resource management and training suffered significant weaknesses, in terms of organizational structure and culture, personnel selection and placement, training and development, job design, and performance appraisal. Recognition of the seriousness of these weaknesses has led to the establishment of management development institutes, some with considerable success such as the Malaysian Institute for Public Administration and the Eastern and Southern Africa Management Institute.
- Many public agencies have also been lax in improving efficiency, delivery, access, timeliness, and customer concern in providing goods and services.
- Part of this failure in delivering goods and services has been traced to lack of coordination, and in some cases outright conflict, among public administration units. Interorganizational and intergovernmental relations needed to be streamlined and clarified, including areas of primary and secondary responsibilities of each level of government and administration and effective horizontal and vertical coordination among agencies (World Bank 1997, 80–85).

The first phase of civil service reforms in developing countries was aimed at improvements within the existing centralized structure to upgrade the skills and knowledge of the personnel, changing the culture of the organization and the process of its internal management, clarifying missions, and improving the quality and timeliness of public services.

These reforms were considered necessary but not sufficient by policymakers and development practitioners to promote national integration and effective development management. Equally important was the need to change administrative and political boundaries and create special administrative units to bypass inherent administrative capacity deficits and overcome inertia within the administrative machinery. Special regional development authorities were established to promote regional development and planning, such as the Narmada Valley Authority in India, the Superintendency of the Northeast in Brazil, and land development authorities in eastern Malaysia. These administrative units were granted considerable autonomy, separate budget lines outside of the ministerial structures, and posting of high-performing officials in high managerial positions. The objectives were both political, to gain greater legitimacy in less developed areas, and administrative, to deliver quick results by creating autonomous decision-making structures that did not have to go through the ponderous standard processes.

Another, more micro-level strategy to bypass the standard administrative machinery of government was the establishment of program management units (PMUs), often with the support of the bilateral donor agencies, the World Bank, and regional development banks. The PMUs were created to implement such programs as the Integrated Rural Development Program and the Urban Management Program. Staff of these PMUs were recruited from among the most qualified personnel from within and outside the civil service and given higher salaries and incentives. In many cases, special procedures were identified to give more discretion and flexibility to these units to make decisions related to program implementation and monitoring. Funding from the donor community was often channeled through these units instead of the conventional administrative machinery of the government. Because of the above factors and external technical support, the PMUs contributed to relatively more effective implementation of the programs.

However, such units had several negative consequences. In some countries such as Sri Lanka, channeling more funds for integrated rural development to selected districts through the special units than through the normal machinery of government weakened the roles and capacities of nation-building departments, such as agriculture and health, and created greater fragmentation of the government machinery. Different patterns of district administration and flow of donor resources through the different procedures used by the PMUs impeded effective coordination at the subnational and local levels.

During the 1980s, the introduction of structural adjustment policies supported by the IMF, the World Bank, and other international financial institutions led to the declining role of the state and its legitimacy in providing access to services to the vast majority of the people. In the late 1990s, however, the confidence in market forces as the engines of growth and means for allocating resources eroded. The frustrations with persistent poverty and environmental degradation contributed to this decline in confidence, and the 2008 financial crisis further undermined the legitimacy of the market leading to demands for greater regulation (Hydén 2011).

A strikingly opposing view is that the motives and actions of the state, shorn of rhetoric, are not directed to determining how to enhance growth, human dignity, or any other public good, but rather to strengthen and reward the state itself. James Scott (1998) attributes state action to the imperative of government leaders to shape and control society. Some functions must be performed—taxation, monitoring of the population, general information gathering, etc.—and these imperatives dictate governance structures. Our perspective is that while power certainly covers one set of objectives of government leaders, many other values—including the commitment to enhance human dignity—cannot be so easily dismissed, especially as democratization has spread despite formidable obstacles.

Thus, the latest phase in the evolution of governance practice emanated from the rapid pace of democratization over the past three decades. Scholars and development practitioners began to assess governance from the perspectives of three interrelated doctrines—responsive and accountable governance; inclusive governance that protects the interests of otherwise marginalized people; and participatory governance that involves citizens in public policy making. They began to examine not only efficiency criteria to assess civil service but also its accountability and transparency, inclusiveness, and degree of citizen participation in articulating local demands and needs, identifying local projects, and monitoring and evaluating government initiatives. Democratic reform programs were introduced to realign powers and resources of different branches of government and relationships among different democratic institutions in the context of the constitution. These political reforms included:

- Establishment of independent electoral management bodies with sufficient autonomy and capacity to spur free and fair elections that are insulated from potential interventions by the executive branch and the bureaucracy to affect electoral outcomes (notably in Argentina, Botswana, Chile, Colombia, Indonesia, India, South Korea, and South Africa);
- Enhanced capacity of the legislature to perform important functions, such as initiating legislation (as opposed to simply endorsing executive initiatives), representation and oversight of the executive branch, including the bureaucracy (notably in Bangladesh, Ghana, and Mexico);
- Decentralization of powers and resources from the central government to subnational and local governments, including deconcentration of ministerial functions, delegation to semiautonomous government organizations, devolution to local governments, and transfer to NGOs (notably in Bangladesh, Brazil, India, and Indonesia); all require competent subnational and local authorities, structures ensuring that local elites do not capture the bulk of the benefits under their control, and restraints on the tendency of some central governments to use deconcentration to penetrate local affairs through the increased presence of officials reporting to the central government (Schou and Haug 2005, 18);[1]
- Reform programs to promote the rule of law, judicial independence, and access to justice in order to ensure checks and balances among different branches of government, check abuses of authority, and take advantage of opportunities provided by the rapid pace of globalization (notably in Malaysia, Nigeria, and the Philippines);
- Programs to enhance the enforcement capacity of government agencies and human rights institutions to make people aware of their rights, enforce national and international legislative frameworks, protect against human rights violations, and promote inclusive development policies that take into consideration the interests of minorities, women, and marginalized groups (notably Chile, India, and South Africa).

Roles and Perspectives of Civil Society

Over the past few decades, civil society engagement in the process of governance practice has expanded considerably (Cheema 2010). This increased role emanated from the inherent weaknesses of government-dominated doctrines as well as changes within developing countries that created demand for greater participation (such as the Arab Spring), inclusiveness, and accountability. These domestic pressures have been reinforced by pressures from the global community, including bilateral donors and multilateral agencies such as UN Development Program and the World Bank, to engage all segments of the society in making decisions that affect people. The work undertaken through the UN Human Development Reports, the rights-based approach to development, and the MDGs hastened the process of CSO engagement in political and economic development processes. Thus, the need for increased space for civil society has been a key component of the evolving democratic governance agenda and practice. However, the simultaneous proliferation and fragmentation of CSOs makes it increasingly difficult to assess their impact or to see them overall as barometer of change.

Definitions of civil society have evolved from two interrelated traditions—one rooted in democracy promotion and the other to development practice (Cohen and Arato 1994, ix; Diamond 1994, 5; White 1994, 379; Carothers 1999; Alagappa 2004, 32; Cheema 2010; Levy 2010). The United Nations, the World Bank, and the Organisation for Economic Cooperation and Development define civil society to include community groups, NGOs, labor unions, indigenous groups, charitable organizations, faith-based organizations, professional associations, and foundations (Cheema 2010, 3).

Social movements and activist groups have been instrumental in the global discourse on the role of civil society in governance (Cheema and Popovski 2010; Hydén and Samuel 2011). Among others, they have emphasized social issues, human rights, fair wages and working conditions, and environmental concerns. People-centered local initiatives in countries such as India (social audit), Brazil (participatory budgeting), and Kenya (citizen tribunal) provide many lessons about the potential roles for CSOs in ensuring that government is responsive and accountable through the active engagement of civil society.

CSOs are playing increasingly important and expanding roles at the local, national, and global levels to improve the quality of democratic governance, provide access to basic services, and protect the rights of minorities and marginalized groups through their advocacy functions (Cheema and Popovski 2010). At the local level, CSOs are now actively engaged in community development, skill improvements for sustainable livelihoods, and access to basic social services. At the national level, CSOs often perform a watchdog function to improve the quality of electoral and parliamentary process, promote access of the poor to justice through paralegal services, and seek the accountability of public officials by informing media about violations by public officials. In Abra, the Philippines, the Concerned Citizens of Abra for Good Governance, an NGO, investigates projects for substandard materials, poor construction techniques, and fraudulent contracting procedures. In Bangalore, India, the Public Affairs Center has established and released Citizens Report Cards that rated and compared agencies on the basis of public satisfaction and responsiveness. CSO accountability functions are not limited to direct monitoring, as shown by the capacity building of the Philippine Centre for Investigative Journalism (PCIJ). The PCIJ is an independent, nonprofit media agency specializing in investigative journalism, which provides training for investigative reporting to full-time reporters, freelance journalists, and academics. The Local Governance Barometer in South Africa was aimed at increasing citizen engagement and at assessing the impact of policy on local issues at the district level. In Vietnam, a governance and public administration index was established to assess the citizen demand for better governance performance. In India, there were many local initiatives led by citizen activists and civic movements, through such participatory instruments as social audits and watch measures, policy score cards, and budget monitoring at the local level. Finally, the Philippines Report Card on Pro-Poor Services, a pilot project, was supported by the World Bank to get feedback of citizens concerning the performance of government services, including basic health, elementary education, housing, potable water, and food distribution (Cheema 2010).

It is worth noting that despite the potential for rivalry between government and civil society in determining societal directions, some governments have actively encouraged a greater governance role for CSOs. The Porto Alegre experiment of participatory budgeting obviously could not have proceeded without government support. In Mongolia, the government led assessments of governance programs involving multiple stakeholders. The African Peer Review Mechanism engaged actors from government and civil society to assess governance programs and strengthen democratic process in the region (Hydén and Samuel 2011).

At the same time, assessing how well civil society undertakes its roles has also been a major task of CSOs themselves, particularly NGOs. CIVICUS in South Africa, the Philippines Rural Reconstruction Movement, the Centre for Budget and Governance Accountability in India, the Centre for Governance at BRAC University in Bangladesh, and the Brazilian Institute of Social Analysis have not only launched civil society initiatives shaping governance processes at the local, national, and global levels but have also focused on the accountability and transparency of CSOs, their capacity, their roles in democratic change, their internal codes of conduct, their advocacy roles on behalf of minorities and marginalized groups, and their role in improving access to basic social services. This introspective aspect of CSOs is key to their evolution.

The roles of CSOs have expanded at the global and regional levels as well. International CSOs are also addressing issues that transcend national boundaries such as human trafficking and cross-boundary water management. CSOs also perform advocacy functions for global public goods, such as debt relief for highly indebted and least developed countries; greater awareness of climate change; and implementation of UN conventions and treaty bodies dealing with civil and political rights, transparency in global governance, and increased foreign assistance from the donor community (Clark 2010). In 1996, the Economic and Social Council of the United Nations launched the accreditation process that enables a number of CSOs to be associated with it and thus be able to influence intergovernmental processes. In 2004, the Panel of Eminent Persons on the UN-Civil Society Relations recommended an enhanced civil society participation in various UN bodies such as UNICEF, UN Development Program, World Food Program, and the UN High Commission for Refugees. It facilitated CSO engagement in the design and monitoring for the MDGs. Today, CSOs are playing an important role in global advocacy, normative intergovernmental processes of the United Nations, and the programs of various UN entities promoting the global public goods including the environment and climate change, human rights protection, and refugees. They pose a challenge to traditional notions of state and national sovereignty with implications for governance and stability (Popovski 2010).

However, understanding the dynamics of civil society engagement in governance and its strengths and weaknesses requires an analysis of their specific contexts, histories, and patterns of growth; the legal framework under which they are established; their capacity to deliver on their mission; and their upward and downward accountability. The pattern of CSO growth in China, for example, reveals the massive influence of political orientation, socioeconomic history, cultural influences, and the forces of globalization on the sphere (Roy 2009). The new but still limited space for civil society in China emanated from the need for a new social safety net as the liquidation of many state-owned enterprises left millions of Chinese unemployed and without health care, while the state's role in providing social welfare shrunk as part of the marketization of the economic system. Chinese who have migrated from their home province are often denied the free or heavily subsidized health care in their new locations (Roy 2009).

The opening for civil society in China also grew out of the slight loosening of restrictions on political discussion, demonstrating the correlation between the degree of permitted political participation and the growth of CSOs. Unlike China, Indonesia has undergone a complete transition from authoritarianism to democracy, with CSOs playing a prominent role in promoting democratic governance. In turn, the democratic government and decentralization reforms have facilitated further growth of CSOs and embedded them in the political and social landscape.

This increased presence of CSOs of different types does not necessarily provide a united front. For example, the efforts of Latin American NGOs to coordinate across the entire region in order to formulate and implement strategies for "transforming the current reality of inequality, injustice and exclusion" (Asociación Latinoamericana de Organizaciones de Promoción al Desarrollo 2010, 5) through a highly publicized, UN-backed series of national dialogues and panregional interactions were boycotted by labor unions. The CSO leaders acknowledged that "[i]t continues to be difficult to coordinate trade unions with other social movements. Although trade unions were invited and a previous global agreement for their participation had been reached, not a single one participated (Asociación Latinoamericana de Organizaciones de Promoción al" Desarrollo 2010, 11).

Legal Framework

Another factor that has influenced the degree of CSO space is the legal basis for their establishment and the legal framework in which they operate. Malaysia is a prime example of a highly regulated legal process for the organization of civil society, such as to provide many avenues for state oversight and control. In a political system that is dominated by a coalition of ethnic-based political parties, and in which race-based affirmative action is well entrenched in the development policymaking and implementation processes, advocacy-based issues such as human rights and transparency are considered highly sensitive. The legislative framework, thus, is not an enabling one, nor is it conducive for advocacy-based CSOs. However, as had been the case of the Republic of Korea, Malaysia's rapid pace of economic development and expanding middle class has been coupled with increased demand for space for civil society (Jayasooria 2010).

Capacity of Civil Society

Experience suggests that the technical capacities of CSOs (fundraising and financial management, information gathering and research techniques, and communication skills to attract the attention of citizens) are essential for them to utilize the available space to engage in political and developmental processes. CSOs need networking skills to develop coordinated advocacy, document and adopt best practices, cultivate professionalism and trust building to educate citizens, and develop a willingness and ability to speak truth to power. Bangladesh, for example, has some world-renowned and highly successful large NGOs, but many others face a host of capacity problems, including inadequate technical and managerial skills, high levels of corruption, and weak monitoring and evaluation systems for project activities. Corruption and nepotism weaken the quality of management capacity. Many NGOs have too loose a structure, often with limited accountability to beneficiaries and are unduly influenced by donors' interests (for further elaboration, see Cheema and Popovski 2010).

Accountability of Civil Society

Many CSOs lack clear, enforceable rules that govern the ways in which their officials relate to beneficiaries. Furthermore, some NGOs are highly dependent on international donors whose program priorities can overlook or undermine the needs and aspirations of their intended beneficiaries. The Malaysian case shows the weakness of civil society in ensuring its own accountability and transparency of funding sources, utilization, and financial management practices. Often they lack published and audited accounts that can be shared publically. Moreover, few CSOs have a common position on external audits or impact assessment of their activities. No accountability or audit panel is available to certify that CSOs observe universally accepted standards of governance practices or are accountable for their funding. In contrast, Philippine CSOs have their own code of conduct to ensure their accountability (Cheema 2010).

Civil Society and Democratic Change

In some countries civil society is now playing a vital role in stimulating democratic change, indeed through multiple ways: CSOs are directly involved in all stages of the electoral process, including voter registration, voter education, and electoral monitoring. In Indonesia and India they are monitoring the performance of human rights institutions; in India they are also holding local government officials accountable for access to services. CSOs are engaged with parliamentarians and local government leaders to communicate the concerns of citizens, such as the Pakistan Institute for Legislative Development and Transparency. They provide paralegal aid and other support mechanisms for access to justice and highlight abuses of power, especially by the local elite, through media and other channels.

CSOs have also influenced democratic governance practice through the protection of rights of minorities and marginalized groups, supporting the independence of the judiciary, and holding local officials accountable to improve access to services. In Latin America a combination of Bolivian, Ecuadorean, and Peruvian CSOs and international NGOs has pushed to strengthen indigenous rights, successfully mobilizing the indigenous identity as distinct from alternative nonethnic identities such as peasant or citizen (Lucero 2008, 91–95, 147). Each of these aspects is crucial to promote and sustain democratic local governance. However, in order to ensure that they remain effective advocates of public goods, CSOs must strengthen their linkages with other organizations, address issues related to their legitimacy, and strengthen their accountability through various measures, without subtracting from their organizational flexibility and improved capacities.

The degree of citizen activism and the role of CSOs in promoting democratic change, however, is strongly influenced by the national political structures that provide them with opportunities to be actively engaged. Countries such as India, the Republic of Korea, Indonesia, South Africa, and Brazil all have relatively stable national-level political institutions that are conducive to CSO engagement at a systemic level. In countries that are still going through democratic transitions, CSOs contribute to mobilization of pressures from below for greater political pluralism (Hydén 2010).

Perspectives of Donors and International Organizations

Over the past two decades, the role of the external partners in promoting governance has expanded due to several factors (Schmitter 1996; Cheema 2005; Hydén and Samuel 2011). First, with the end of the Cold War and the failure of many of the top-down policies, developing countries have been more willing to seek the support and advice of external partners to improve state capacity and effective functioning of electoral systems, parliaments, human rights institutions, local government systems, the judiciary, and public sector management.

Second, pressures for democratic governance have emerged from within developing countries, with the Arab Spring as the latest example. With increasing access to information and economic opportunities, there is a growing citizen demand for transparent and effective governance to achieve national objectives.

Third, the United Nations and all of its associated organizations have been important global advocates of human rights, a participatory approach to development, an increased role of nonstate actors, and respect for fundamental freedoms as reflected in the Secretary General's Agenda for Peace and Agenda for Development (United Nations 1997, 66). The UN conferences and summits on key issues of global concern, including the environment, human rights, human settlements, social development, the status of women and children, and financing for development, have emphasized the central role of governance systems and institutions in promoting sustainable and human development. Specifically, the UN Millennium Summit was a landmark event leading to the UN Millennium Declaration and the MDGs. The central role of core governance functions continues to be seen as a critical component of the Post-2015 Development Agenda of the United Nations. Each of these intergovernmental processes emphasized the role of governance in achieving development objectives.

Finally, CSOs are assuming the role of advocate, watchdog, and interlocutor between the government and the people in the accountability and transparency of governance process. Through their global membership, they provide direct support to developing countries and also monitor and evaluate the donor-assisted democratic governance programs.

Despite their expanding support for governance, however, donor support has often been accompanied by the conditionalities, mentioned in earlier chapters, requiring social, political, and economic policy reforms that may not be acceptable to the government or CSOs. Donors have advanced three arguments in favor of conditionalities: development effectiveness requires efficient and transparent governance systems and process; democratic governance is based on a set of universally recognized values even though their practice could vary from one country to another; and donors need to show results to their own constituents and tax payers, which requires monitoring to show effective use of assistance (Cheema 2005).

Donors and international development agencies, however, are not monolithic. While bilateral donors and the World Bank typically explicitly frame the governance reform requirements as conditionalities, the UN Development Program and other UN programs frame their relationships with governments as partnerships with their country-level counterparts and strengthening their capacity to design and implement governance reform. In practice, however, the distinction is less sharp, because in both cases there must be a meeting of the minds in terms of mutual expectations and obligations.

The various international development organizations have anchored their programs somewhere along the continuum between effectiveness and political legitimacy depending upon their mandates and missions (Hydén 2011). The World Bank and other predominantly economically focused entities largely focus their governance programs on such issues as fighting corruption, promoting transparency in government agencies, and reduction of transaction costs. For private firms, compliance with anticorruption laws in their headquarter countries motivates reform efforts to reduce demands for bribes and other corrupt practices. For the World Bank and the regional development banks, the degree of democratic practice is one component of the country evaluations and by extension a factor in determining the magnitude of support for each recipient government, but this macro-governance focus is less relevant in making funding decisions on development projects and programs. Many other bilateral and multilateral agencies, in contrast, emphasize the legitimacy dimension of governance practice including political pluralism, inclusiveness, accountability and transparency, and engagement of civil society. The UN Development Program takes a two-pronged approach: strengthening of political institutions such as election commissions and parliaments; and improving the administration and management capacities of the government at various levels to effectively design and implement development policies and programs.

World Bank

Governance as a concept and decision-making criterion was not explicitly prominent at the World Bank prior to the late 1980s, basically because of the stricture in the World Bank Articles of Agreement prohibiting it from taking political considerations into account in its loan and grant decisions. The quality of governance was, of course, taken into account indirectly in the annual decisions on the total volume of loans and grants designated for each recipient country, in terms of the absorptive capacity of the country to use World Bank funding constructively. However, the elevation of governance as an explicit criterion was first announced in 1987 (Shihata 1997), then applied to specific contexts (e.g., a 1989 World Bank study "Sub-Saharan Africa: From Crisis to Sustainable Growth"). This study argued that institutional reform toward a more efficient public sector (a key dimension of governance) could support the right economic policies (meaning structural adjustment programs) in bringing about economic growth and development. In that study, governance was defined as "the exercise of political power to manage a nation's affairs."

At birth, governance practice was largely technocratic because the goal of the World Bank was and has been economic growth and development, even though many World Bank staff were committed to understanding deeper causes of development failures. Governance was assumed to be one of the key means to achieve this aim. As a means, governance has been mostly concerned about design, sequencing, implementation, or enforcement of institutional reform. As such, mechanisms of anticorruption, decentralization, transparency, performance-based view of accountability, and rule of law have been used to complement and strengthen the efficiency of market institutions (Maldonado 2010).

With time, this instrumental/technocratic understanding of governance practice gave way to more substantive/societal insights. In 1994, the World Bank expanded the scope of governance practice to include a transparent process of policymaking, a bureaucracy imbued with professionalism and internal codes of conduct, an accountable executive branch of government, and a strong civil society participating in public affairs. In 1997, the World Bank's *World Development Report* embraced more elements in its governance definition, including the role of state and citizen participation, thereby moving closer to a more political understanding of governance (Doornbos 2004). A recent World Bank report on governance approached the concept from the problem-driven governance and political economy (PGPE) approach, emphasizing the good-enough-governance perspective for feasibility purposes. Accordingly, a PGPE analysis comprises three layers: (1) identifying the problem, opportunity, or vulnerability to be addressed; (2) mapping out the institutional and governance arrangements and weaknesses; and (3) drilling down to the political economy drivers, both to identify obstacles to progressive change and to understand where a drive for positive change could emerge from (Fritz, Kaiser, and Levy 2009, ix).

Thus, governance reform is now at the forefront of the World Bank's agenda. Over the years, the World Bank has developed its internal capacity and partnerships in the public sector and structure of government arena, including fiscal decentralization, administrative and civil service reform, public expenditure analysis, anticorruption measures, legal and judicial reform, and institution building in such sectors as education, health, and water. As one of the key players in international development, its role in promoting transparency in government and combating corruption has significantly influenced the governance practice in developing countries.

The World Bank's position on when it can invoke human rights violations as explicit reasons for denying financial support to a government, however, has been limited and sketchy, even though the importance attached to context and culture has increased. The World Bank contends that it takes human rights into consideration in making decisions about the projects it funds if (i) the borrowing country asks; (ii) human right violations have economic consequences; and (iii) the human right violations in question would lead to a breach of the World Bank's international obligations, such as those created under the binding decisions of the UN Security Council (Daniño 2006; Sarfaty 2012).

Regional Development Banks

Regional development banks, most prominently the Asian Development Bank, the African Development Bank, and the Inter-American Development Bank, have been supporting their member states to strengthen their governance capacities and promote economic development, facilitate service delivery, and improve infrastructure. For example, Asian Development Bank's (2010) *Strategy 2020* has identified governance as one of the "drivers of change." Asian Development Bank funding in governance and public sector management is provided to achieve six governance outcomes: accountability for economic performance; effectiveness of policy formulation and implementation; effective use of public resources; participation of the poor; predictability and transparency of regulations and decisions on cross-border issues; and combating corruption.

United Nations

The UN system, in particular the UN Development Program, has played a significant role in making good governance more humane, while putting it on equal footing with economic development rather than simply an appendage to it. The 1995 Social Summit in Copenhagen and the creation of UNDP's *Human Development Reports* and Human Development Index have revolutionized the developmental logic that economic well-being is not equal to human progress and that the content of domestic policies is crucial to both.

To varying degrees, all parts of the UN system are involved in supporting the member states in strengthening governance capacity at the local, national, regional, and global levels. The UN Department of Economic and Social Affairs leads the UN system in supporting the normative, intergovernmental processes of the UN system with a focus on public sector management and civil service reform. Specialized agencies including UNICEF, UN-HABITAT, UNIDO, and FAO examine governance issues in the context of their respective missions and mandates. The UN regional commissions focus on cross-border governance issues such as water management, migration, and trade that require cooperation of the countries in the region and globally.

The United Nations has followed a two-pronged approach to governance practice: a sectoral governance approach to promote processes and mechanisms in such sectors as health (World Health Organization) and education (UNESCO), and a holistic governance that examines systemic processes and mechanisms. For example, UNDP's holistic approach to governance defined it in terms of administrative, economic, and political governance, usually within a democratic framework (UNDP 2011).

UNDP is recognized by the member states as the central entity of the UN system with the mandate of expanding governance capacity. In recent years, grant funding from UNDP to strengthen governance in developing countries has materialized in three distinctive phases or generations. The first generation of governance projects (from the 1960s to the 1980s) addressed the need for the improvement of public sector capacity in policymaking, implementation, and evaluation at both systemic and sectoral levels. UNDP played a key role in supporting developing countries to strengthen their public administration capacities. This included support to independent central banks and regulatory commissions to facilitate the political insulation doctrine for economic policymaking institutions, overcome the skill deficits of the mainline bureaucracy, and establish permanent public administration reform entities within the central government apparatus. During this period, UNDP also supported the creation of administrative units with separate budget lines and administrative autonomy to undertake such tasks as integrated rural and urban development programs, enhancing the capacity of the planning bodies, and providing grant funding for the preparation of feasibility studies for infrastructure projects to elicit investments from the World Bank, donors, and the private sector.

In the second generation of governance projects, as mentioned in chapter 4, a shift occurred in UNDP grant funding that is best encapsulated as from government to governance. To act on the recognition of the increasingly overlapping spheres of interest of government, civil society, and the private sector and to find new methods of encouraging participatory and transparent governance, the UNDP took the lead for the UN system. It prepared and mainstreamed its first policy papers on governance for sustainable human development in 1997 and the integration of human rights with human development in 1998. The 2002 report *Deepening Democracy in a Fragmented World* (UNDP 2002) set the stage for a dramatic increase in UNDP grant funding to strengthen parliaments, electoral management bodies, and electoral processes; engage CSOs; improve local governance and decentralization processes; promote access to justice; and strengthen accountability and transparency.

The third generation of governance projects represents a natural progression in the understanding of ways to improve governance in line with changing global and regional conditions and takes into account the lessons learned during the previous generations. The examples of these are cross-border issues in Asia and other regions of the world such as human trafficking, water management, and infectious disease and response. Rapid globalization has led to increased flows of goods, services, capital, ideas, information, and people between countries. Furthermore, for the first time in history, an overwhelming majority of countries are electoral democracies. Within this context, two dimensions of governance have assumed prominence: governance infused with the universally recognized principles and values of democracy, and cross-border governance issues that can be addressed more successfully through regional and global cooperation.

To bridge the gap between theory and practice, UNDP identified several service lines or entry points to support different aspects of democratic governance. Table 5.1 shows examples of these entry points.

Table 5.1 Generic examples of UNDP grants through the service lines/entry points of democratic governance

Entry points	*Examples of assistance*
Electoral and parliamentary process	To strengthen electoral and parliamentary processes through voter registration and education, election monitoring, support to electoral bodies, legislative drafting, committee systems, constituent relations, etc.
Public administration and civil service reform	Programs to reform the civil service into an effective, transparent, and accountable institution that helps to promote growth and private sector development including policies and procedures for recruitment, compensation, promotion, human resource development, and motivation and performance management
	Programs to strengthen capacities of central government ministries and departments to support policy analysis and development
Judicial and legal reform	Training for judges and prosecutors to strengthen capacities to restructure and reform legal and judicial systems; undertake legal drafting in commercial, tax, trade, investment, labor, and human rights law; and create and maintain independent lawyers' associations
	Programs to support legal reforms that ensure the security of foreign investments, and legal and regulatory frameworks for the engagement of CSOs and paralegal groups to increase citizen access to constitutional justice
Accountability and transparency	To strengthen the capacity of supreme audit institutions and related national financial control and management institutions
	Support to community awareness and education programs to combat corruption and strengthen the capacity and resources of national anticorruption bodies
Decentralization and local governance	Programs to support the devolution of financial and political authority and local government capacities for participatory planning and management; strengthen data and information systems; improve people's participation in the development process; and raise awareness among legislators and policymakers on the importance of local-level planning
Civil society organizations and media	Programs that support the engagement of civil society in free and fair elections through voter education and registration, holding the members of the parliament accountable, supporting local-level development projects, and advocating access to information and transparency of governance process
	Programs to develop the capacity of women to participate in political processes and national governance at all levels
Human rights	To develop human rights capacity in institutions of governance and in institutions specially devoted to the promotion of human rights, ombudsman offices, police, etc.

Source: Adapted from Cheema (2005).

The United Nations has complemented and broadened the World Bank's conception of governance: (1) good governance encompasses all structures and processes for determining the use of resources for public goods; (2) it has, as such, emphasized the sociopolitical nature of governance as well as its public welfare aspect, beyond democratic symbols. In addition to the emphases on transparent public agencies, accountable public officials, local governance, and citizen engagement in public policies and decisions, these theoretical additions have meant in practice the inclusion of the universal protection of human rights and nondiscriminatory laws into the concept and practice of governance. UNDP country offices and their counterparts increasingly pilot-tested project-level modalities and approaches to promote "governance for human development and poverty alleviation," that is, governance that is inclusive and responsive to the full range of needs and aspirations of the vast majority of the poor and marginalized groups.

European Union

The governance practice of the European Union has been diverse and opaque. Three main perspectives have dominated European Union's governance support to developing countries: (i) the main accent on well-functioning government institutions and transparent and accountable mechanisms; (ii) the main focus on enforcing human rights, particularly through well-functioning judicial systems and democratic structures; and (iii) the governance approach with a concentration on poverty reduction and citizen engagement (Hoebink 2006, 155). The European Union does not clearly distinguish between good and democratic governance and uses the two terms interchangeably. However, the European Union conditionalities for governance support are clearly specified in two articles of the Cotonou Agreement that governs the relationships between the European Union and developing countries (Slocum-Bradley and Bradley 2010, 37).

US Agency for International Development

The US Agency for International Development (USAID) is one of the largest donors of democracy promotion and governance projects in developing countries. Its "Democracy, Human Rights, and Governance Strategy" consists of five areas of concentration: rule of law and respect for human rights, promoting competitive elections and political processes, strengthening a politically active civil society, transparent and accountable governance, and promoting free and independent media. In practice, USAID programs have pursued five strategic goals: (1) anticorruption activities such as civic education and advocacy for the reform of laws and practice to promote transparent and accountable governance institutions, processes, and policies; (2) democratic governance of the security sector through such activities as public sector reform and public management, strategic planning, and oversight; (3) decentralization and democratic local governance to improve subnational public administration, citizen participation, and fiscal performance and economic growth; (4) strengthening legislative function and processes focused on the quality and effectiveness of laws and regulations, ensuring the accountability of the legislative and executive branches and overseeing the government budget and laws; and (5) policy reform through strengthened executive and public sector performance, including capacity development of executive offices, ministries, and government bodies, to facilitate the implementation and enforcement of laws and policies, civil service reforms, public-private partnerships, and linkages between and among branches, levels, and functions of government (USAID 2013).

Nordic Countries

Denmark, Finland, Norway, and Sweden have over the years been the strongest advocates of democratic governance. A unique feature of their funding of governance support is that most of their assistance is channeled through such multilateral entities as the World Bank, UNDP, regional development banks, and the European Union. Often their democratic governance assistance is not guided by economic and military interests. The areas of their support include capacity development of political parties, legal education, electoral and parliamentary process, civil society, and promotion of human rights.

All of the Nordic countries have prepared policy papers on good governance practice. They played an important role in the creation of the International Institute for Democracy and Electoral Assistance, which focuses on the consolidation of democracy at the country level by providing a forum to academic institutions, governments, and NGOs. Norway has been playing an active role in supporting conflict resolution initiatives such as the case of Sri Lanka. UNDP and the Government of Norway jointly established the Oslo Center on Governance to support the exchange and sharing of experiences and good practices in governance, including access to justice.

As table 5.2 shows, various bilateral and multilateral agencies have defined governance practice in order to undertake their respective mandates and missions and fund governance capacity development programs.

The perspectives of the international organizations on governance, although nowhere near monolithic, have suffered from comparable biases as in the academic field. For instance, they all have conceived governance at the macro-level, applying to the entire political, administrative, and economic systems and having the state as their main interlocutor. This has meant a managerial and top-down approach as the dominant rationale of the donor community. Likewise, formal institutions of rule have been prioritized over the underlying forces, such as informal formations and societal forces that often determine implementation and outcomes.

Other issues with the donor perspectives on good governance have included: (1) the assumption that things are broken and need fixing in the Global South and the underestimation of the capacity of existing local institutions; (2) the generalization that entire sectors need fixing, implying broad civil service reforms, legal sector overhauling, and local government makeovers; (3) the presumption that good governance will bring development or that bad governance will impede it, thus leading to the imposition of institutions sometimes out of touch with the socioeconomic realities in the field; (4) the inclination toward quick fixes and results-based approaches as opposed to the perceived needs and rights; (5) the preference for large data sets of governance indicators (such as the World Bank's World Governance Indicators), often based on perceptions, to the detriment of specific policy issues and diagnostic

Table 5.2 Governance perspectives of donors and international organizations

United Nations—Good governance is ensuring respect for human rights and the rule of law; strengthening democracy; and promoting transparency and capacity in public administration.
World Bank (1992, 1994, 1997)—Governance is the manner in which power is exercised in the management of a country's economic and social resources for development. The World Bank has identified three distinct aspects of governance: (i) the form of political regime, (ii) the process by which authority is exercised in the management of a country's economic and social resources for development, and (iii) the capacity of governments to design, formulate, and implement policies and discharge functions.
Asian Development Bank (1997)—"Good governance... is synonymous with sound development management."
IMF (2002)—The term governance encompasses all aspects of the way a country is governed, including its economic policies and regulatory framework.
UNDP (1997)—Governance is viewed as the exercise of economic, political, and administrative authority to manage a country's affairs at all levels. It comprises mechanisms, processes, and institutions through which citizens and groups articulate their interests, exercise their legal rights, meet their obligations, and mediate their differences.
Organisation for Economic Cooperation and Development (1995)—Governance is the process or method by which society is governed. It denotes the use of political authority and exercise of control in a society in relation to the management of its resources for social and economic development. This broad definition encompasses the role of public authorities in establishing the environment in which economic operators function and in determining the distribution of benefits and the nature of the relationship between the ruler and the ruled.
Commission on Global Governance (1995)—Governance is the sum of the many ways individuals and institutions, public and private, manage their common affairs. It is a continuing process through which conflicting or diverse interests may be accommodated and cooperative action may be taken. It includes formal institutions and regimes empowered to enforce compliance, as well as informal arrangements that people and institutions either have agreed to or perceive to be in their interest.
International Institute of Administrative Sciences (Rampersad and Saleh 2014)—Governance refers to the process whereby elements in society wield power and authority and influence and enact policies and decisions concerning public life and socioeconomic development.

Note: Compiled from Singh (2003), Weiss (2000), Cheema (2005), and Hydén (2011).

tools drawing on local practices; (6) the lack of effective, reliable, uniform, and evidence based measurement, monitoring, and assessment tools for many sensitive governance issues, such as human rights violations; and (7) the tendency to plug Western institutions into the local rules and practices, thus leading to analogies being drawn with colonialism or the development administration movement of the 1960s.

Emerging Trends in Governance Practice

Recently, there has been a clear path in governance practice toward adopting a rights-based approach, centered particularly on human rights. This has been termed humane governance—governance that supports the creation of participatory, responsive, accountable polities embedded in a competitive, nondiscriminatory, yet equitable economies (Ul Haq 1999). Components of humane governance were enumerated as broad ownership, decency, and accountability; its objectives are freedom, genuine participation, and sustainable human development.

Over the past few decades, governance has become one of the core themes of the global discourse on political and economic development. Governments have identified different strategies and doctrines based on the prevailing development paradigms and their country's peculiarities. Donors and international organizations have played an important role in shaping governance practice, though there have been significant differences among their respective approaches. The most visible change in the discourse on governance practice has been the increasing roles of civil society and activists in promoting citizen engagement in the process of governance at the local, national, and global levels. Cumulatively, each of these sets of actors has contributed to the state of present governance practice.

Presently, several new trends are discernible in the governance practice in developing countries.

Quest for Legitimacy

During the early phases of governance practice, the focus was on managerial effectiveness in order to cope with specific development problems. Now the quest for legitimacy has become one of the core goals of governance. In most countries, legitimacy is sought through democratic governance that includes systems of checks and balances among different branches of government, the rule of law and independence of the judiciary, and free, fair, and regular elections. Thus, the shift has taken place from managerial to political aspects of governance. It is widely recognized by policymakers and development practitioners that governance is political in the sense that it deals with politically sensitive issues such as the distribution of power and the allocation of resources to different segments of society and regions of the country. Regarding political legitimacy, although political pluralism is still the predominant model in the world today, some countries such as China and Vietnam are attempting to promote legitimacy though government programs that increase the income levels of citizens and provide services to them effectively. This, in turn, has enhanced the legitimacy[2] of the government. The citizen trust in the government in China, for example, is higher than in most other countries in Asia, which is largely attributed to the ability of the Chinese government to reduce poverty on a massive scale and improve the standard of living of the majority of the people.

Decentralizing Governance

There is an emerging consensus that governance requires not only building of national-level political institutions, such as parliaments and electoral management bodies, and administrative structures and processes to effectively implement policies and programs, but also the devolution of powers and resources to subnational and local units of government and administration (Cheema and Rondinelli 2007). Administrative capacity is needed at the local levels to implement projects. The UN

Capital Development Fund (2005) has, through a set of case studies in developing countries, shown that where local government plays an active role, it becomes easier to engage local actors in development processes and more effectively coordinate local economic development. In order to capture the shift from top-down to bottom-up approach, a new set of tools has been developed to examine the perspectives of different sets of governance actors at the national, subregional, and local levels. These include UNDP's governance assessment methodology that includes the perspectives of different stakeholders at multilevels, Afro- or Latino-barometers focused on a broad range of issues and institutions, and the Africa Peer Review Mechanism that focuses on self-assessment (Hydén and Samuel 2011).

Delivering Services

It is widely recognized that effective access to such services as education, shelter, and health is a true test of governance effectiveness. In contrast with the first generation of democratic governance initiatives focusing on building institutions of democracy, the focus during the present phase of democratic governance reforms is on democratic dividend, especially coping with the deficits in access to services for the majority of the rural and urban poor.

Promoting Country-Driven Strategies

The global community, including bilateral donors and international organizations, played an important role in the general advocacy for democratic local governance, delineation of global norms and standards, and, in some cases, general blueprints. This has now changed to the need for a country focus, recognizing that democratic governance is inherently a political process and thus is determined by various political interests in the society. Context does matter a great deal in terms of institutional reform options that countries effectively design and implement. The World Bank's Problem-Driven Governance and Political Economy Analysis, the UK Department for International Development's Drivers for Change methodology, and the Swedish International Development Agency's Power Analysis approach are some of the examples of governance practice frameworks promoted by the international development community to ensure that governance interventions take into consideration such factors as history, social and cultural norms, and political power structures and processes. Because each society has distinctive characteristics of underdevelopment and governance deficits, it is logical to identify institutional reform options based on a clear understanding of these characteristics and deficits. Furthermore, it is essential to ensure that reform processes are country-owned and -driven, with external development partners providing technical and financial support. The 2005 Paris Declaration on Aid Effectiveness was an attempt by donor countries, multilateral donors, and recipient or partner countries to create new international norms of cooperation. Regarding country ownership, the Declaration highlighted the need for partner countries to exercise effective leadership over their development policies and strategies and to coordinate development actions, It also called for the donor community to base its overall support on partner countries' national development strategies, institutions, and procedures.

Enhancing Citizen Monitoring of Government

Representation though elections at various levels of government continues to be one of the core elements of democratic governance practice. Increasingly, however, citizens seek greater participation in the process of policymaking and program implementation. New opportunities for civil society entail institutional mechanisms beyond the formal representative institutions to enable citizens to monitor the implementation of government policies and projects. The monitory democracy utilizes mechanisms that go beyond the traditional institutions of accountability, such as anticorruption commissions, codes of conduct, offices of the auditor general, the conventional media, and parliamentary oversight of the executive. Examples of extraparliamentary power-monitoring institutions are workplace tribunals, youth parliaments, associations of minorities, citizens' assemblies, independent public inquiries, expert reports, participatory budgeting, and new forms of media oversight such as blogging (Hydén and Samuel 2011).

Focusing on Process and Qualitative Data

With the ascendency of the global discourse on governance and the rapid increase in external funding for governance reform programs, a series of quantitative indicators to measure governance was identified. These include the World Bank's World Governance Indicators, Transparency International's Corruption Perceptions Index, Freedom House's State of Freedom in the World, the Ibrahim Index of African Governance, and the Millennium Challenge Corporation's three democracy indicators. Each of these indicators serves useful purposes such as global advocacy, mobilizing country-level support, and undertaking comparative analysis. They serve the growing demand for indicators to cope with corruption and transparency issues. Quantitative indicators, however, have been criticized by academics and development practitioners and challenged for "normative, legitimacy and methodological reasons" (Hydén 2011). Some developing countries perceive these to be biased toward market-oriented liberalization. Others contend that quantitative indicators fail to show the complexity of social, economic, and political processes in the country. These concerns have given rise to the development of qualitative indicators. Qualitative governance assessments focus on process as the entry point to empower local stakeholders. Hydén has identified three characteristics of the new phase of governance assessments: Bottom up, citizen-initiated approaches that emphasize dialogue and social accountability; the application of basic principles of the rights-based approach (especially strengthening accountability relationships); and the influence of the aid effectiveness agenda with its emphasis on national ownership of development and capacity development (Hydén 2011). Examples of the process-focused approaches are the State of Democracy assessment initiated by the International Institute for Electoral and Democracy Assistance, the Urban Governance Index supported by the UN HABITAT, and UNDP's Global Program on Democratic Governance Assessments.

Conclusion

From the 1960s until the 1980s, central state administration based on legal-rational bureaucracies was believed to be the engine of growth and prosperity. The decade from the mid-1980s to the mid-1990s entailed more citizen participation, but maintained the emphasis on outcomes than on process; on assessing what people want and need than on full authority of the citizenry to determine government policies; on relying on market liberalization than on state intervention. Starting with the mid-1990s, a gradual shift away from this mentality to sensitive governance areas like human rights, legislative support, judicial reform, and corruption started to occur. This could also be attributed in significant ways to changes within developing countries such as increased access to information, expanding urbanization, and pressures to reduce economic disparities. It could also be attributed to changes in the donor approaches and strategies to reflect changing social, political, and economic contexts of developing countries. Good governance, at both turning points, acted as the tipping point. It reversed the state-dominated socioeconomic development patterns in the 1980s and softened the dominance of the markets in the mid-1990s. Then, it gradually coupled its socioeconomic agenda with political leadership and democracy, followed by a focus on human rights and accountability. Governance practice has thus taken on humanistic qualities, increasingly abandoning its more technical face in government policy and aid conditionality.

CHAPTER 6

Evolving Roles of NGOs in Developing Countries

Introduction

Expanding on the governance focus on the emergence of civil society institutions, this chapter examines the evolution of the theories and doctrines that have guided NGOs operating in developing countries and how, in light of this evolution, NGO practices have themselves evolved. The roles and influences of NGOs have increased considerably since the end of WWII. Their sheer number is huge and probably still growing, but what that number might be is basically unknown and probably even unknowable (Salamon, Sokolowski, and Associates 2004, 15–17). Despite the confusing terminology of nongovernmental organization, nonprofit organization, private voluntary organization, etc., it is clear that entities that are not part of government, business firms, or families number well above fifty thousand. For the purpose of this chapter, it is useful to confine our definition and inquiry to such organizations that are ostensibly dedicated to serving the public good in some way. Therefore, labor unions, business groups, and grassroots organizations that unite people explicitly to pursue their own interests are not within the purview of this chapter.

NGOs existed well before the United Nations reportedly coined the term in 1945.[1] Although conventional wisdom holds that NGOs is a relatively recent phenomenon (Stokke 2009, Part I), Thomas Davis argues persuasively that "international NGOs have a long and turbulent history, which has often placed these actors at the center of key transformations shaping international society over the last *two centuries*" (2013; emphasis added).[2] The transformative and turbulent constants that characterize this long history are notable, persistent, and guide what follows in this chapter. The approach taken here focuses on the core or prevalent development theories evident at different times and locations and how these theories are applied to achieve ends or goals. These means are represented by the realistic practices that do not necessarily conform precisely to the theoretical expectations driving them. Exceptions, changes, outliers, and failures are as interesting as the successes sought. Indeed, learning from collisions of theoretical expectations and doctrinal wishes with realities more complex than can be imagined is an additional goal of this chapter and the volume more generally.

The Evolving Context of NGO Operations

Eras of the Analysis

To summarize the context in which NGOs have operated over the past seven decades, it is helpful to segment this period to better grasp major events and the developmental trends thus created. Korten (1990) and, subsequently, Lewis and Kanji (2009) offer their own perspectives of the early eras of NGOs as agents of development (table 6.1).

Thus, NGOs can serve a host of functions, including provision of basic services, improving access through local accountability mechanisms, and advocacy functions usually for the poor and marginalized and global public goods (Salamon and Anheier 1992a, 1992b). The theory of global public goods triggered both the evolution of various development theories and the role of NGOs at higher levels as well. These include debt relief, anticorruption movements, human rights, gender equality, and climate change issues.

Era 1 covers the immediate post-WWII period of humanitarian relief, relocation, reconstruction, the Marshall Plan, and the settling in of the Cold War antagonisms with various food, development, and military assistance programs meant to combat Communism around the world. While much of this effort was governmental, NGOs, particularly those based in the United States, played a major role in assisting refugees and restoring services in war-ravaged areas (McCleary and Barro 2008). The shift in

Table 6.1 Eras of NGO emphases

	Era 1 (1945–1963)	*Era 2 (1963–1980)*	*Era 3 (1980–2000)*	*Era 4 (2000–2015)*	*Era 5 (2015–)*
	Relief/basics	*Community development*	*Technology/ fall of iron curtain*	*Overreach/ Push Back*	*Social impact/ balanced systems*
Focus	Specific	Local/specific	Regional/ national	National/ global	Ecosystem/ mega-cities
Time frame	Immediate	Project life	10–20 years	Open-ended	Variable
Scope	Individual	Neighborhood/ village	Region/ nation	Nation	Ecosystems (natural/ constructed)
Participants	NGO members	NGO/ communities	"Everyone"	Networks	Super networks
NGO role	Primary/ central	Mobilize/direct	Catalyze/ innovate	Active/direct involved	Educational/ tech support

Sources: Adapted and extended from Korten (1990) and Lewis and Kanji (2009).

focus from recovery to development, and the emergence of Kennedy-era development programs, ends this era and opens up many new possibilities for NGOs.

Era 2 begins with newly elected President John F. Kennedy's stirring call to action to Americans and also to citizens of the world. The familiar line—"And so, my fellow Americans: ask not what your country can do for you—ask what you can for your country"—is followed by another less familiar but still important one: "My fellow citizens of the world: ask not what America will do for you, but what together we can do for the freedom of man."[3]

Local development, specific projects, areas targeted around the world, and innovative organizational forms all characterize and help define this era (Lasswell 1965). However, early in Era 2 the roles and influences of NGOs were marginal as compared to the dominance of governments as the Cold War took greater hold. The Peace Corps, the US Agency for International Development (USAID), the Alliance for Progress with Latin America, and financial means such as the World Bank, United Nations, and others focusing on regional development enable a variety of new programs and possibilities to emerge in this era. Indeed, development activities routinely supported military objectives. Various insurgencies, wars of national liberation, coups, revolutions, and the underlying specter of communist aggression defined the early phase of this era. Winding down Vietnam in the late 1970s was for the United States a time when many military instruments were replaced by nongovernmental ones.

Era 2 in retrospect foreshadows contemporary environmental movements—sources of much NGO activity. Rachel Carson's *Silent Spring* (1962) and the 1972 Stockholm Conference on the Human Environment, along with concerns about feeding and supporting a burgeoning world population, signaled this emergence (Bickel 1974; Cullather 2010).

The 20-year span of Era 3 began in 1980 and is characterized by increases in the number of NGOs and in various missions they undertook. In nations freeing themselves of Soviet domination, such as Hungary, Czechoslovakia, and Poland, NGOs sprouted and flourished as an accepted form of political opposition. Human rights, the environment, good government, and other specific mission-driven organizations came into being just before and after the fall of the Berlin Wall in 1989. Their roles varied as widely as their putative missions, although a general constancy of purposes—to catalyze, innovate, and generally transform moribund and failing states—pertained.

By Era 3 the division of labor between official organizations and NGOs had become apparent. As chapter 7 on official foreign assistance notes, governments and multilateral development institutions such as the World Bank and the regional development banks were committed to large-scale projects such as physical infrastructure that would move substantial amounts of funds. These organizations eventually acknowledged the importance of social services and microfinance, but have had no comparative advantage in establishing small-scale interactions with local communities or mobilizing other elements of civil society. Governmental and multilateral entities (owned by the governments in the nations in which they work, as well as by donor governments) are generally less prepared to side with community groups against government policy. Therefore, several niches have been open for NGOs, even if some of their funding is from governmental or intergovernmental sources.

The environmental agenda took more concrete form with the 1992 UN Conference on Environment and Development. The Rio Conference or "Earth Summit" stimulated the subsequent launch of hundreds of NGOs committed to environmental improvement and what lately has been labeled sustainable development.[4]

Early and promising successes of the Green Revolution to expand food production through technological means, such as genetically modified crops, synthetic forms of fertilizer, selective breeding, and many other innovations, did much to feed a growing world population and also to sustain an increasing number of consumers whose positive impacts on growing economies made a substantial difference in millions of lives. Much of the impetus behind the Green Revolution came from the planning and financial support of the Rockefeller Foundation and the Ford Foundation (Jain 2010).

The importance of institutions, especially in terms of the managerial and technical competences required to operate effectively, emerged with growing urgency in the early 1980s and continues to this day in numerous NGO efforts for capacity building. (Esman and Uphoff 1984; Austin and Ickis 1986; Ickis, de Jesus, and Rushikesh 1986; Korten 1987).

Era 3 also encompasses the so-called Washington Consensus, an effort in 1989 to come to terms with the increased complexity of development programs and policies around the world. Much debate surrounds this effort to simplify and focus economic theoretical ideas in the interest of development practice (Williamson 1990; Stiglitz 2002). Many NGO leaders, believing that they were advocating on behalf of the poor, railed against this development approach. Usually lost in the technical and personal back and forth is the fact that most of the empirical referents were from Latin America, and claims to be general and global were seldom made (Kuczynski and Williamson 2003).

The consensus itself is usually summarized by the principles reviewed in chapter 2,[5] which include market-freeing prescriptions based largely on neoclassical economic concepts and practices.

The complexity of possible causal links between these liberalization reforms and the trends of growth and equity has resulted in enormous ambiguity as to their merits, especially because, as chapter 3 reviews, they were only partially implemented in most developing countries—but in many cases when economic conditions had become so adverse that austerity measures were required alongside of structural reforms. Many NGO leaders focused on the painful austerity measures rather than on the structural reforms

Several other important trends became more evident in retrospect. The Washington Consensus is best seen as an effort to simplify a complex collection of circumstances and activities careening out of the familiar control decision makers relied on in the 1945–1980 eras. The specter of Everyone being able to participate during the 1980–2000 stretch of Era 3, often in the form of numerous, independent NGOs, is indicative.

Even if an economy is stripped of the special privileges enjoyed by the already wealthy and powerful, assets are still the major determinants of economic returns. Therefore, an essential challenge to poverty alleviation is to provide assets to the poor through multiple channels: security of physical capital, access to financial capital, and opportunities to increase human capital through education, job skills, and health. Williamson (2004, 13) argues that "The solution is not to abolish the market economy, which was tried in the communist countries for 70 years and proved a disastrous dead end, but to give the poor access to assets that will enable them to make and sell things that others will pay to buy. That means: Education. Titling Programs. Land Reform. Microcredit." Each of these initiatives provides fertile ground for NGOs to operate worldwide. The creative linkage of financial institutions and means with philanthropic sources in creative ways is of special interest as the world moves forward into Era 5 and beyond.

The relationships between state authorities and civil society became more complex and confusing in the late 1990s and into the 2000–2015 (Era 4). Globalization, sheer increases in the numbers of NGOs, and several other developments all contributed to what Indian author and political commenter Arundhati Roy (2002) termed the "NGO-isation" of politics.[6]

> In India, for instance, the funded NGO boom began in the late 1980s and 1990s. It coincided with the opening of India's markets to neoliberalism. At the same time, the Indian state, in keeping with the requirements of structural adjustment, was withdrawing funding from rural development, agriculture, energy, transport, and public health. As the state abdicated its traditional role, NGOs moved in to work in these very areas. The difference, of course, is that the funds available to them are a miniscule fraction of the actual cut in public spending. Most large-funded NGOs are financed and patronized by aid and development agencies, which are, in turn, funded by Western governments, the World Bank, the UN and some multinational corporations.
>
> ...They have become the arbitrators, the interpreters, the facilitators. In the long run, NGOs are accountable to their funders, not to the people they work among...It's almost as though the greater the devastation caused by neoliberalism, the greater the outbreak of NGOs.

This was also the period in which a paradigm shift took place in the way development was defined. The UN HDR, first published in 1990, led by Amartya Sen and Mahboob ul Haq, presented the HDI to move the focus beyond from economic growth. The need for expanded space for NGOs and other civil society organizations was a critical part of this. The HDRs, now published annually in over 150 countries, transformed the perspectives on state-civil society relations by emphasizing more dimensions of welfare. In addition, as outlined in chapters 4 and 5, development took on a much stronger emphasis on governance and administration. In concert with the various UN agencies, NGOs took on a greater role in assisting local grassroots organizations to strengthen their institutions.

By Era 5, NGOs had to face intense criticism regarding their relation to the state, as the quote above by Roy vividly illustrates, and concerns that they were overreaching. These concerns were already articulated by concerned academics and practitioners in the late 1980s and 1990s (Clark 1991; Tandon 1992; Edwards and Hulme 1996), but they culminated in strong pushback during Era 5. Many criticisms of the NGOs fell into five general categories: effectiveness, accountability, autonomy, commercialization, and ideological/political objections to their rising influence (Reimann 2005). Many of these specific matters emphasize *credibility* as a collective concept (Gourevitch, Lake, and Gross Stein 2012). The concern over NGO effectiveness has been exacerbated by the fact that NGO performance evaluation typically involves unclear or nonexistent measures of effectiveness. In a world where much more than a simple profit-and-loss statement exists to keep score, it is little wonder that questions about effectiveness would occur. Is the NGO promoting democracy, development, innovation, popular participation, poverty alleviation, peace, or what? Assessment of NGO efforts to promote sustainability or political empowerment is often overwhelmed by the lack of any commonly accepted and operational definitions of those objectives. Even when goals are relatively clear, such as providing humanitarian assistance or ending internal conflicts, questions remain as to whether NGOs are doing "more harm than good" (de Waal 1997; Maren 1997; Anderson 1999; Katz 2013).[7]

NGOs have also been under scrutiny over concerns of accountability, ranging from keeping track of fund flows to obeying regulations at different operational levels. Transparency and legitimacy are also at issue (Edwards and Hulme 1996). Questions about what proportion of an NGO's funds went toward program goals versus paying big salaries and supporting lavish lifestyles arise, and often for good reasons (Buffett 2013). NGOs have responded to these concerns about accountability through various mechanisms such as the codes of conduct, the Global Chapter of CSO Accountability, and so on.

The worry is that some NGOs become virtual extensions of their primary donor or donors, and in so doing they lose their independence and—some believe—their legitimacy as well. Dependence on official governmental sources, as is often the case for development NGOs, may open them up to criticisms of donor capture or, for example, becoming agents of imperialism. Criticisms of commercialization occur when the NGO becomes more and more like a private business whose main objectives are to raise funds and generate surpluses—a euphemism for profit in the nonprofit world. This issue becomes more problematic with the increasing size of a given NGO and its operations. Paradoxically, efforts to improve the professionalism and effectiveness of NGOs by building their capacity may work in some sense but at the cost of lost focus on an organization's core mission and as it seeks commercial gains and economic efficiencies. Powell and DiMaggio (1991) first identified this problem in the late 1980s and labeled it institutional isomorphism. In the Era 4 world of increasing numbers of large NGOs, commercialization is, if anything, even more significant and troubling (Ritzer 2000; Cooley and Ron 2002).

Ideological and political pushbacks run the gamut from the far left to the far right on the political spectrum. Long-running fears among the US conservatives, among others, focus on the specter of potential domination of the United Nations through global governance. Others see NGOs as an arm of US imperialism, or the real intentions of those promoting social responsibility, population control, or food security. Or even the inevitable disasters looming somewhere down the road related to technologies, such as genetically modified organisms, biotechnology, nuclear power, and so on.

Whatever the specific circumstances, the Era of Overreach and Pushback took increasing hold as we entered the twenty-first century.[8] The four eras spanning the decades from 1945 until 2015 have been anchored in and conditioned by the past, even though they bring us up to the present (and a couple of years beyond). The question for the twenty-first century is whether new roles for NGOs can be both imagined and put into practice. We return to some of the possibilities after discussing core theoretical models that help one understand the construction and operations of NGOs between 1945 and 2015.

Theories and Doctrines

There are no hard-and-fast rules to rely on in summarizing and constructing theoretical models of NGOs in the last 70 years. Nor is there much to be gained by refining or sharpening this or that theoretical point for its own sake. We offer what follows mostly as a guide to help one understand—and thus appreciate—NGOs as they operate in developing countries. These theories must also be considered from the temporal perspectives of the four eras just characterized. None of the theories operates independently of context, most specifically the spatial and temporal, cultural, and unique individual and institutional settings where development occurs.

Theory of Beneficent Charity

The simplest explanation of NGOs is that they are motivated by beneficent charity. As a doctrine it means that organizations comprising people with no direct personal stake in the outcomes provide assistance to the people affected by adverse conditions, where other sources and institutions are unable to do so without the collaboration of NGOs. Beneficent charity may be engaged in temporary crises: the typical events prompting beneficent charity are both natural and human-caused disasters resulting in starvation, displacement, destroyed homes and communities, and refugees. The main issues are related to finding food, medical supplies, and shelter and then figuring out ways to deliver them to desperate people. Hurricanes, earthquakes, and displacements caused by war are obvious examples.

However, the impetus of beneficent charity may be conjoined with other motives. A vivid example from the immediate post-WWII period is the fabled Berlin Airlift, where military and NGOs combined to save Berliners from a drastic Soviet power play (Shlaim 1983). This is also an early example of ways in which military and humanitarian objectives could combine for mutual benefit. Many challenging logistical problems were confronted, thus forming a valuable collection of procedural and scheduling tools for subsequent humanitarian crises. The focus of beneficent charity was and continues to be straightforward, immediate, and relatively simple help and humanitarian relief

For NGO assistance motivated by beneficent charity that addresses more chronic problems, such as endemic malnutrition or long-term refugee camps, deeper explanations are needed to account for why problems deserving charity persist.

Theory of Community Development

The importance of unmet needs of individuals and the communities in which they live and work gains prominence in theories of community development.[9] The theory is that local capabilities are necessary for improvements to be sustainable; institutions and skills, and perhaps attitudes as well, must be transformed to break out of stagnant patterns. Widespread poverty and alarming population growth, especially in the world's poorest countries and regions, have stimulated efforts to work at the community level to improve health, education, food supplies and security, and other basic human needs and requirements. A focus on the community as a collective with medium to longer term goals, rather than on individuals with immediate needs, is noteworthy. Thus, the main premise of this theory is that communities are unable to provide for themselves in important ways, such as education, simple infrastructure provision, communal self-help, and boot-strap operations.

The orientation of the doctrine, therefore, is very much from the bottom-up to develop local and specific neighborhoods and villages, project by project. Departing from the focus on overall national development invites NGOs to design projects, whether or not in collaboration with other types of institutions, on a scale they can manage and for which they can also take responsibility and credit.

The doctrine offers up bite-sized and specifically grounded projects that are often ideal tasks for NGOs to take on. There is a long history of NGOs being called upon to implement the community development projects that the major international development agencies, such as USAID and the World Bank, have funded.

However, another noteworthy premise of the theory is the rejection of the hypothesis that what to the outsider appear to be abject conditions may in fact serve essential societal interests that outsiders may not adequately appreciate. The practice of rural women walking substantial distances to wells may seem like a failure to provide potable water directly to small villages, but it may also provide the women with an otherwise lacking opportunity to socialize beyond the few residents of their villages. It may be that piped water would indeed turn out to be a boon for all, but the question is whether the NGOs have the capacity to determine whether this is the case.

Theory of System Failure

The premise behind this theory is that the complexities of a nation's systems are beyond the constituted authority's coping abilities. Corruption is often a central feature underlying the diagnosis. Doctrines based on the system failure theory call for clean, intelligent, transparent, well-meaning, and morally superior agencies—all best provided by something other than the formally constituted authorities. In other words, NGOs to the rescue! Capacity-building prescriptions—to improve the managerial abilities of the NGO's clients such as other civil society institutions and, on occasion, the public institutions—are sometimes encountered in circumstances diagnosed as system failure.

Failure to innovate and otherwise transform whole societies to make them more transparent, more efficient, less corrupt, legitimate, more effective, more democratic—the list of rationales accounting for system failure is long and often very creative—usually underpins both doctrine and practice when theories of system failure are invoked. If a disaster explanation or story is part of the failure theory, then the doors open wide for a collection of NGOs to surface and conspire to replace the ineffective, corrupt, inefficient institutions that are most liable and responsible for the failures. Of course, replacement may not mean that NGOs are better or best able to deliver the goods and services.

The recent dramatic rise of private sources of funding accelerates the clash of extant political power and authority with outside and alien forces—mainly cloaked as simple and benign NGOs. The William and Melinda Gates Foundation, the Soros Foundation, and the Billionaire Challenge are illustrative (*Christian Science Monitor* 2010). The magnitude of the resources and the global reputation of success of the individuals who literally personify many of these NGOs elevate their ambition.

With even modest success, initial practices based on the system failure theory and doctrines may result in community-level successes being promoted and taken to higher, regional, national, and international arenas. However, the premises accounting for subnational failures are often different than those pertaining to national problems, if only because the local initiatives are constrained by national regulations, legal recognition, and other factors. In addition, a typical limitation of local capacity is the scarcity of personnel, whereas the typical national problem is the bureaucratic bloat. These differences challenge the presumption that capitalizing on initial smaller scale successes and trying to recreate them in other circumstances, or to scale them up in greater numbers and larger sizes, proves often to be problematic (Dees and Anderson 2004).

Finally, system failures open up the time horizons of both problems and expected measures to deal with them well beyond the normal ones embraced in project contracts or even public policies and programs. Now, with the specter of failure looming large, prescriptions and solutions often emerge unbounded with completion dates or other management constraints. System failures, in effect, require endless open-ended solutions.

Because the system failure diagnosis puts some blame on national governments, local governments, or both, there is a risk that governments will be cut out too much. This is a serious problem not only because partnerships between NGOs and governments are often essential, as each has a different set of capabilities, but also because NGOs that shun government involvement frequently provoke reprisals by the government. Increased space for NGOs is an important need, but there are some things that governments must do, including establishing the legal status and regulatory framework of the NGOs themselves.

Theory of Social Enterprises/Social Impact

The newly emergent and rapidly evolving fields variously identified as social entrepreneurship/social enterprise/social impact provide another set of theories and related doctrines that affect development practices—practices that often rely on NGO and NGO-like organizations. As is the case with emerging fields, basic definitions are hard to come by (Martin and Osberg 2007). Nonetheless, it is still possible to identify and roughly characterize many core aspects, and by doing so, doctrinal precepts begin to emerge:

- Unmet needs, especially those felt and expressed by individuals who are disadvantaged, exist and persist in large part because of extreme poverty and scant political influence, in addition to the failure to properly define the problem as lack of opportunities for greater productivity for the poor.[10]
- Opportunities to innovate and even transform exist but require a creative approach that includes value creation in other dimensions than just economic ones (Brooks 2008).

- Beyond value creation, social enterprises are concerned with creating a more equitable sharing of these new values.
- Ultimately, the social entrepreneur is in the business of shaping and sharing a range of human values in the interest of improving human dignity—a summary idea defined in terms of distinctive human needs for respect, well-being, affection, enlightenment, skill, rectitude, wealth, and power (Lasswell and Kaplan 1950).

Social entrepreneurship as a practical matter is where most activity occurs, with social assistance combined with business initiatives directly involving the poor: microenterprises, savings groups, community infrastructure self-help programs, and so on. In a most optimistic formulation of this point, Muhammad Yunus, Nobel Prize winner for his creative microfinancial Grameen Bank, describes it through the title of one of his publications: *Building Social Businesses: The New Kind of Capitalism that Serves Humanity's Most Pressing Needs* (2010).

Yet the theoretical underpinnings and certainty about the future trajectory of the social entrepreneurship movement are slight, because of the blurring of distinctions between seeking profit and providing for social needs.[11]

It is very hard to know with any certainty exactly how social enterprises and social impact activities will eventually play out in the development realm. We do know that many different kinds of organizational forms are involved and that these range essentially from traditional for-profit operations all the way to traditional nonprofit ones. It is the space between these anchoring extremes that creates both excitement and confusions. NGOs figure in somehow, but at the moment it is unclear just exactly how and where the NGO form will contribute and offer a comparative advantage over alternatives. Perhaps that is just the point: NGOs are so highly diverse and variable that they readily become tools for many different social enterprises.

To simplify, we offer the following simple sketch of the organizational spectrum to situate social enterprises and variants. The inspiration for this comes from an article by Dees (2011) on using different social ventures as learning laboratories, which combines theory and practice so as to learn by doing (table 6.2).

Table 6.2 NGO mission and organizational spectrum

Traditional nonprofit	Nonprofit with income generating activities	Nonprofit practicing social innovation	Social enterprise	Socially responsible business	Corporation with strong corporate social responsibility program	Traditional for-profit

Dimensions

Mission motive	Profit motive
• Stakeholder accountability	• Shareholder accountability
• Income reinvested in social programs or operational costs	• Profit redistributed

Sources: Adapted from Dees (2009) by Isadora Tang and Tony Sheldon, Yale School of Management.

According to the analysis by Tang and Sheldon, different organizations can be located along this spectrum. Nonprofits with income-generating activities are represented by many museums, arts, and cultural organizations. But they can also be collective enterprises for the poor in standard business ventures, with auxiliary social services, training, and community-building efforts. Those practicing social innovation to enhance social services include Teach for America and Kiva. Social enterprises in a very clear form are represented by the Acumen Fund, SELCO, and the Grameen Bank. Stonyfield Farms, striving to keep as much of its dairy products from organic milk producers, and Seventh Generation, with its adherence to biodegradable and hydrocarbon-free cleaning and other household products, are socially responsible businesses. These have a stronger social commitment and presence than corporations with a strong Corporate Social Responsibility program such as Ben and Jerry's, Timberland, and Burt's Bees.[12]

Theory of Global Necessity

The prospect of addressing the mounting global challenges, such as climate change, exceeds the governance capacities of conventional and constituted authority. The argument is that these globally rooted challenges exceed the governance capacities of conventional and constituted authority. Major, even revolutionary, reorganizations and reallocations are deemed essential to even have a chance of managing the risks presented by these challenges. Much of the thinking on needed reorganization has focused on climate change (National Research Council 2010a; 2010b; Field et al. 2012; Revkin 2013), but other global dislocations may also call for radical changes in nongovernmental structures and their interactions with governments.

To address these challenges, a global necessity doctrine would posit that technologically informed and expert scientific sources of information will be in special demand, and these demands will be global even for problems that manifest locally simply because of the huge scale and numbers of humans involved. National and lower level public authorities are simply overwhelmed by the massiveness of the demands created and leveled at them. Direct challenges to all constituted authority, in the face of their inability to deliver even rudimentary services, will increasingly become the norm (Moss et al. 2013). The doctrine calls for high levels of integration of nongovernmental and governmental institutions across developed and developing countries. As the emerging Era 5 looms large and most challenging, question is whether institutions and organizations, including nongovernmental ones, will be sufficiently adaptive to be able to cope.

The evolving theories and practices of governance and development over the past 50 years have had the most critical impact on the state of theories and practice of civil society, including NGOs today. A paradigm shift has taken place from government (with monopoly) to governance as a process co-produced by three sets of actors: those from the state, civil society and the private sector (see chapters 4 and 5 on governance).

Another important point is the role of the United Nations in providing legitimacy to NGOs in the developing world in several ways—by making NGOs active participants in global norm-setting process; by restructuring internal mechanisms in the United Nations that provide formal space to NGOs registered with Economic and Social Council; by the Secretary General's Panel on Civil Society led by President Cardoso, which enabled the United Nations to engage NGOs (traditionally done only through intergovernmental process), and other changes accepted by the General Assembly; by the ECOSOC accreditation process, and so on.

Key Considerations and Conditioning Factors

Considerations and conditioning factors focus attention on activities and processes within the context of the problem or considered subject that will, most likely, influence or impact the flow of outcomes and events. No one can say with certainty which of these will matter and by how much. The bet is that, other things being equal, these are the issues that loom large.

Measurement/Keeping Score

Measures of effectiveness and other means to figure out how an NGO functions have become fashionable requirements used by philanthropic and donor agencies to evaluate NGO performance. In the words of one prominent philanthropist, Gordon Moore, it is finally possible to make "a huge push toward measurability" (Hechinger and Golden 2006, A-1). This may seem simple, but it *decidedly is not* when one tries to evaluate an NGO. Well-meaning intentions to add metrics in order to improve performance can be challenging, even in some cases counterproductive. NGOs are not simple profit-making machines, where a bottom-line score card exists to measure performance. Efficiency and effectiveness are not equivalent, a truism reflected in efforts to create Double and Triple Bottom Lines, rather than the standard one based mainly on economic efficiency (Elkington 1994; Hacking and Guthrie 2008; Savitz 2013).

Efficiency-based assessments dominate the private sector, and profit is what matters. The often unexamined premise is that when producers compete for a consumer's dollars, the most efficient survive and drive out inefficient competitors. Market inefficiencies attract new producers and encourage them to enter and compete. The driving force propelling and sustaining globalization is understandable in these terms. Producers are gauged according to their profits, and efficiency of operations is a key determinant of profitability.

The nonprofit sector, like the public sector, differs from the private one in efficiency terms. The public sector manager is not legally allowed to maximize profits or to capture residuals, so profit maximization is converted to budget maximization. Efficiency in this case may become a negative incentive because economizing could lead to a subsequent reduction in one's budget and, as a consequence, less ability to deliver a public good or service. For instance, donors put great emphasis on impact measures of effectiveness, but in practice we know how difficult it is to define or measure either impacts or effects: "Mission impossible," in the words of one seasoned professional.[13] We also know that adaptability and innovativeness are crucial in any organization, and perhaps even especially so for NGOs.

The nonprofit and nongovernmental sectors are also not well tuned to efficiency's discipline, although the reasons why differ from the public sector case. Nonprofit/

nongovernmental organizations are mission-driven and goal-directed. That is, they exist and are driven to accomplish a mission—usually stated in general, strategic terms—and operate according to goals set as means to direct and accomplish that mission. Efficiency may not matter nearly as much, in specific circumstances, as getting the job done at any cost, or fulfilling the mission. In matters of life and death, where this distinguishing characteristic is most evident, little concern may be accorded bean counting or comparison shopping for less expensive ways of doing the job.

Innovativeness or being adaptable to changing circumstances—the rise of both opportunities and problems—is an essential criterion for nonprofit management. However, telling someone to innovate and adapt or trying to teach them how to do either is itself chancy. Still, innovativeness clearly counts as a measure of an individual and an organization's performance and effectiveness.

Effectiveness may be defined technically and with relative ease if the context does not change substantially: It is a ratio measure relating observed output to planned output over some time period. Because both planned and realized outputs are measured in the same units, the ratio expresses the percentage of effectiveness for the period. Unless two programs are nearly identical in the mission or goals set for them, and the resources allocated to accomplish these goals, comparisons of effectiveness between programs will not make much sense. Many foundations and granting agencies fail to appreciate this elemental point. In addition, the measure may create perverse incentives. An ambitious program that is also efficient in its production can look bad in effectiveness terms when compared with a conservative program that is inefficient. Or, setting minimal goals or ambiguous objectives may result in assessments of high effectiveness as contrasted with programs having clear and expansive goals that somehow fail in the end to achieve everything promised or sought. Keep expectations low so that no one is disappointed is one way of describing this.

Agenda Expansion/Mission Creep

A pervasive theme throughout this book is that the agenda for development has expanded enormously since the end of WWII. It has also become more complex. Moving beyond obvious and basic human survival objectives, it now includes a multitude of other goals: human rights, women's rights, environment, education, food security, sustainability—the list goes on and on. Likewise, simple, clear statements of an NGO's mission are becoming increasingly rare, often as a consequence of something known as mission creep, a term and concept borrowed from military operations and activities, including those where initial military objectives open up to and give way to civilian, humanitarian, nation-building ones (Siegel 1998). The melding and blurring of military and humanitarian missions has long been with us, as the Berlin Blockade and Airlift long ago demonstrated (Shlaim 1983).

Mission creep for NGOs represents a somewhat special case distinctive from the more commonly encountered military one (which we discuss in chapter 9). In the American setting, the creation of a nonprofit requires a clear statement of the proposed organization's mission from its founders and original funders. While there are several different provisions of the tax code that possibly legitimate a nonprofit or an NGO, a common one is referred to as a 501(c) 3 in which the tasks and prospective benefits of the organization must be specified. For most NGOs operating as nonprofits, the first day of their existence is perhaps the last day when mission clarity is so evident, for reasons suggested by Wilson (2011).

> Over time, however, new programs are added. Perhaps because of the lure of new efforts and new needs, earlier programs may wither. The original, stated mission may become de-emphasized, while new, unstated missions take over. When assets and resources are allocated to efforts not related to the organization's mission—as well-intentioned as the efforts may be—the original mission of the nonprofit may be obscured.

Mission creep is not necessarily all bad. Goals may be attained. Other organizations, including public sector ones, may assume responsibility for service provision. More pressing problems may rise up, demanding attentions and resources. More typically, however, mission creep occurs as the tastes and priorities of funders change, sometimes to such an extent that the original mission hardly conforms any more. Creep in this case also can mean compromising goals and objectives merely to ensure the organization's survival. This is an era where individual donors expect "engagement beyond writing a check; and our organizations aren't used to that kind of engagement." The traditional pattern—"They give it, and trust you, and they go away until the next time you come"—is over (Silicon Valley Business Journal Staff 2012). Donors increasingly want specific outcomes, many of which will not conform to the organization's mission statement and goals. Mission creep to accommodate donor wishes is a frequent result. Another reason for mission creep is that in the globalized and interconnected world, new global public goods and norms have emerged. Human rights, gender equality, inclusiveness and minority rights, debt relief, etc., become NGO advocacy missions. In addition to the services provided directly by the NGOs, the expanded agenda entails pressuring governments to incorporate these global public goods in policy and program interventions.

NGOs as Agents to Secure Other Strategic and Political Goals

This transformation of many NGOs into political agents, in which usual humanitarian and charitable ends are replaced with more overt political ones, is a feature of present-day experience. It represents, in a sense, a perverse instance of mission creep. That this transformation may not end well for the NGO can be observed in places like Egypt, Pakistan, Russia, and China where NGOs are increasingly attacked by constituted authorities—and ideological crazies, as in Pakistan where polio technicians were killed by the Taliban simply for giving polio vaccine to children (Murphy 2013). Of course, the politicization of NGOs depends on the nature of each political system and social structure. One need not go so far afield, however, to find evidence of this politicization, as the recent proliferation of tax-exempt organizations in the United States doing overtly political activities attests. The blurring and consequent confusion attending nonprofit/nongovernmental organizations accepting or bending to donors' political goals and agendas have been around for some time, although the consequences were not as evident as they have been in recent years. It is also reflected in the nonstop confusion about the differences between NGOs, nonprofits, advocacy groups, and almost every other kind of organization other than government and conventional business firms. In India, for example, NGOs have so frequently taken critical positions against the government that some NGOs operate virtually as political opposition groups, even if they do not run their own candidates for elective office.

Formal Legal Constraints and Opportunities

Different countries' legal traditions and standards create different incentives, constraints, and opportunities for the creation and operation of NGOs. Opposite trends can be observed in developed and developing countries. Current unsettling changes in US law—generated by the Supreme Court cases and decisions diminishing transparency and broadening the scope of acceptable NGO activities, for example, the *Citizens United* case and several others—stimulated a rush to create so-called 501(c) 4 Public Benefit Corporations, which are essentially politically enabled NGOs.[14] In many developing countries, the ability of NGOs even to survive is challenged by governments' capacity to shut them down through a host of measures. One approach has been to deny the NGO the legal status needed to own property or censoring the NGO with the threat of doing so. Another, increasing common approach is to restrict NGOs from receiving external funding. For example, a 2009 Ethiopian law prohibited civil society organizations from accepting more than 10 percent of their funding from external sources. In Russia, a 2012 law requires NGOs engaging in political activities and receiving any amount of foreign funding to register as foreign agents, which, Human Rights Watch (2014a) notes, is "a term generally understood in Russia to mean 'traitor' or 'spy.'" The even more draconian approach is simply to arrest, or threaten to arrest, NGO members for alleged crimes. In Zimbabwe, the Mugabe regime has long practiced arbitrary arrests under a series of vague laws against defamation or incitement (Human Rights Watch 2014b). Ironically, heavy-handed government actions can have the same result of politicization.

Also in contrast to the distinctively supportive American approach to beneficial institutions, governments ranging from the developed nations of Europe and East Asia to the developing world have either been unwilling to downgrade the dominance of the state in supplying public goods, or deeply skeptical of the challenges that NGOs pose to their authority. European governments do little to encourage private contributions to various socially beneficial institutions. Rather, the state itself is expected to supply these services and resources; the same holds for the relatively statist governments of Japan and South Korea. In Latin America, the statism that has long been invoked to defend government dominance in the economy is reinforced by the parallel doctrine that the economically powerful state should be responsible for social services as well. The problem is that the generally weak tax effort of Latin American governments has left major gaps where, if permitted, NGOs have a major role to play.

Social, Educational, Humanitarian Goals—Or What?

The wide and changing array of goals pursued by the major international NGOs dedicated to development issues add to the many challenges of trying to understand

and improve NGO performance. Perhaps the best example is provided by the Ford Foundation, the largest US foundation until the mid-1990s.[15] It has been a significant and important development change agent for many decades now. Established in 1936 by Henry Ford, primarily as a means to avoid paying federal estate taxes, the vagueness of its original mission statement—"for scientific, educational and charitable purposes, all for the public welfare"—provides scant guidance about what might constitute success in any of its programs, much less in whether it was achieving its overall mission. The personal goals of the Fords to hold onto the family fortune finally began to get clarified in the mid-1950s after Henry Ford II took the company public and professionalized the Ford Foundation's board and staff. He resigned as chairman of the board of the Foundation in 1956 and then as a board trustee in 1976, saying that the Ford Foundation no longer had much to do with the sources of the money that created it.[16] He had already created the Ford Motor Company Fund (FMCF) as a nonprofit controlled by the company to promote local initiatives and projects favorable to corporate ends.

The main point is to call attention to the wide and changing array of goals to which any particular NGO or nonprofit may aspire and adhere. Such goals also include those officially stated and those unofficially but importantly sought. Such goals are also linked to those of the donor or funding agencies in often direct and telling ways. Thus, the Ford Foundation was roundly criticized in the mid-2000s for funding Middle Eastern NGOs that the Foundation claimed were dedicated to human rights and dialogue but others castigated as bent on the destruction of Israel. Whether this funding was intended to serve either educational or charitable purposes, the contribution to the public welfare has been hotly debated,

Sustainability as a Theory—Put into Practice

NGOs are turning their attention to sustainable development around the globe. Conservation NGOs that began in developed countries have greatly expanded their operations into developing countries. For example, the Nature Conservancy, the world's largest conservation NGO with US$6 billion of assets, made its first conservancy purchase in 1955, but only began its International Conservation Program in 1980, focusing on Latin America, with its first expansion beyond Latin America in 1990. It now operates in 35 countries. Policy advocacy has impelled the Natural Resource Defense Council, a US-centered sustainability NGO, to begin work in China, and now has 30 staff members in Beijing, focusing largely on climate change and air pollution.

Domestic sustainability NGOs in developing countries have also proliferated, some within international confederations.

Many of these efforts go beyond influencing governments to include businesses and the private sector generally. Multinational firms have been targeted, and these activities can be observed at the regional and national levels as well. The current status of the NGO/business engagement is far from clear. Collaborations with the private sector are also controversial, especially in cases where supposed conflicts of interest, forthrightness, and even duplicity are open to question and debate (IISD 2013).

The practical implications can and will be revealed through the actions of main development and nongovernmental organizations, such as the World Bank, UN

Environment Programme, the UN Development Programme, regional banks, and unique entities such as pension and sovereign wealth funds. The niche positioning of NGOs such as CERES (Coalition for Environmentally Responsible Economics) as a corporate conscience and of CALPERS (California Public Employees Retirement System) as a public aggregator for political and financial pressure on corporations needs to be monitored in the years to come.

The growing movement to include social/cultural and environmental activities and measures along with conventional economic ones is especially significant. The so-called Double and Triple Bottom Lines attempt to give equal weight to cost/benefit accounting in each of the two or three separate (but related) arenas: economic, social/cultural, and environmental (Elkington 1994, 1997; Savitz 2013; Hacking and Guthrie 2008). When NGOs in developing countries focus on cultural preservation, human rights, and the environment, they almost inevitably engage in policy advocacy to change government policies that contribute to the problems. The focus on poverty alleviation often raises the parallel critique of government policies that exacerbate poverty and inequality. It is no wonder, then, that the expansion of NGO activities beyond simply providing goods and services has dramatically politicized the relationship between NGOs and governments.

These considerations and conditioning factors are admittedly selective and at best offer only a few of the numerous possibilities. Nonetheless, each of them—alone and in different combinations—will likely help shape and direct events in the coming era, which for the present purposes we have labeled Social Impact/Balanced Systems.

Exploring the Next Era of Development NGOs

The historical progression from small and simple NGOs to extremely complex and numerous creates a sense that it may be time to pause, take stock, and rethink or reimagine the entire NGO phenomenon (Lehmann 2013). While there are occasional critiques, usually delivered in anecdotal form, of some specific and not altogether successful operation or organization, or as a proposal for a framework to measure some aspect of NGO performance (Ebrahim and Rangan 2010), there is nothing like a full-scale, comprehensive assessment of NGO experience in the development realm. Despite the paucity of rigorous assessment, NGOs continue to hold the public trust around the world, but with important exceptions. According to the Edelman Trust Barometer (2013), a survey organization that tracks trust and confidence in a range of institutions,

> For the fifth year in a row, NGOs are the most trusted institution in the world, and in 16 of the 25 countries surveyed, more trusted than business. Trust in NGOs has reached a record high of 79 percent in China among 35 to 64-year olds . . . The growth in NGO trust, a by-product of becoming the world's second-largest economy, also indicates that China's people and media outlets are breaking long-standing traditions and now relying more heavily on non-traditional sources for information. Since 2009, trust in NGOs has surged in India to 68 percent among 35–64-year olds.

> But some countries weren't as trusting of NGOs. In markets that dealt with crises and scandals such as Brazil (down 31 points), Japan (21 points), and Russia (14 points), NGOs suffered severe drop-offs in trust.

Doing a thorough appraisal of the NGO considered as a social experiment would be a monumental undertaking to be sure, but it could help clarify many of the issues, conflicts, and considerations we have touched upon so far in this chapter. More importantly, taking the measure of the 70-year-old social experiment that the NGO institutional form presents could result in its refocusing and realignment to deal better with emergent problems of the twenty-first century. Even without such an ambitious project, there are plenty of immediate, basic, and practical questions to answer that have become part of an ongoing discourse:

- What roles and purposes could and should NGOs fulfill?
- How could/should they be funded?
- What are appropriate ways and means to appraise NGOs? And to what extent is public trust and confidence in them subjective and emotional, rather than based on solid evidence of the sort determined in assessments?
- What does effectiveness mean—in general and in specific, important settings where NGOs operate?
- What does it mean to hold an NGO accountable, and if one is found to be deficient, what sanctions exist to rectify matters?
- How is globalization affecting the way NGOs emerge?
- How are the advocacy functions of NGOs affecting their other missions?
- What are the patterns of their relationship with the formal government institutions?
- What type of pressures do NGOs exert on governments to respond to the revolution of rising expectations?

There is one additional, possibly transformational, question to ask:

Is Development Even the Right Purpose?

Poverty continues around the world, of course, but significant progress in some areas is evident in reducing the numbers of human beings subjected to its most extreme forms. According to different sources, about a billion people worldwide have been lifted out of extreme poverty in the last generation. This quote from the *Economist* (2013c, 11) is representative: "Between 1990 and 2010, their number fell by half as a share of the total population in developing countries, from 43% to 21%—a reduction of almost 1 billion people." However, over 70 percent of the poverty reduction in the world was in China, and the situation in some African and South Asian nations is getting worse.

As new MDGs are set to replace by 2015 those first established in 2000, the degree of dominance of poverty reduction as a discrete focus that guided NGOs since the end of WWII, namely providing goods and services directly to the poor, is open for debate. Poverty alleviation will and should remain a major focus: the post-2015 development agenda of the United Nations emphasizes poverty eradication and inclusiveness. The World Bank's very recent strategy aims to limit its programs to the least developed countries in order to alleviate poverty. Indeed, poverty alleviation in the developing and developed world is going to be more prominent as far as the global agenda is concerned. However, the growing awareness of the interconnections between poverty and overall economic policy, environmental justice, social exclusion, property rights regimes, political rights, and governance has meant that for NGOs to contribute most effectively to poverty alleviation, they must sort out which of the many potential roles that they should play: providing goods and services funded through their own efforts vis-à-vis private donors, delivering either domestic government-funded or internationally funded goods and services, or advocacy on any of the issues mentioned above.

The specific contexts and government orientations obviously make a huge difference in how NGOs can address poverty and connected challenges. Getting a billion people out of abject poverty happened for many different reasons. Consider China, whose economic success has created the wherewithal to lift more than 680 million out of extreme poverty, "more than the entire current population of Latin America . . . China alone accounts for around three-quarters of the world's total decline in extreme poverty over the last 30 years" (*Economist* 2013c, 23). The contributions of NGOs to this remarkable achievement have been mixed and certainly are not as important as the force of capitalism and a global market-based economic system (Studwell 2013). NGOs have played an important role in filling gaps in social safety nets after many of the state enterprises were abolished with the introduction of socialist market economy. This is reflected in the number of NGOs over the past few decades that have responded to natural disasters such as floods and earthquakes. Clearly, NGOs did not have the space to play any political and even advocacy role.

In contrast, in many nations of Asia, Africa, and Latin America, the NGOs have more space to go beyond the provision of uncontroversial goods and services. Indeed, even without going into advocacy and direct criticism of policies and practices, the provision of what may seem like straightforward goods and services become controversial. NGOs may provide conventionally Western education in countries where the government or others regard this form of education as cultural imperialism or sacrilege. NGO-supported health clinics often provide controversial family planning services. NGOs involved in providing health services in the Pakistan's Taliban-dominated Swat Valley have been banned, and polio vaccinations halted.

Yet, the central controversy pits the bulk of developing country NGOs against the governments that oversee economic growth *and* poverty alleviation through authoritarian rule. The Chinese example points to a painful dilemma for progressive NGOs committed not only to poverty alleviation but to democracy and also the defense of property rights. NGOs cannot quarrel with the pace of poverty alleviation in China, but the remarkable transition has been accomplished through suppression of civil rights and abrogation of property rights. China's industrialization has required huge inputs of raw materials and energy, prompting major Han incursions into the Uighur regions of western China, undermining Uighur property rights. By the same token, the energy demands have driven the enormous programs of hydroelectric dam construction, displacing millions. And while China is the most prominent example of authoritarian-led poverty alleviation, it certainly is not the only one. South Korea's remarkable growth and poverty alleviation occurred under authoritarian governments; Chile's economic reforms that laid the groundwork for making it the most successful Latin American nation were undertaken under a brutal military regime.

The most complex paradox is that, as the case of China demonstrates, poverty alleviation can go hand in hand with growth income inequality. This is not surprising in light of the fact that the dynamism that brings previously impoverished people into better-paying jobs also, and to a greater degree, benefits those with the greatest assets to take advantage of the dynamic growth opportunities. To be sure, sometimes the rich grow richer at the expense of the poor, through mechanisms of exclusion, such as keeping agricultural prices low so that factory owners can pay lower wages, expropriating the user rights of low-income resource extractors, or using tax revenues to subsidize capital made available to wealthy business owners As long as NGOs are fixated on income inequality per se rather than economic exclusion, they will find themselves opposing some government policies and actions that propel poverty-alleviating economic growth.

Where domestic NGOs have the space to choose among roles, they have to decide what to emphasize: Provide goods and services, whether controversial or not. Keep a low profile to strengthen other civil society organizations. Offer critical advocacy. But, of course, none of these functions is monopolized by NGOs. Government agencies, political parties, religious institutions, international organizations, and foreign governments all provide different means. A crucial opportunity exists to reconsider and rebalance the contributions of private, public, and nonprofit (including NGO) participants in decision making around the world. The roles that *international* NGOs ought to play are conditioned by the decline in NGO funding.

World economic considerations, beginning with the global financial crisis of 2008/2009, are now taking center stage as emerging markets slow down markedly—"Hitting a Wall," "The Great Deceleration," and "When Giants Slow Down," in the colorful terms of three recent reckonings (Cowen 2015; *Economist* 2013a, 2013d). Figuring out why there is a slowdown matters, but for NGOs an obvious and immediate consequence is a drying up of funds. Ironically, the reductions in funds are most evident in countries and in programs where NGOs have been successful. "[I]nternational government contributions are drying up. Between 2005 and 2008, contributions to NGOs in Latin America grew an average of 37 percent. From 2008 to 2011, the donations grew just 3 percent, well below the average regional inflation of 6.1 percent for 2012" (Brodzinsky 2013).

Cutting back funds for NGOs means setting priorities and deciding what to support and what to terminate. In countries where some measure of economic development success exists, this presents serious threats because traditional donors are now made to refocus on hard and basket cases while reducing or even ending traditional support for their relative successes. To a substantial degree, this will be shaped by the governments of wealthy countries. Consider the US context: it is not just the US NGOs that determine the magnitudes and contents of NGO assistance to developing countries; the USAID is a major funder and motivator for many development NGOs. Much of the USAID on-the-ground work is conducted or at least managed by US NGOs. USAID has been a mainstay, but it has never achieved high levels of government support, at least as compared to equivalent foreign aid organizations in other developed countries around the world. Sweden, for instance, aims for a 1 percent of GDP allocation to foreign aid and so would have spent some $5.8 billion to this end in FY 2014. Were the United States to spend an equivalent proportion or fraction in foreign aid, the $156 billion result would far surpass the actual $20.4 billion requested in the FY 2014 budget.[17] Sweden obviously is not fighting multiple wars around the world, nor does it merge military and development goals and objectives in the ways that the United States has done since WWII.[18] Nonetheless, as the US public loses its taste for foreign military adventures, the relative importance and role for development programs could be reassessed and increased.

A large number of possible pathways exist for NGOs as one peers out into the next era. No one should dare to know and predict what will happen, although many will undoubtedly try as in the past—but with little or no success.[19]

So how can we at least explore the future and perhaps identify problems that might be minimized or averted in the first instance and discover opportunities that could be maximized or at least taken advantage of in the second? One way is to create scenarios, many of which start with plausible extensions of past trends and conditions and then extend these forward. For shorter term and stable systems, such scenarios are often helpful (Brewer 2007). For the longer term and in dynamic and unstable situations—such as what appears to be the case for NGOs at the moment—simple extrapolations will not yield much of value. More creative uses of scenarios exist and include Best and Worst Case exercises and also explicitly normative ones where desired end-state conditions at some specified time in the future are postulated and then actions to get from the current "here" to the desired "future" are explored.

The following brief discussion sketches out four broadly drawn scenarios that are constructed in the spirit of exploring future possibilities. It is very important to keep in mind various past theories, doctrines, and practices because these will continue well into the future if, for no other reason, than the sheer inertia that they exhibit. The fabled "sunk costs" that economists tell us to ignore are in reality often with us institutionally long after they should be.

To illustrate, we imagine and present four different theories of NGOs in a scenario format. These include Economics/Growth, Resiliency, Sustainability, and International Rapid Response.

Economics/Growth

This is built upon the assumption of continuing economic development of the sort realized in the last 25 to 30 years. The social safety net component of the Washington Consensus—present from the beginning as the international development institutions' doctrine but often ignored by governments—will continue to grow in importance, but for many governments, growth is likely to remain a higher priority than direct redistribution. The pairing of liberalization and democracy is challenged by the so-called Beijing Consensus of state capitalism and authoritarianism (*Economist* 2013b, 10). The attractiveness of this model will wax or wane depending on the outcomes of China's slowing growth and disruption. Population growth—sheer size—matters as it naturally generates its own demand in the form of additional consumers. Environment, equity of wealth and other value distributions, and many other longer term outcomes and effects are temporarily subordinated to wealth accumulation in the form of high annual growth rates.

The roles for NGOs within this set of scenarios continue to be marginalized or contested. China, one can see, fits this picture. What of the other so-called BRIC (Brazil, Russia, India, China) nations that have received so much attention as the rising economic powerhouses? Brazil and India may be seen as promising models of growth with democracy, but both are likely to continue to put faith in and so rely on growth to a considerable extent. Both have high-profile affirmative action plans (for Brazil's Afro-Brazilians; for India's dalits [untouchables], adivasi [tribals], and Other Backward Castes), and Brazil has a prominent conditional cash transfer program for low-income families. However, the income opportunities and affordable social services are still low for the very poor in both countries. Moreover, Brazil's economy is slowing down and the political consequences are encouraging politicians to renew efforts to stimulate growth. India is the world's largest democracy, but it is still unable to meet basic economic and many other needs of its citizens. This contradiction is forcing important policy conversations there about stimulating business growth as a way to solve chronic problems. The increasingly free market economies such as India and Brazil will continue to provide more space to NGOs in influencing economic and political decisions in the future. Fobbing off many public services on NGOs has not worked, for many of the reasons Arundhati Roy (2002) anticipated years ago. Nor is simply being a democracy the same as simultaneously having an efficient or effective public sector (Drèze and Sen 2013). Thus, one of the potential challenges is that as NGOs continue to expand their influence, government and public sector capacities are being weakened. It would lead to more failed and failing states, but with strong civil society in the future.

The longer term concerns neglected in the pursuit of growth will intrude, often dramatically, in the form of extreme pollution events, worker abuses, or climate-related natural disasters. The traditional way of dealing with such events is on a one-to-one basis and in such a way so as not to impede economic growth. The rise of mega-cities presents a very different longer term concern that will not be so readily addressed. The persistent patterns and problems of a single-minded pursuit of growth, and the neglect of the social/cultural and environmental costs and consequences of it, will mount. There are hints that these pressures open up chances for NGOs to operate more broadly in contexts like China:

> Society is becoming too complex for the old structures to handle. Hence the [Chinese] government's decision to allow the development of what it calls "social organizations." In essence these are NGOs. The party dislikes the idea of anything non-governmental and has long regarded NGOs as a Trojan horse for Western political ideas and subversion, but it is coming to realize that they could solve some of its problems—caring for the sick, elderly and poor, for instance. The growth of civil society is not just important in itself. It is also the bridge to the future, linking today's economic reforms to whatever putative future political reform might come. (*Economist* 2013a)

In this case, the Overreach and Push Back of what we have labeled the end of Era 4 could be yielding to a more balanced systemic one, in which many social functions and needs are met by social organizations whose impacts are enhanced by technologies and super networks of social media and scientific support.

Resiliency

The concept of resiliency owes much to engineering and the ecological sciences, and it is the latter version that will dominate in coming years as climate change forces societies around the world to adapt and otherwise face an array of challenging outcomes and effects (National Research Council 2009, 2010a, 2010b). The idea of resiliency is not the same as the one of stability, as clarified by noted ecologist C. S. Holling (1973, 14), the creator of resiliency theory:

> It is useful to distinguish between two kinds of behavior. One can be termed stability, which represents the ability of a system to return to an equilibrium state after a temporary disturbance; the more readily it returns and the less it fluctuates, the more stable it would be. But there is another property, termed resilience that is a measure of the persistence of systems and of their ability to absorb change and disturbance and still maintain the same relationships between populations or state variables.

Climate stability is at risk as a consequence of human-caused perturbations; the resilience of ecosystems is thus becoming a matter of increasing concern. The joining of natural and human systems is a defining characteristic of the resiliency scenario, which draws heavily on the Theory of Global Necessity discussed previously.[20] Among other fears are potential violent conflicts that may result from climate change sources (Scheffran et al. 2012).

Short-run natural events trigger basic humanitarian relief measures of the sort where NGOs have long performed. Disaster relief and taking care of refugees—in this case climate refugees displaced by hurricanes/typhoons, flash floods, and prolonged heat waves, and other natural events such as tsunamis and fires—will figure prominently. Longer term natural events such as droughts, crop failures, land submergence due to sea level rise, water shortages, and resultant conflicts will also demand and stress civil society's full array of resources.

Technological and scientific demands are also considerable and include monitoring and measurement systems, long-range forecasting, computational requirements, satellites, computer modeling, early warning systems, and so forth. The information technologies required are formidable; many probably have not been conceived. Lessons being learned in a cascade of natural disasters in recent years will eventually refine responsibilities for all three institutional sectors—public, private, and, most especially, nonprofit/nongovernmental ones.[21]

Developing all these means and then linking them into effective super networks capable of monitoring, measuring, and serving decision makers and the public around the globe will demand far more from NGOs than they have ever been asked to provide.

Sustainability

As discussed previously, sustainability is a strong conditioning factor for many potential developments in Era 5. Sustainability, taken together with many of the core ideas and practices in the social enterprise/social impact fields, is already actively involving NGOs in new and interesting ways (IISD 2013). The Triple Bottom Line (TBL) approach to sustainability forces businesses to include many others in addition to their shareholders and employees. Innovations from the social enterprise/social impact fields come at the TBL from a different angle as they start by identifying unmet social needs and then devise ways to address and meet them. Governments are involved in many different ways, as they set and enforce environmental goals and expectations and also oversee and monitor a range of social and cultural activities commonly used to describe the social/cultural dimensions of the TBL.

A very practical list of what needs to be measured in each of three components of the TBL reveals opportunities for NGOs to participate constructively in scenarios based on the theory and practices of sustainability—particularly with respect to environmental and social/cultural elements (Slaper and Hall 2011).

> Environmental variables should represent measurements of natural resources and reflect potential influences to its viability. It could incorporate air and water quality, energy consumption, natural resources, solid and toxic waste, and land use/land cover. Ideally, having long-range trends available for each of the environmental variables would help organizations identify the impacts a project or policy would have on the area.
>
> Social variables refer to social dimensions of a community or region and could include measurements of education, equity, and access to social resources, health and well-being, quality of life, and social capital.

Many of the specific variables required in the TBL regime are the consequence of activities done by NGOs and governments and so present opportunities to work in partnership with businesses to elaborate and enrich the strong economic variables they have to offer.

A rapprochement between neoclassical and ecological economics is a strong possibility and would be most welcome theoretically to redirect and guide economic and business practices around the world. Various new movements and fields, such as industrial ecology, urban and regional symbiosis, the Natural Step, Cradle to Cradle, the Blue Economy, the Circular Economy, systems thinking, ecosystems, food systems, and watersheds, will be available to help puzzle through, balance, and manage human and natural systems. Partnerships, networks, and closer involvement of educational and scientific disciplines and practitioners are the expected model characteristics in the sustainability scenario for Era 5—Social Impact/Balanced Systems.

International Rapid Response

The number, scale, and devastation of natural disasters call for extraordinary measures in response. War-like images abound in the aftermath of typhoons, hurricanes, floods, tsunamis, forest fires, droughts, earthquakes, volcanoes, and other disasters.

Increasing concerns about human health threats on a large scale present similarly huge challenges to conventional organizations. In many of these situations, war-like images are reinforced by the presence of regular military forces bringing their own talents and resources to bear for humanitarian objectives. Good and humane intentions aside, the blending of military and humanitarian cultures, people, and organizations is seldom easily done.

The core idea of the international rapid response scenario is to capitalize on the comparative advantages of military, nongovernmental, and private organizations in the interests of responding to natural and human-caused disasters. For instance, strategies and doctrines based on past lessons learned, preplanning, prepositioning materiel, field exercises, C^3I (communications, command, control, and intelligence), and other routine military activities can all be applied to humanitarian and disaster situations. The venerable idea of problem-specific military task forces is almost ideally suited to the needs of challenging, very context-specific humanitarian and disaster needs. "What is the problem at hand and what needs to be done?" is the first question to ask. Next it is necessary to find the personnel and other resources that are likely to manage and solve the problem. This is the basic task force. Next, one needs to formulate an operational plan that pulls together and creates the processes and procedures required to blend forces and resources and provides the steps needed to deal with the problem.

What distinguishes the Rapid Response scenario from the ad hoc ways disasters have typically been handled is that the military is treated as if it were an essential partner with the full range of the usual NGO and other organizations we rely on. For the military, this means creating, redefining, and then committing to a new doctrine and a new way of doing business. It will not be easy, as the peripatetic experiences over the years with counterinsurgency as a doctrine to win hearts and minds readily attest. Nor will it be simple or easy for traditional lead NGOs such as the Red Cross, Doctors without Borders, Save the Children, Oxfam, Direct Relief, and many others to accept the military as an equal partner for the long haul.

The Intellectual Challenge Ahead: Assessing the Future of NGOs

NGOs play many different roles and assume a wide array of forms and functions. The past seven decades since the end of WWII present a complex story and a collection of experiences that reveal successes, failures, and everything in between. This story has not been adequately told, as our suggestion to mount a full-scale assessment of the NGO as social experiment suggests. Now is the time for such reflection and redirection for NGOs, and we hope that the present review will encourage many others to embrace the future with confidence.

CHAPTER 7

Evolution of Foreign Assistance Theories and Doctrines

This chapter traces out how foreign assistance has been conceived and rethought throughout the post-WWII period. Whether in the hands of the US government, the Soviet Union, Western European nations, the East Asian Tigers, or international organizations, foreign assistance has served—but also disserved—both donors and recipients.

The optimistic premise of foreign assistance as a remarkable boon to nations in need of development or reconstruction has often been borne out. From the US Marshall Plan that restored much of post-WWII Europe to the remarkable success of heavily aided Tigers, foreign assistance has had some impressive successes. Major recipients such as Japan and South Korea have become major foreign assistance donors themselves. Yet the evolution of theories and doctrines must respond to the negative experiences of foreign assistance as well, when the impacts have been ineffective, politically disruptive, supportive of undemocratic governments, environmentally destructive, and regressive in terms of income distribution. Critics of foreign assistance have also accused both donor governments and multilateral foreign assistance agencies of distorting the economies of recipient countries, through policy conditions or trade commitments that limit economic opportunities. There are also complex questions as to whether foreign assistance funds end up funding the military of other purposes beyond humanitarian or developmental assistance.[1]

Thus, in this chapter we review the evolution of theories that have established, and then altered, foreign assistance doctrines. Some of these theories link development to democracy; others see foreign assistance as the linchpin to the maintenance of international alliances or to induce peace among otherwise belligerent nations; yet others relate foreign assistance to strengthening the donor country's economy. And, of course, many of the personnel of foreign assistance agencies ascribe to the belief that their role is to aid the residents of the recipient country; different theories have been embraced to guide the foreign assistance strategies to accomplish this.

It is important to keep in mind that although US bilateral foreign assistance dominated for several post-WWII decades and resumed its leading role when Japanese assistance plummeted,[2] foreign assistance from other nations and from multilateral institutions has been important in both volume and orientation. Foreign assistance from the Soviet Union, Japan, EU countries, other developed nations, and now China has had major impacts politically and economically and has shaped US assistance in reaction to the assistance of others. Multilateral assistance, particularly from the World Bank and the regional development banks, has had parallel impacts.

Theories

Despite the eventual relevance of foreign assistance from countries other than the United States and the Soviet Union, for at least the first three post-WWII decades the primary arenas of theorizing about the purposes and modalities of foreign assistance were within the governments of these Cold War rivals.[3] For US policymakers, the remarkably successful reconstruction of Western Europe seemed to hold the keys to the development of the Third World: a large infusion of foreign assistance, *if* properly deployed, could spur economic growth and in turn support democracy and positive relations with the United States. The European Recovery Program, popularly known as the Marshall Plan, was remarkable in its economic support of the vanquished Axis countries.

The Marshall Plan reveals two early theoretical perspectives on rationales for US foreign assistance. Foreign assistance would foster democracy, and it would be an essential instrument in the rivalry with the Soviet Union.

Foreign Aid and Democracy

The disastrous precedent of the treatment of Germany following WWI reinforced the logic that avoiding economic adversity would reduce the chances of political instability, and political instability runs the risk of undermining the democratic progress. Prosperity was seen as necessary to safeguard the democratic constitutions resumed by, or imposed on, Western European countries, justifying the infusion of foreign assistance.

For developing countries, a more complex theory emerged. So-called modernization theorists[4] argued that developing country societies were mired in low expectations, limited political participation, elite dominance, and rigid structures offering very limited socioeconomic mobility. Democracy was essential both for its own sake and to inspire confidence that broadening the voice of nonelites would raise expectations and spur economic mobility. Despite this diagnosis of deep economic, political, and social problems, the prognosis was optimistic, at least in the pronouncements of public officials and modernization theorists—if the big push could be mounted alongside of proper direction. Progress in social mobility and standards of living would reinforce the preference for democracy, associated with these positive economic patterns because of the tutelage of the United States. This would be supported by US funding, calculated according to each country's absorptive capacity based on targeted investment and growth rates. As mentioned in chapter 2, the volume of capital to achieve a given rate of economic growth could be calculated—assuming that the capital would be effectively used. To escape the low-level equilibrium traps, social and economic restructuring was needed, but governments had to be induced to engage in this restructuring.[5] This required recipient countries to engage in comprehensive development planning with heavy input from US officials and development experts. This was in stark contrast with the approach of the Marshall Plan and subsequent Europe-oriented foreign assistance programs[6] that permitted European officials to take the lead in devising the recovery packages. The Soviet Union was not the only world power that could offer ambitious, highly visible plans—so could the United States, despite the prevailing American disdain for central planning.

The theory-driven differences between the approaches to European reconstruction and Third World development are worth noting. The Marshall Plan and subsequent Europe-oriented foreign assistance programs were remarkable in the lead that the European officials were permitted to take in devising the recovery packages; developing countries required externally imposed conditions to overcome the imbalances that the policy elites were otherwise unwilling to address. A long-lasting consequence of this diagnosis has been the highly directive nature of US interactions with developing countries in establishing assistance-supported programs. The Kennedy administration reinforced the strongly directive approach by insisting that for Latin American countries to qualify for foreign assistance through the major Alliance for Progress program, they would have to file medium-term (e.g., five-year) national economic plans.

The striking contrast between the US approach and the Japanese foreign assistance approach that emerged later, once Japanese transfers shifted from reparations to aid to other Asian nations, growing dramatically in the 1980s is worth noting.[7] The Japanese approach was to accept projects as long as they were within the allocation for that country and met technical quality standards. Fairly strong signals were sent that physical infrastructure projects would be welcomed, which probably induced recipient governments to request funding for such projects. Japan's strong heavy equipment sector and engineering expertise reinforced this orientation, but within this sectoral emphasis, Japanese assistance was far more passive in accepting the recipient government's project selection.

Foreign Aid and the Cold War Rivalry

The rationale of foreign assistance as a Cold War instrument was based on the premises that strategic countries may be more favorable to the more generous Superpower and that allies required support to fend off clients of the other Superpower. In this competition over hearts and minds, the logic was that assistance ought to go where the security threats were greatest; where support in bodies such as the United Nations could be most useful (Rai 1980); and, if the ally had been successful in direct or surrogate wars, much assistance had to go into postwar reconstruction to consolidate the victory. Some developing countries permitted US military bases on their soil as a deterrent, but foreign assistance has also been an allure for agreeing to host US and allied troops, even if increasing the security risk of the host country.

In addition, even today some of the economic development funds may be directly or indirectly channeled into military assistance. USAID, the primary American foreign assistance agency since 1961, has sometimes commingled development assistance and military assistance. In addition, if foreign assistance covers investments that the government would have made with its own funds, the freed-up funds can be used for other purposes. This fungibility is the development practitioner's nightmare but can be the defense policymaker's boon. Insofar as officials of recipient countries can forgo spending the nation's foreign exchange on development projects, these funds can be devoted to defense materiel and other military-related spending. However, fungibility can cut both ways. If a recipient country's government leaders feel compelled to devote a given amount to security, military and other forms of security assistance can free up budget resources for nonsecurity purposes. In Colombia, for example, US military support against guerrilla groups, inextricably commingled with the antinarcotics initiative, received nearly US$2 billion from 2000 to 2013 directly from the US Department of Defense, in addition to more than US$4 billion coming from the US government's Andean Counterdrug Initiative/Andean Counterdrug Program. The total US funding of Plan Colombia, the omnibus plan to combat the drug trade and the guerrillas, amounted to over US$9 billion for this period. To complicate the situation even further, some of the monies that would appear to be for the antidrug program actually go to USAID.[8]

The objectives of Soviet foreign assistance were more limited, in part because the issue of international trade was less important for the internal trade among Russia, Ukraine, the Baltic republics, and many other republics within the Soviet Union. Trade with the other Warsaw Pact countries did not need the inducement of foreign assistance. Therefore, the Soviet foreign assistance was driven more strictly by geopolitical concerns, split between support for the most acute US-Soviet conflict areas, particularly Vietnam and Cuba, and overtures to major nonaligned countries, particularly Egypt and India. Some of the financing of Soviet foreign assistance was accomplished through the Council for Mutual Economic Assistance (Comecon), a consortium of Warsaw Pact countries, Cuba, Mongolia, and Vietnam.[9] In a sense, therefore, Eastern Bloc foreign assistance was multilateral, though with clear dominance by the Soviet Union in contrast to the relatively less influence of any single national over the World Bank and major regional development banks (African Development Bank, Asian Development Bank, and the Inter-American Development Bank). Because of the relative wealth of the Eastern European countries, they bore a disproportionate burden in providing assistance to recipient countries: Cuba, Vietnam, and, though to a lesser extent, Mongolia. Foreign assistance to Cuba was reciprocated by Cuban military activities in Latin America and Sub-Saharan Africa; assistance to Vietnam was but one facet of Soviet rivalry, first with the United States and then with China.

Comecon foreign assistance beyond the Comecon countries themselves was spread very thinly (62 developing countries in 1970, over 100 in 1985 [Goodrich 1989]), clearly intended to provide at least a presence and basis for government-government relationship in so many countries, but the bulk of the funds going beyond the Comecon countries were focused on swaying hitherto nonaligned governments (e.g., Egypt received more Soviet aid in the 1960s than any other nation beyond the Soviet bloc [El Beblawi 2008, 21]) or supporting successful leftist regimes (e.g., the MPLA in Angola, the Derg regime in Ethiopia, and the FRELIMO in Mozambique).

Post-Cold War Geopolitical Objectives

The demise of the Soviet Union certainly did not put an end to the geopolitical and defense rationales of foreign assistance, for the United States and other NATO members, resurgent Russia, emerging East Asian nations, and oil-rich Middle Eastern nations. The post-9/11 concern with terrorism gripped both East and West. The US war on drugs extended from Latin America to Afghanistan. Direct and proxy wars in Iraq, Afghanistan, and Syria, and the chronic confrontations in Egypt and Israel, have prompted development and military assistance from NATO members, Russia, and the Middle Eastern oil exporters. The confrontation over the South China Sea is likely to drive the foreign assistance calculations by China, Japan, and South Korea vis-à-vis the Southeast Asian nations. In short, the post-Cold War peace dividend has not reduced the relevance of geopolitical considerations in foreign assistance.

Economic Theories

While political and security concerns dominated Cold War bilateral foreign assistance approaches, and still do in significant respects, theories concerning the economic impacts on both recipient and donor countries have also been important in shaping the targets and modalities of assistance.

Dominance of Physical Infrastructure Projects

If large volumes of capital were to be transferred, what sectors should be favored? The Japanese were not unique in placing heavy emphasis on physical infrastructure. Physical infrastructure could remove the bottlenecks holding back productive activity by enabling greater transportation, energy, and communications capacity. In principle, physical infrastructure can enhance social services (e.g., through greater access to rural schools and health clinics) as well as manufacturing, agriculture, extractive industries, and banking. Because of the lack of engineering expertise in many developing countries, aid agencies had expertise as well as money to contribute.

For the donor, physical infrastructure projects meant business for its construction companies, heavy equipment manufacturers, and engineering consulting companies. Reconstruction assistance even more clearly called for physical infrastructure assistance to restore what the conflicts had destroyed. Donors with heavy machinery industries would have the greatest capacity to maximize ties through construction projects as opposed to other sectors within the recipient country. Thus, Soviet assistance also concentrated on physical infrastructure, demonstrating Soviet engineering prowess and providing outlets for heavy machinery and engineers. Initially the industries to benefit both from infrastructure expansion and direct industrial support were involved in import substitution industrialization; in the 1970s and 1980s, the emphasis shifted to export-oriented industries (Goodrich 1989).

Industry Emphasis Giving Way to Agricultural and Social Sector Concerns

As reviewed in chapters 2 and 3, early post-WWII development thinking strongly emphasized manufacturing and other aspects of industrial development, on the grounds that industry had more promise than agriculture, with its limits on land area. Therefore, the concern over agriculture was largely confined to restructuring land holdings rather than directing money to improve agricultural productivity. In fact, through price ceilings on food that allowed industrialists to offer lower wages, capital was drained out of agriculture. The initial bias in favor of industry was reinforced by the attractiveness of infrastructure, inasmuch as many projects such as roads, power plants, and ports can be directed to industrial development. In addition, on the assumption that industrial jobs would come with training, little external assistance was addressed to education. Development assistance for health care has also been largely neglected and disorganized, leaving coordination to the poorly funded World Health Organization and its regional affiliates.

The rethinking of sectoral theory was driven by greater recognition of the multiple problems of declining agricultural productivity, famine, rural poverty, inefficient protected industry, overcrowding of cities due to flight from the countryside, and the loss of foreign exchange because of food imports. The new theory was that in light of the unwillingness of national leaders to focus adequately on agricultural development, foreign assistance was needed to supplement agricultural investment and to influence governments to reorient their proindustrial strategy.

In addition, the disappointing progress in human resource development in many developing countries, linked to industrial stagnation, prompted rethinking of whether education and health improvements would come naturally through economic growth. The impressive economic performances of Japan, South Korea, and Taiwan highlighted the strong commitment of all three countries to education and health care—whether or not these countries, with their distinctive nature and histories, would be good parallels to other developing nations. When the World Bank under Robert McNamara (1968–1981) and some of the regional development banks adopted the poverty alleviation doctrine, reinforced (as noted in chapter 2) by the basic needs theory, the multilateral foreign assistance agencies faced a difficult theoretical quandary. On the one hand, comparative advantage theory argued that each institution ought to concentrate on doing what it did best. On the other hand, insofar as these institutions served as exemplars and thought leaders on how development ought to be pursued, neglecting particular aspects of the growing agenda of development summarized in chapter 2 might send a signal that neglected aspects ought to be neglected by other institutions, including the recipient country governments. With the partial exception of the Asian Development Bank, dominated by Japanese funding and maintaining a predominantly infrastructure focus, the development banks have greatly broadened their foci. They diversified their project staffs beyond the prior near monopoly of economists and engineers, to include agronomists, education specialists, anthropologists, sociologists, environmental experts, and so on.

One consequence of the expanded range of the sectors targeted by foreign assistance is the greater difficulty of assessing the degree of project or program success. As reviewed in chapter 2, the technical means of gauging the societal rate of return for a particular development initiative is more problematic for social sector efforts (Jimenez and Patrinos 2008). While the use of rate-of-return analysis (i.e., benefit-cost analysis) has been declining across all sectors, it has declined much more precipitously for education and health projects.[10] This means that foreign assistance officials are less accountable from a technical perspective, but it also means that they lack the opportunity to invoke benefit-cost analysis to defend their efforts.

Land Reform

One of the major aspects of attempted leverage through foreign assistance was to urge governments of countries with high levels of land concentration and landlessness to engage in significant land reform. The Japanese, Taiwanese, and South Korean economic successes were accompanied by major land redistribution; this seemed to be a way of assisting the rural poor in other countries without the need to raise agricultural prices. For Latin American countries, the qualification for USAID Alliance for Progress funding had to mount land reform efforts. And although some governments passed laws but were lax in implementation (Feder 1965), land reform initiatives provoked strong resistance, and even civil wars (Lipton 2009).

Political Economy Theories

The political economy considerations faced by foreign assistance donors and recipients center on different aspects of influence, reciprocity, cooperation, and autonomy. Thus, the conditions that assistance providers require—or at least attempt to require—of recipients; the actions of recipient governments to protect their autonomy; and the tradeoffs between gaining influence through nation-nation assistance vs. multidonor collaboration have all strongly shaped foreign assistance doctrines.

Fundamental Exchanges

The most fundamental political economy premise of foreign assistance is that governments of donor countries often exchange economic support, through bilateral or multilateral channels, for some combinations of (1) favored status with the government and population of the recipient country, (2) greater support from constituents within the donor country, and (3) favored status with other donor governments. Much paper has been wasted by efforts to determine which objective *generally* dominates the motivation of the donor. It should be clear from the fact that both donors and recipients face great variation in context that the donor motivations vary as well: trade, diplomatic support, security alliance, drug interdiction, and/or access to raw materials from the recipient country; political popularity within the donor country and/or support for companies exporting to the recipient country; stronger alliances with other donor countries, and so on.

However, some of these exchanges may backfire. The favorability with the recipient government may be jeopardized if the aid is perceived as excessively manipulative from the perspective of the recipient government or other actors within that country. Strengthening military alliances may exacerbate tensions with other countries.

Constituents within the donor country may resent the fact that their taxes are going to other countries, or they may object to the support of governments of which they disapprove.

Rationales of Multilateral Assistance

Because the grants and concessional loans channeled through multilateral foreign assistance come predominantly from the funds or guarantees of individual developed countries, it may seem irrational for these governments to forgo taking direct credit for their transfers. However, collective action theory clarifies that governments interested in maximizing the total volume of funds for development have an incentive to enter into agreements that bind multiple countries, with defection either infeasible or costly. A government considering defecting may face criticism from both developed and developing countries. It is also possible that a government wishing a recipient nation to receive a given level of foreign assistance would be deterred from free riding if that meant that the multilateral collective initiative would collapse as other donors defect. For example, when the US government moved to withhold its contributions to the World Bank Group's soft-loan International Development Association (IDA) facility, other governments enacted triggering mechanisms to reduce their contributions, jeopardizing the entire IDA initiative until the US government largely relented (Kapur, Lewis, and Webb 1997).

The second rationale for contributing to multilateral assistance at the expense of losing bilateral credit is that direct bilateral relations may not be as effective as the indirect impact through multilateral assistance and influence. Influence from multilateral institutions may be more palatable politically within the recipient country than from a single nation vulnerable to accusations of neoimperialism. By the same token, donor country leaders who wish to see a recipient country receive more assistance may need to circumvent opposition to direct bilateral assistance within the donor country. It may be politically contentious within a donor country to provide direct assistance to problematic government of a particular recipient country. In short, multilateral assistance channels can depoliticize this assistance. The obvious tradeoff is that the donor country government loses some leverage based on the threat to cut off assistance.

NGO Assistance

As chapter 6 notes, international NGOs devoted to development assistance have proliferated, and the financial resources that they command have grown dramatically. For the United States, at least, this is a direct consequence of the broad policy doctrine of promoting private voluntary contributions through tax exemptions for both domestic and international assistance, making the total of US development assistance highly distinct in comparison with the typical state-centered assistance of other donor nations.[11] Davis and Gelpern (2010, 1211) concluded that "Foreign assistance from private sources is estimated to have reached $49 billion in 2007—just short of half of its official counterpart, which stood at nearly $105 billion. By some estimates, private aid for development is approaching the level of bilateral official development assistance." This reflects the combined effect of US bilateral aid dominance in the total volume of assistance and the high proportion of US assistance from private donors. While this does not mean that US development assistance is more generous on either a per capita basis or as proportion of GDP, it does mean that the emphasis of US NGOs on the social sectors has greatly increased the overall contributions coming from the United States.

Aid Coordination

Similar premises and tradeoffs hold for the coordination of aid among bilateral agencies, multilateral intergovernmental institutions, and nongovernmental institutions. The dilemma for a bilateral foreign assistance agency begins with the risk that the credit to the donor country may be diminished if its aid is not distinctive from that of other donors. There is also the administrative burden of coordination. On the other hand, uncoordinated aid is often less effective, with fragmented efforts lacking the best scale for maximum effectiveness. In addition, the effort at coordination can reinforce the collective action logic, as joint initiatives may entail highly visible commitments.

Aid coordination has had to include nongovernmental assistance providers as well. Moreover, the UN Charter recognizes NGOs as an integral participant in UN activities. In addition, emerging theories of aid effectiveness recognize the utility of working through civil society, which includes NGOs within the recipient country as well as grassroots organizations. Bilateral aid agencies, particularly of Western European countries, have long endorsed the premise that working through NGOs, both international and those within the recipient country, strengthens civil society while also increasing the likelihood that assistance will be responsive to local needs and wants. In terms of the multilateral agencies, the UNDP, which has lacked the huge financial resources requiring channeling through large infrastructure projects under government control, has had a long history of NGO collaboration. When James Wolfensohn took over as president of the World Bank in 1995, he gave international NGOs far more presence in World Bank deliberations.

However, the overall coordination imperative has added to the bureaucratic overhead of foreign assistance. On some issues, such as food security, new multilateral entities have been established primarily to coordinate, yet the bureaucratic burden and the lack of clear benefit to their efforts have led to their dissolution. For example, the World Food Council was disbanded in 1993 as not worth the expense of a separate organization.[12]

Theories Regarding NGOs

The enthusiasm of foreign assistance agencies toward NGOs has somewhat cooled because of growing awareness of the divergence of the interests between NGOs and grassroots organizations.[13] Even if leaders and members of international or domestic NGOs truly believe that they are acting in the best interests of the societies they serve, the time scales may differ dramatically. For example, environmental NGOs typically emphasize conservation, which often clashes with the short-term interests of local people. Many domestic NGOs have strongly antigovernment orientations that local groups may not share, and the NGO's confrontations with government may enmesh others in dangerous clashes. For bilateral or multilateral assistance agencies, collaborating with NGOs can sour relations with the government of the recipient country. Therefore, the emerging doctrine is to work more directly with grassroots organizations and to work with NGOs that are less contentious.

Another consideration regarding working through NGOs is that the political credit that a donor government can gain with the government of the recipient country may well be lower unless the flows are government-government.

Theories of Conditionality

The controversies over the bilateral or multilateral conditionalities that impose policy reform requirements on recipient countries rest on different political economy theories that have normative as well as practical implications. No other aspect of the relationship between donors and recipients has been more contentious than this issue.

Defenders of conditionality rely on five premises. First, they must presume that the reforms are appropriate. For assistance to be effective, government economic policies must be sound (Burnside and Dollar 2000); therefore, the precious foreign assistance resources would be wasted unless recipient governments are pressured into adopting sounder policies. Second, if government officials do not want to act in the best interests of the country, conditionalities can pressure these leaders to accept appropriate reforms. It is worth noting that during the debt crisis decade of the 1980s, when many developing country governments faced difficult economic and political conditions that discouraged adopting painful economic policy reforms, the conditionalities became far more specific (Kapur and Webb 2000, 2).

The third premise is that if appropriate conditionalities are not honored, the government officials may be punished by the failure to secure the assistance. This presumes that failure to accept or maintain appropriate conditionalities would result in withdrawal or reduction of the foreign assistance.

Fourth, conditionality communicates to other nations what the foreign assistance experts regard as appropriate policy reforms. The validity of this premise depends not only on whether the conditionalities are sound but also whether the criticism mounted against them erodes their credibility.

Fifth, from the perspective of the recipient country, the positive aspect of conditionality is that external pressure by international organizations can reduce the political costs that government officials would face in enacting appropriate reforms that otherwise may be too controversial to enact. "The World Bank (or the IMF) made us do it" is a common refrain.

Yet government officials of the recipient countries also recognize that foreign assistance agencies face their own pressures to expend the resources available to them and to fulfill their mandates to promote development. As mentioned in chapter 3, the international entities requiring conditionalities are often loathe to terminate an agreement even if the conditionalities are not met; the collapse of an agreement has costs to both the people of the country and the conditionality imposer: without an agreement, the shortage of borrowing capacity can seriously damage the economy, and the relationship between the external entity and the government is likely to be damaged. Government officials of countries under conditionality constraints know that under some circumstances the risk of losing the support because of noncompliance is low. Many agreements over the decades have simply been renegotiated, with different, often weaker, conditionalities (Mikesell 1983; Boughton 2003).

The critiques of conditionality vary. One argument is that conditionalities are likely to be inappropriate. This may rest on the assumption that the recipient government's actions, without the pressure from assistance providers, would be more appropriate, reflecting a deeper presumption that the recipient government acts in the interests of the country. This is often paired with the skepticism as to whether the donor has the interests of the recipient nation at heart. In addition, a common premise of those who are skeptical of the soundness of conditionalities is that external actors, even if well-intentioned, do not know enough about the political, economic, and social context to judge what is best.

If the conditionalities are regarded as appropriate, imposing them may still be regarded as counterproductive. From the perspective of the donor government, conditionality may reduce the credit that the recipient government and population will grant to the donor government and country, insofar as conditionality is perceived as heavy-handed and self-serving. Moreover, as briefly mentioned in chapter 2, even if the conditionalities are regarded as appropriate, the question of whether the recipient government has the political will or capacity to enact the conditionalities it has formally accepted has led to skepticism toward conditionalities that are not owned by the recipient government.

This last concern has given rise to two somewhat new doctrines. One is government ownership: the government must embrace the reforms enshrined in conditionalities. For the international financial institutions, this came into prominence in the early 2000s. Boughton (2003, 3) of the IMF staff reported,

> When the IMF embarked on a reexamination of it policies on conditionality in the millennium year 2000, a key objective was to promote national ownership of policy adjustments and structural reforms. It was clear from experience and from formal studies...that the main reason for failure of Fund-supported programs to achieve their objectives was that governments too often did not implement policies to which they had committed. Whatever could be done to deepen and strengthen commitment was likely to improve implementation and raise the success rate.

Boughton acknowledges that the phrase national ownership is difficult to define "with sufficient precision to make it operational"; however, the IMF developed a working definition: "Ownership is a willing assumption of responsibility for an agreed program of policies, by officials in a borrowing country who have the responsibility to formulate and carry out those policies, based on an understanding that the program is achievable and is in the country's own interest" (Boughton 2003, 3).

There are, however, several serious complications. For one thing, governments are not monolithic, making the concept of government ownership at best ambiguous. The ownership by the government can never be totally absent or fully complete.

Typically, those government officials mandated to maintain economic stability have already sided with the IMF, World Bank, or other international financial institutions advocating policy reforms. In contrast, the government's political leaders, and the officials of the spending ministries, may have different sets of incentives than the government's financial managers. The political leaders within the government may prefer the reforms were it not for the political consequences, but if ownership means making an explicit commitment to them despite the political costs, the public stance of criticizing the conditionalities, and perhaps reneging later, will result if these costs are regarded as too high. Top government policymakers often state opposition to conditionalities, in order to reduce accountability for the unpopular aspects of the reforms, even if they welcome being forced to accept them. For their part, the leaders of spending ministries typically wish to pursue their mandates by spending more to promote activities within their sectors. Thus, by the time agreements have been reached, *some* government officials typically have already been collaborating with the multilateral institutions, while others have not. Second, the owned reforms must still be acceptable to the international financial institution; the difference between the preexisting doctrine and the ownership doctrine is far less than one might imagine. The requirement of government ownership could result in the failure to reach agreement, if the relevant government officials are unwilling to own an agreement that the assistance institution cannot accept. On the other hand, government leaders who wish to qualify for support may claim ownership, but renege later.

These difficulties have led to proposals, largely targeted to the IMF, to ensure country ownership (in the sense of firm commitment) by requiring prequalification for eligibility to receive a loan or loan guarantee. The prequalification would require meeting specified policy reform criteria (Meltzer 2000). In principle, if this were a sufficiently credible approach, it could strengthen the hand of the government officials in favor of IMF-favored reforms vis-à-vis government officials who would otherwise oppose these reforms were it not for the possibility that the international financial institutions would refuse to serve as lender of last resort if a financial crisis occurred. The question is whether the international financial institutions would refuse to support a country that had not prequalified, in light of the damage to the people of that country and potentially to the international economy, if this support is not forthcoming.

Governance Conditionality

Much of the disappointment of the economic stagnation in many developing countries came to be attributed to the realization that capital in itself is not enough; how it is deployed is crucial, and, as chapters 4 and 5 address, this depends on the strength of institutions of government, markets, and society. As mentioned in chapter 2, Rostow (1956, 25) had recognized this; development "is likely to require political, social and institutional changes which will both perpetuate an initial increase in the scale of investment and result in the regular acceptance and absorption of innovations," yet the early foreign assistance magnitudes were not very responsive to the institutional constraints. Currently, admonitions and conditionality regarding governance are prominent, but in some quarters, the governance constraints are still disregarded in terms of absorptive capacity. Matthew Auer cogently derides the initiative urged by Jeffrey Sachs to launch a new big push even where, as in Sub-Saharan Africa, institutions are acknowledged by all to be weak.[14] Auer (2007, 180) observes,

> Sachs... presents the problem as a dichotomous choice—either institutional development or social investment. From an institutional perspective, this is a maddening prescription. Vaccinations are not acquired, distributed, or administered without institutions, nor is soil conserved, nor are roads built. Consider the latter. Who designates which roads are to be paved and with reference to which rules and regulations? Who takes charge of procurement for road building and what procedures are invoked? Who audits the bidding and disbursement of road building funds? Who maintains the roads, and who trains the professionals who perform these functions? Answers to these questions must have institutional reference points.

If assistance without adequate institutions cannot reach its potential, assistance that requires and accomplishes institutional improvements—that is, governance conditionality—could be a major gain. However, governance conditionality is even more contentious than economic policy conditionality. As Kaufmann's research cited in chapter 2 has demonstrated, economic policy conditionality can be straightforwardly justified by the premise that the loans and grants can be put to good use. Governance conditionality, which more directly challenges national sovereignty but it also has a powerful but indirect link to economic development, requires a more complex supportive theory.

Governance conditionality is also premised on the theory that better governance enhances both aggregate economic performance and equity. De Janvry and Dethier (2012, 6) posit that in poor countries, large-scale financial transfers are unlikely to be effective given the limited absorptive capacity of these countries. So what can be done? Broadly, the answer likely includes an intense focus on capacity building, combined with more direct delivery of human development services and humanitarian assistance. Donors should focus on institution building, capacity building, and knowledge transfer to facilitate change. Thus, strong governance, as defined by broad participation, low corruption, efficient administration, and the rule of law (see chapters 4 and 5 on governance), is presumed to enhance the success of projects and programs that foreign assistance supports and lead to more coherent and effective policies that provide benefits broadly to the nation's residents.

For multilateral agencies, the argument that sovereignty shields governments from external pressure on internal policy issues has seriously eroded in terms of international norms, through the extension of the interventionist principle for the most serious internal human rights violations, such as genocide, to a much broader range of concerns over democratization and human rights in general.[15] Therefore, just as US and Western European foreign assistance has long exercised strong though informal governance conditionality for recipient countries (though US allies against first Communism and currently terrorism have often been exempt from governance conditionality), multilateral foreign assistance agencies (e.g., World Bank, regional development banks, UNDP) have imposed both formal and informal governance conditionality, despite the fact that the recipient nations share ownership of these organizations, with the premise that the pressure is necessary to get the governments to do more. In some circumstances, external conditionality is also useful if the relevant government officials within the country desire the governance reforms to go through despite opposition from others within the country (Kapur and Webb 2000, 8).

Conditionality does not have to be formal; bilateral foreign assistance, in particular, provides an avenue for influence on issues of governance, repression, corruption, and conflict. The implicit risk to recipient governments that resist these pressures is that aid might to reduced or completely cut off. However, the interests of the assistance-providing nation in maintaining good relations with the recipient nation may limit the pressure. For example, Feasel (2013, 222) notes the limitations of Japanese pressure on the Vietnamese government:

> Japan has largely played a hands-off role when it comes to the internal decisions and workings of the government. It has suggested direction and provided aid and expertise, but has not used the potential cessation of assistance as a stick to punish lack of speed in implementing market-oriented reforms or government actions restricting certain freedoms... Other donors and multinational institutions have played a more active role in pushing the GoV to increase the pace of reforms.

Feasel concludes that "Japan has not placed any conditions on aid based on the government's treatment of ethnic minority and religious groups" (225). However, neither have the multilateral agencies.

Tied Aid

Physical infrastructure, insofar as it calls for imported construction equipment and other imported components, had fed into the doctrine of tied aid as an element of the donor country economic rationale. If a nation is transferring funds to another nation, it might seem only fair, and not a significant cost to the recipient nation, for foreign inputs—both equipment and personnel—to be of the donor nation. However, the noncompetitive nature of some procurements seriously reduces the value of the benefit insofar as the goods and services are overpriced, inappropriate, or both. Hans Singer (1968, 54) criticized the limitations of tied aid to the recipient country and pointed out that while each donor nation would gain from excluding others, each would also be excluded from providing inputs funded by the other countries. Jepma judged that "tied aid represents only a small percentage of the donor countries' total exports. Thus, it is improbable that aid tying provides significant macroeconomic benefits to any donor's domestic employment or balance of payments aggregates. The case for tying is therefore essentially political rather than macro-economic" (1991, 13). Petermann (2013) extends Jepma's assessment into the present. The theory of tied aid, then, has to focus on the political acceptability of foreign assistance. The premise is that resistance to foreign assistance in the donor country budget process would be great enough to reduce the assistance program unless the enticement of providing benefits for domestic firms is present.

Collective action theory calls for donor nations to collaborate on limited degree to which their assistance is tied aid. The Organisation for Economic Cooperation and Development (OECD) has succeeded remarkably in this respect, countering the domestic political pressures, at least among OECD donors. Through its Development Assistance Committee (DAC), encompassing the major Western European donors, Australia, Canada, Japan, New Zealand, South Korea, and the United States, the OECD organized several consultations that resulted in accords in 2001 and 2012 to reduce tied aid. Before the 2001 accord, total DAC untied aid was at roughly 45 percent; it is now over 75 percent (OECD 2013a), with much of the remaining tied aid due to technical assistance that donor nation governments use to design and monitor the assistance. In contrast, China, now a significant foreign assistance donor through concessional loans, still utilizes tied aid broadly. Wolf, Wang, and Warner (2013, 7) note that "Chinese concessional loans also include a stipulation that at least 50 percent of the loan is tied to the purchase of Chinese goods."

Conclusions

The theories underlying the doctrines of foreign assistance for developing countries in the post-WWII period have reflected a host of objectives and premises. Some of the tradeoffs have created severe challenges, mitigated or exacerbated to varying degrees, which have led to changes in theories and doctrines.

Several theoretical advances have had rather impressive impacts on foreign assistance doctrines. The fundamental premise that development requires capital is, of course, still acknowledged, but the overoptimism of the Big Push has been tempered by recognition of the importance of institutions, even if the idea that capital can be do wonders still resurfaces periodically.

The theory that tied aid would redound to the benefit of individual donor countries has been undermined sufficiently to diminish the prevalence of this wasteful practice, at least among OECD donors. Recognition of the power of collective action has generated more aid coordination and increased multilateral assistance as well. However, multilateral assistance is no less contested than when its conditionality was introduced in the 1980s, and the government ownership doctrine is largely hollow.

With the expansion of the sectoral scope of multilateral assistance, and the emphasis of NGO assistance on the previously neglected social sectors, development assistance is more capable of addressing the complex web of factors that determine the pace and quality of development.

In sum, it is fair to say that as the premises behind foreign assistance have been tested through experience, the doctrines have matured. Yet further improvements are still required.

CHAPTER 8

International Development in the American Grain: From Point Four to the Present

Introduction

This chapter recounts the changing institutional and political context of US foreign assistance. This evolution is unique, not only in its dominance over the past 70 years but also in its major role in the global development assistance effort. The search for a stable US foreign assistance institutional structure has largely been accomplished, but the broader institutional arrangements have placed development assistance into a chronically precarious position. We must also acknowledge that even with a stable organizational structure, no matter what the organizational chart looks like, the funding for USAID and smaller assistance agencies is far too small to make major inroads in economic assistance. In FY 2014, total US foreign assistance was roughly $34 billion, of which $8.5 billion was for international security assistance and $3 billion for multilateral institutions. USAID's economic assistance was only $19.3 billion, as other US government agencies absorbed $1.3 billion of the rest of the foreign assistance budget (US Department of State 2014, 62). The FY 2015 request is to cut the USAID foreign assistance budget to less than $18 billion. To put this into perspective, the newest US aircraft carrier, the USS Gerald Ford, cost $14 billion.

The Current Dilemma

In early June 2008, six months before the election of US President Barack Obama, the Modernizing Foreign Assistance Network (MFAN), a private organization of foreign policy specialists counting among its members Larry Diamond and Francis Fukuyama, issued a plea to "restore America's reputation abroad" by combining and strengthening defense and diplomatic tools with "equally robust tools of development" (MFAN 2008, 1, 3). It was an ironic request. Historically, calls to separate military assistance from development aid fell on deaf ears, even when advocated by John

F. Kennedy, for example. MFAN also pushed, more pertinently, for a Cabinet-level presence and a "whole of government" approach to global development. The historical record notwithstanding, strong evidence suggests that President Obama and his advisors heard the MFAN. Responding directly to the "drip-by-drip erosion of USAID" but also to forge closer strategic and operational ties among defense, diplomacy, and development, the 2010 Presidential Policy Directive on Global Development was a sweeping endorsement of the MFAN recommendations (Hyman 2010) and the USAID Alumni Association recommendations to revive the Agency to reverse its "progressive deterioration" (USAID Alumni Association 2008, 1). Secretary of State Hilary Clinton agreed, "I think it's fair to say that USAID... has been decimated. It has half the staff it used to have. It's turned into more of a contracting agency than an operational agency with the ability to deliver."[1] President Obama declared, "We are rebuilding the United States Agency for International Development as the world's premier development agency." USAID would regain control of "policy, budget, planning, and evaluation capabilities"; recoup its leadership "in the formulation of country and sector development strategies"; resume the task of educating the next generation of development professionals; and once again become "the U.S. government's lead development agency" (White House Office of the Press Secretary 2010a, 1). According to the White House Fact Sheet that followed the pronouncement, the Directive "charts a course for development, diplomacy, and defense to mutually reinforce and complement one another in an integrated comprehensive approach to national security" (White House Office of the Press Secretary 2010b, 3). Not since President Kennedy, speaking before the UN General Assembly in 1960, called for a Decade of Development has a sitting president elevated development to such a central place.[2]

This begs a number of questions: What had changed between Kennedy's declaration in 1960 and Obama's in 2010? What continuities and discontinuities—in development theory, doctrine, and practice—characterize the evolution of US development policy? How do we put into context the nearly stagnant real dollar levels of foreign aid, as revealed in table 8.1, and the precipitous decline in the proportion of US GDP devoted to foreign aid since the end of WWII? What problems and challenges seemingly unique to the postwar world nevertheless survived into the present and in what form? What new theories and doctrines would replace them?

The fundamental if unavoidable weakness in US aid policy is its vulnerability to partisan political demands; to various forms of bureaucratic, interagency infighting; and, perhaps most important of all, to changing circumstances on the ground in those countries either receiving or designated to receive American aid. The leadership of USAID has stabilized since the earlier chronic institutional turnover, but is still subject to fairly dramatic shifts in emphasis and relationships with other agencies as international crises erupt and domestic political winds change. A system of congressional oversight and annual appropriations that has institutionalized gridlock, sacrificing long-term program sustainability to the demand for immediate results—and short-term political interests to a long-term economic vision for the countries that are most in need—has created "an output-oriented process favouring short-term results over long-term impact" (OECD 2011, 53).

Clearly, it is a vastly different world today than in 1960, with new dangers, new vistas, and new opportunity costs for American foreign aid policy. It is not simply

Table 8.1 US foreign aid, 1946–2012

Fiscal year	*Billion in constant 2012 dollars*	*As percentage of GDP*
1946	33.97	1.4
1950	57.59	2.2
1955	32.59	1.0
1960	36.34	1.0
1965	35.00	0.8
1970	35.06	0.6
1975	27.31	0.4
1980	26.30	0.4
1985	38.09	0.5
1990	25.56	0.3
1995	20.42	0.2
2000	23.06	0.2
2005	27.99	0.2
2010	41.63	0.3
2014 (est.)	34.2	0.2

Source: US Overseas Loans and Grants (Greenbook); Office of Management and Budget Historic Budget Tables, FY 2011; Congressional Research Service (CRS) appropriations reports and CRS calculations.

that 9/11 and the War on Terror permanently changed the political landscape; what are now transnational problems call loudly for transnational solutions.

Notwithstanding Obama's call for a new "division of labor"—governments, multilateral agencies, and NGOs "all working together" with a "sense of common purpose"—bilateral US Official Development Assistance (ODA) must account for the bulk of total US foreign assistance (94% in 2005, 87% in 2009, and 83% in 2012) (White House Office of the Press Secretary 2010a, 3; US ODA 2014).[3] The major multilateral institutions with significant financial resources (the World Bank and regional development banks) have to limit their total capital to maintain the creditworthiness of their bonds.

However, bilateral aid has its virtues. A 2012 Development Assistance Committee peer review asserted that US bilateral aid is delivered "more effectively, efficiently and professionally" than at any time in its history, especially to those countries most in need—Sub-Saharan Africa, for example (OECD 2011, 41). Bilateral aid also appears to be the way most Americans want it.[4] And despite lofty sentiments,[5] bilateral aid is widely believed to serve the donor nation: President Kennedy stated in 1963 that "an early exposure to American goods, American skills, and the American way of doing things" was critical to "forming the tastes and desires and customs of these newly emerging nations which must be our markets for the future."[6] Some of the costs of providing foreign assistance are offset when the goods and services are procured from the United States, though less so now that the United States, like other OECD DAC members, has greatly reduced the tied aid practice (see chapter 7). Nevertheless, food stuffs and their transshipment are purchased wholly from American-owned farms and freight carriers.[7] Most military aid finances equipment purchases and training personnel in the United States, or by tying aid to exclusive trade agreements (Tarnoff and Nowels 2004, 19–20).

Since the end of WWII, the effective and efficient delivery of US foreign aid has suffered as a result of initially rapid institutional change, increasing complexity of its functions and accounting, the public indifference to assisting developing countries, and the convoluted relationship between security and development considerations.[8] The late twentieth century downsizing of USAID, the major source of bilateral US ODA, is a good case in point. The Year 1996 saw a major reduction in the Agency's workforce, an attrition in human resources that continued throughout the 1990s (Clinton-Gore Administration History Project 2000, 71). By the end of 2000, USAID was no longer able to meet its compliance requirements under the Chief Financial Officers Act of 1990, the Government Management Reform Act of 1994, and the Federal Financial Management Improvement Act of 1996, legislation designed to provide Congress with reliable, timely, and consistent information about the management of federal agencies (71–73). USAID "managers could not be sure program objectives were being met; resources adequately safeguarded; reliable financial and performance data obtained, managed, and reported; and activities complied with laws and regulations" (78–79). The Obama administration's rehabilitation of USAID, while restoring to the Agency its policy, strategic planning, and budget management capabilities, has done little to overcome the fragmented aid system. That system today includes as many as 27 different public institutions, hampering "a more strategic and consolidated approach for U.S. development co-operation" (Development Assistance Committee 2011, 54).

A second factor in the evolution of US development doctrine is the reordering of the relative weight given to specific countries and regions as well as to the five main categories of US foreign aid—Bilateral Economic Assistance, including an account for political and security assistance; Multilateral Economic Assistance; Humanitarian Assistance; Military Assistance; and Law Enforcement Assistance, a new category previously buried in the USAID budget under the Office of Public Safety. Civil society organizations and the business sector have each in their own way played a part in helping to set the agenda for US foreign aid. This is also true of public opinion generally, which can be a barometer of what are often very subtle shifts of policy and political emphasis within the halls of Congress. While difficult to measure and assess, some accounting of these forces is essential to our understanding of the progress of development thinking and practice.

A history of both overt and covert linkages between military assistance and economic aid suggests a third factor in the evolution of development theory and doctrine, one that is dealt with at some length in a subsequent chapter of this book, drawing on the example of US involvement in Chile, South Korea, and Vietnam.

Finally, the activities of other foreign assistance providers, including developing nations themselves, have had and continue to have a major impact on how Americans think of and respond to the needs of development. Not only other donor nations but international finance agencies like the World Bank, the IMF, and the Asian Development Bank, as well as intergovernmental bodies like the United Nations run programs and tout policies that both parallel and intersect with vital American interests.

These four factors—institutional turnover and paralysis, changes in the basic architecture of US aid, a complementarity doctrine of economic and military assistance, and a growing competition for influence between bilateral and multilateral donors as well as between the former and such newly industrialized countries (NICs) as Turkey and Brazil—all complicate considerably the task of analyzing the large and growing literature of economic aid and development and extracting from it lessons for future policymakers. Although researchers have touched on one or more of these factors to varying degrees, there is no single comprehensive study of development that has taken full advantage of both the primary and secondary sources available to the modern development scholar.[9]

The Revolving Door: Continuity and Discontinuity in the Planning and Administration of US Foreign Assistance

Before being reorganized into the USAID in 1961, the US foreign assistance program was a diversified yet piecemeal affair carried out by an alphabet soup of evanescent administrative entities: the Economic Cooperation Agency (ECA) (1948–1951), which implemented the European Recovery Program (ERP), popularly known as the Marshall Plan; its successor the Mutual Security Agency (MSA) (1951–1953), established by an Act of Congress in 1951 and extending assistance to Asia and South America; the Technical Cooperation Administration (TCA) (1950–1952) established within the State Department and charged with implementing and managing technical assistance; the Foreign Operations Agency (FOA) (1953–1955), which superseded the MSA, followed by the establishment of the International Cooperation Administration (1955–1961) and its subsidiary Development Loan Fund (DLF) in 1955. The DLF, signaling a general shift in US foreign aid policy from grants and soft-loans to hard-loans payable in dollars, gained institutional independence in 1959. The Foreign Assistance Act of 1961 gave birth if not to the last to the sturdiest link in this chain, the USAID.

Whether US foreign assistance lost its organizational focus and direction after the recovery of Europe and the termination of the Marshall Plan is an important question. Certainly it is true that by the 1960s, "criticism," as James Hagen and Vernon Ruttan have pointed out, "became more focused on the administration of aid than on its existence" (1987, 7). This was a sea change. President Truman's new aid initiative, announced in 1949 fresh in the wake of the Truman Doctrine of containment and culminating in so-called Point Four, was nothing if not ambitious in scope and design.

First, said Truman (1949), "we will continue to give unfaltering support to the United Nations and related agencies, and we will continue to search for ways to strengthen their authority and increase their effectiveness." The theory here was that collective action would serve simultaneously to leverage US development funds and to free them up politically for the business of containment. Second, he said, "we will continue our programs for world economic recovery." This was a reference to the Marshall Plan but more generally to the logic of US direct involvement in development. Third, "we will strengthen freedom-loving nations against the dangers of aggression," a major strategy in the doctrine of containment. Fourth, said Truman, "we must embark on a bold new program for making the benefits of our scientific advances and industrial progress available for the improvement and growth of underdeveloped areas." Science and technology transfer would in theory provide the necessary basis for infrastructure development, industrial growth, and agricultural productivity with benefits redounding both to recipient countries and the United States.

In the intervening years leading to the creation of USAID, technical assistance, development loans tied to investor incentives and Buy American requirements, and manpower support—not direct economic aid and assistance—were the preferred forms of US intervention in developing countries. These and other programs, notably the food aid program, Public Law 480 passed by Congress in 1954, "not only channeled security assistance to U.S. allies and potential friends in the developing world," argues Dennis Rondinelli (1987, 2), "but offered countries help in increasing their food production, educating their citizens, and industrializing their economies." In spite of such gains, development remained, as Rondinelli suggests, a mixed bag, catering to American self-interest no less than to the needs of the developing world (as government officials understood them), but prey also to isolationist doubts about the value of foreign aid of any kind, including US support of the United Nations. The Byzantine administration of aid—in 1951, 23 federal agencies were conducting some foreign economic operations—exacerbated these conditions while paving the way for structural reforms, especially the creation of USAID (Partners in Progress 1951, 4).

Based upon new models to control waste and inefficiency in government, the drive was on to consolidate foreign aid into a single all-encompassing agency with broad and inclusive powers. Two aid reports commissioned by President Truman—the Gray Report, chaired by the Secretary of the Army Gordon Gray, and Partners in Progress, a report of the president's International Advisory Board chaired by Nelson A. Rockefeller, the former coordinator of the Institute of Inter-American Affairs, a forerunner of the TCA—recommended the creation of "one overall agency" that would be independent both of the State Department and the ECA (Partners in Progress 1951, 2). In Rockefeller's proposal, a new Overseas Economic Administration would be responsible directly to the president; its charge—the centralization of all US foreign aid operations, including military assistance. The recent outbreak of war on the Korean peninsula made the need for greater unification of aid administration especially "acute," argued Partners in Progress (4). In addition to bringing military assistance and economic development under one roof, reducing duplication while increasing fungibility, the new agency would take a more proactive approach to working with private enterprise as well as with civil society organizations and voluntary groups already engaged in the business of development. In its current configuration, US foreign economic policy lacked "consistency and continuity," focus and flexibility, essential qualities that only a central, unifying institution could provide (5). Still another rationale for the centralization of aid administration, one with growing relevance up to the present, was the corresponding lack of coordination in recipient countries, where government agencies, civil society organizations, and the private sector often operated independently and at cross purposes with one another. Democracy would never follow development, a basic contention of modernization theory, without inclusive, broad-based development administration, of which the United States would have to become the model, but an efficient one.

The Gray and Rockefeller reports recommended a sweeping reorganization of the American foreign aid program, including new responsibilities for the oversight of such operations as foreign procurement (traditionally the sole province of the General Services Administration), export licensing (under the Department of Commerce), foreign mineral imports (Department of the Interior), Export-Import Bank and World Bank loans, and UN development agencies. Released within a year of one another, both reports garnered considerable attention, if not universal approbation, in the national media. The *New Orleans Times-Picayune* had its reservations about the need for a new super-agency, seeing it as an "oblique criticism of the State Department's present coordination of foreign economic policies" (US Department of State 1950, 4). Truman would ultimately come down on the side of this opinion, rejecting the recommendations of Gray, Rockefeller, Hoffman, and others, making it clear in a letter of April 5, 1951, to then ECA administrator William Foster that "The Secretary of State, under my direction, is the Cabinet officer responsible for the formulation of foreign policy and the conduct of foreign relations, and will provide leadership and coordination among the executive agencies in carrying out foreign policies and programs" (Congressional Quarterly Research 1951, 2). This is another example of how presidential politics, never as straightforward and unambiguous as a good theory and the doctrines it produces, can subvert the latter at any stage in the development process.

Geopolitical considerations can also play a role in how successfully even the most rational expectations can be helped to see the light of day. State Department officials objected vigorously to the Gray and Rockefeller recommendations to take the TCA out of it. Security threats arising in Korea and South East Asia made it increasingly difficult to disaggregate economic aid, technical assistance, and military preparedness much less to diminish the role of the State Department in the supervision of all foreign policy operations. The Korean War and the Cold War paved the way for the Mutual Security Act of 1951, which now made political considerations—the willingness of aid recipients (i.e., friendly nations) to enlist in the common struggle against international communism and accede to the demands of US security—a litmus test for receiving assistance.[10]

Although the Mutual Security Act provided for the organization a unified program of military, economic, and technical assistance under the coordination and supervision of a "single person," a director within the Executive Office of the President, it also made abundantly clear that "Nothing contained in this Act shall be construed to infringe upon the powers or functions of the Secretary of State" (Mutual Security Act of 1951, sec. 501a, 505). For Averill Harriman, the new director of the MSA, "coordination" was the operative term (Gordon 1975, sec. 162). He kept the MSA staff purposefully small, preferring to leave the day-to-day business of development assistance to individuals and agencies directly involved in the work or in bringing to the Hill annual requests for appropriations. Nearly omitted from the Mutual Security Bill but for some frantic back-channel negotiations, TCA survived with a $100 million appropriation in FY 1951–1952 for a program that was to be "an attack at the grassroot level on the three basic needs of mankind: food, health, and education" (Andrews 1970, 30). The development doctrine of the TCA was first and foremost: no big projects, no major new industrial plants, no national airports equipped with state-of-the-art airplanes that no one could fly or knew how to repair, no "building shit houses that you can see instead of slow educational work" (31). Andrews and his predecessor, Henry G. Bennett, envisioned putting in small demonstration plants or providing funds for feasibility studies and impact statements for the kinds of resource development ideas, even the big ones, which the Third World economies they visited (some 29 in two years) had in mind.

Harold Stassen, a former governor of Minnesota and twice a candidate for the presidency, was appointed by President-elect Eisenhower at first to lead the MSA and then in 1953 to administer a new agency replacing the MSA, the FOA. Two years later, TCA and FOA were consolidated into the International Cooperation Administration also under Stassen, straight out of the take-off and big-push school of modernization theory. However, over the course of the Mutual Security Act period (1953–1961), the average annual total of all economic assistance declined by nearly half compared to the preceding Marshall Plan period (1949–1952), from $36.1 billion to $17.6 billion in constant 2012 dollars (USAID Greenbook 2014). The overall reduction of the foreign aid budget combined with a relative increase in security concerns and large-scale loans and grants for economic assistance meant that the kinds of technical assistance programs that Andrews describes (and Stassen abjured) had little or no chance of survival (Ruttan 1996, 71–72; USAID 2005, 1).

The organizational problem continued to return to haunt the US aid program throughout the 1950s as one new federal agency metamorphosed into another. Jurisdictional confusion persisted over the respective roles of the Office of the Executive, the State, Commerce, and Agriculture departments, and the Export-Import Bank. It was very easy to conclude, as did the President's Advisory Committee on Government Organization, that development policy was adrift. No longer was a "better coordination of existing responsibilities" sufficient to fix the problem, when the problem lay instead in the "suitability" of the existing machinery for implementing such policy (President's Advisory Committee on Government Organization 1954, 1–2). Urgently needed was a study of ways in which to shore up the Office of the Secretary of State to assume full responsibility for all policy considerations and for relating foreign economic conditions and US foreign and domestic policies. Under the current chaotic state of affairs, "a maximum of overall responsibility is coupled with a minimum of direct line authority," emasculating the authority of the president and the State Department (Dodge 1954, 4). In response to these concerns, in 1957 Congress gave undersecretary status to a new central administrative unit located within the State Department and responsible directly to the Secretary of State. This Undersecretary for Economic Affairs was put in charge of the Mutual Security Program now housed within the International Cooperation Agency, made chairman of the Board of Directors of the DLF, represented the United States on the governing boards of the World Bank and the IMF, served as the State Department's liaison to the Export-Import Bank, provided guidance and direction to Public Law 480, and represented State Department on the president's National Advisory Council. Congress had solved the problem of decentralization not by rebalancing the many forces at work in foreign aid policy and practice, thereby strengthening the hand of the president and the State Department, but by once again housing administrative oversight of these programs in a single, all-encompassing authority (Dodge 1954, 4).

For a young John F. Kennedy (1960, 52) surveying the landscape, anything was acceptable but the status quo.

> The alternative [to reform] is chaos, not economy—a continuing of *ad hoc* crisis expenditures—a further diffusion and dilution of our efforts—a series of special cases and political loans—an over-reliance on inflexible, hard loans through the Import-Export and World Bank, with fixed dollar repayment schedules that retard instead of stimulate economic development—a lack of confidence and effort in the underdeveloped world—and a general pyramiding of overlapping, standardless, incentiveless, inefficient aid programs.

In the simple organization and administration of aid, there was a deplorable "lack of coherence," according to Kennedy's Task Force on Foreign Economic Policy (1960). "Today it consists of bits and pieces of policy developed at different times by different people to meet particular situations" (10–11). Kennedy's answer, the creation of USAID, a single overarching agency led by an administrator with undersecretary-level access to the president, hardly put an end to "coordination issues" (Task Force on International Development 1970, 34). The 1969 Rockefeller Report and the Peterson Task Force on International Development in 1970 recommended abolishing USAID and replacing it with one or more alternative institutions. USAID has managed over the years to survive such onslaughts but not without a price: tighter congressional controls and oversight, new and more onerous reporting requirements, periodic internal reorganizations and staff reductions, half of its permanent staff since 1975, and progressive outsourcing of the design and implementation of its field operations (Ruttan 1996, 96–97; *USA Today* 2009).

From the Nixon through the Obama administration, calls for the reorganization of development assistance have remained the one constant in an otherwise fluctuating world of strategic giving. Such calls have ranged from demands for greater efficiency and less waste to cries for centralization one day, decentralization the next, more or less coordination versus more or less authority, to the subordination of project design and approval to implementation. In the next section, we examine the evolution of doctrine pertaining to the appropriation of aid, funding priorities that have shifted dramatically over the years with new alignments in foreign and domestic politics, with changing geostrategic forces, with increasingly interdependent private-public sector relations, and with changes in popular opinion.

Foreign and Domestic Doctrine in the Allocation and Distribution of Aid: Who Gets What, When, and Why?

The ERP was America's first large-scale bilateral foreign aid operation. Although it had contributed far more than any other member government (some $2.66 billion) to the UN Relief and Rehabilitation Administration (UNRRA) established in 1943 to provide food, fuel, shelter, and other basic necessities to the victims of WWII, nothing could prepare the world for the massive infusion of American dollars—$13.3 billion (94.3 billion in constant or real dollars) in a little over three years between 1949 and 1952—to secure the postwar economic recovery and integration of Europe (UNRRA 1950, 4; Orme 1995, 15). Sixteen countries, including occupied West Germany, formed themselves into the Organization for European Economic Cooperation, a precursor to the European Community and today's European Union (Orme 1995, 16). The United States channeled aid into this organization through the new ECA, which over the course of the Marshall Plan period poured an additional $15.3 billion for a total of $28.6 billion into other countries around the world where it deemed itself to have "a vital interest"—China, Korea, and Southeast Asia in addition to Western Europe (Orme 1995, 25; USAID 2005, 1).

According to John Orme, the Marshall Plan was the child of synergistic and conflicting forces in the early evolution of foreign aid: "the continuation of postwar prosperity in the United States; the relief of distress abroad; the containment of communism; and the reconstruction of Germany at an acceptable cost to U.S. taxpayers" (cited in Orme 1995, 35). With the Chinese Revolution of 1948 and the onset of the Korean War in 1950, and as Europe itself began to reap the benefits of American investment, the work of the International Cooperation Agency devolved to a new agency—the MSA—that would increasingly blur the lines between economic and military assistance to places then called the Third World (to distinguish them from the First World of developed capitalist countries and the Second World of advanced communist ones). Military security and economic stability were interchangeable parts of a unified theory of democracy building during the heyday of US foreign aid and development when obligations and loan authorizations swelled from $43.3 billion during the period of the Mutual Security Act (1953–1961) to $196.7 billion under the Foreign Assistance Act and its amendments (1962–1985), leading up to the break-up of the Soviet Union in 1989 (USAID 2005, 1; Tarnoff and Lawson 2011, 29–30). Although for record-keeping purposes USAID separates economic assistance from military assistance, in practice the Military Assistance Program and USAID funding for development, including a catchall Economic Support Fund (ESF), were coterminous—especially in trouble spots where these forms of assistance tended to pool anyway: Taiwan and Korea in the 1950s; Laos, Indonesia, and Vietnam, as well as such nascent, politically unstable Latin American republics as Chile and Guatemala in the 1960s; the Middle East generally and Israel and Egypt in particular from the 1970s to today; and, of course, such recent hotspots in the War on Terror as Iraq, Afghanistan, and Sudan. In these and other areas of the world, especially in those countries posing US security threats or facing their own challenges of poverty, disease, and instability, humanitarian impulses combine with national security objectives to enforce a doctrine of development rooted in the belief that political stability, social mobility, and higher standards of living, as well as greater political participation would reinforce the preference for democracies friendly to the United States.

From the funding priorities of the immediate postwar period, when this trend was just getting under way, to the current level and distribution of foreign aid in our post-Cold War, post-9/11 world of roughly $50 billion of annual US foreign aid operations, who gets what, when, and why has involved a complex desideratum of the following doctrines, which we examine in their turn:

- US foreign policy entails a response to changing geostrategic conditions in the world, determined by the State Department working closely with the president.
- Congressional decision-making powers are exercised through authorities and appropriations under the Foreign Assistance Act of 1961, as amended and after 1986 through so-called continuing resolutions (CRs) (a euphemism for the current practice of congressional earmarking).
- Domestic as well as international public opinion counts in this process.
- National security, a complex calculus of economic, social, and political considerations including but not limited to questions of military strength, democracy building, and economic growth and stability at home and abroad, is the penultimate driver of the US foreign assistance program.

It should come as no surprise that foreign aid reached an all-time postwar low in 1988, just a year before the fall of the Berlin Wall and the collapse of the Soviet Union (Gimlin 1988, 1). Until then, fears of Soviet expansionism had dominated US foreign policy and, with several notable exceptions, dictated the flow and distribution of US military and economic support first to the developed world in war-torn areas of Europe—emergency aid to the tottering governments of allied Greece and Turkey being perhaps the prime example—and then later to large swaths of potential social and political influences across the developing world. The dogma of anticommunism, the doctrine of containment, and the policy of the "falling domino" first introduced by President Eisenhower in 1954 became the filters through which "the United States interpreted events in the rest of the world" (Ruttan 1996, 278). Although subsequent events have discredited the domino theory, American policymakers remained under its thrall for a generation.[11] They worried with George Kennan (1947, 576, 581), author of the doctrine of containment, that unless resisted at every turn, unless "contained by the adroit and vigilant application of counter-force at a series of constantly shifting geographical and political points, corresponding to the shifts and maneuvers of Soviet policy," communism would gradually but ineluctably metastasize to every cell of the free-world body politic, "a cautious, persistent pressure toward the disruption and weakening of all rival influence and rival power."

Yet as the British historian Eric Hobsbawm (1994, 358) has noted, while Soviet-American relations may have dominated the interactions between states, "It did not entirely control them." Other forces were at work in the world, more salient for the purposes of development than the bipolar competition for influence between the United States and the Soviet Union (Walters 1970, 97–98).[12] One of them was a population explosion in the Third World, which was placing profound stress and strain on already limited physical and human resources. By 1968 most of these nations had only completed the first leg of a demographic transition to low birth and death rates. Average life expectancy was increasing (thanks in no small part to both US and Soviet scientific and technological assistance), but birth rates remained the same or rose only gradually. The pressures of population growth came to bear directly on economic development. By the 1970s, family planning programs were getting the same public attention, if not funding, as the building of roads and dams and other more traditional forms of donor investment (Hobsbawm 1994, Chapter 12; Ruttan 1996, 125).[13]

No less vexing for development officials was the multiplication of new states, 200 by 1990, many of them former European colonies unschooled in self-government and prey to the blandishments less of communist or democratic forms of governance than of naked military rule (Hobsbawm 1994, 347–349). Providing economic and military assistance to states prone to political corruption, cronyism, and authoritarianism in the hope of making friends for your point of view or in the larger illusion that strengthening foreign military capabilities would serve as a deterrent to either internal subversion or external invasion was a "serious misconception," according to Max Millikan, an old development hand during the Kennedy era. Reproducing the colonial dependencies of the past in the guise of development would only leave poorly educated and poorly performing countries in "recurrent states of chaos and confusion" (Millikan 1955, 11), making the establishment of "milieu goals" (Walters 1970, 47)—environmental conditions favorable to the social and political ideals, not to mention the investment needs, of the major powers—exceedingly difficult, if not impossible. By the 1960s, under increasing pressures to explain why the fond hopes of the 1950s had not been achieved, development practitioners turned the blame onto the recipient countries; development administration (see chapters 4 and 5) focused on correcting the mismanagement of aid in recipient countries. USAID came to believe that institutional development was "key to promoting sustainable economic growth and social progress in poor countries" (Rondinelli 1987, 4).

The Nonaligned Movement

Another mitigating effect on the organization of foreign policy around Cold War exigencies, including challenges to sovereignty that may or may not have the support of local elites, has been the presence of a small but growing number of nonaligned, neutralist countries committed to political self-determination and to protecting from external interference their new-found independence. Meeting in April 1955 in Bandung, Indonesia, delegates from 29 countries in Africa, Asia, and the Middle East issued a Final Communique calling for greater economic and cultural cooperation, human rights and self-determination, and the "promotion of world peace and cooperation." The latter would entail three "abstentions" from the "big powers": abstention from "intervention or interference in the internal affairs of another country"; abstention from "the use of arrangements of collective defense" to advance the interests of the First and Second Worlds; and abstention by *any country* from "exerting pressures" on any other country, an implicit criticism of the foreign influence peddling, otherwise known as "presence aid" (aid in return for establishing a political presence), that had become common in the developing world (Asian-African Conference 1955, 7–9). This is not to say that the countries subscribing to Bandung stopped accepting aid from or trading with either the United States or the Soviet Union or both in the case of India, Egypt, and Turkey. Contentious states like India and Afghanistan became adept at playing the great powers against each other, taking their money and receiving their technical assistance while retaining the right to control and develop their own natural and human resources for themselves. At the height of the Cold War between 1946 and 1968, India received the bulk of its economic assistance from the United States ($7.8 billion), but it also received over a shorter period of time, 1954–1968, an additional $1.6 billion from the USSR (Walters 1970, 84).

The Land Reform Initiative: Neglect of Political Realities

With the exception of a few NICs, for the majority of people living within the developing world, agriculture is still the primary means of earning a living. And for those whose livelihood was tied to the land, land reform took precedence over the state-planned industrialization and large-scale structural adjustments favored by donors like the United States and Russia. Land reform, however, could mean many things, ranging from the outright seizure and redistribution of large estates to milder, more gradualist measures such as rent reduction and tenancy reform. As Hobsbawm (1994, 356) has written, "what the modernizers saw in [land reform] was not what it meant to the peasants...whose demand for land was not based on general principle (communist or otherwise) but on specific claims." The Green Revolution of the late 1960s—touted by American officials as an alternative to a Red Revolution of the Soviet-kind or a White Revolution of the Iranian-kind—was a mixed blessing for farmers. The new agriculture had demonstrable long-term benefits in the form of higher crop yields per hectare of land, but the initial investments were often beyond the reach of small farmers.[14]

Emergent Doctrines

As countries around the world turned to urban industrialization, the agricultural sector contracted. As globalization and integration of markets proceeded, foreign policymakers had to admit that Kennan's bipolar world was simply not a reality. Bipolarity was no longer a sound basis for making decisions about who gets what, when, and why in land, industrial goods, or investment capital.

As this realization began to strike the people at USAID and the State Department, Congress took matters into its own hands, moving development in New Directions after a disastrous war in Vietnam. In 1973 Congress passed New Directions Legislation (NDL) designed to shift its aid package from Southeast Asia to the Middle East and from large-scale, capital-intensive infrastructure and investment projects that focused on growth and productivity to a new basic human needs agenda that put the stress on health, education, family and population planning, and poverty reduction (Ruttan 1996, Chapter 6). Asia by this time had commenced its economic miracle, the combined result of US foreign assistance, grassroots patterns of saving and investment, and years of human capital deepening. Taking this lesson to heart, NDL encouraged popular participation by aid recipients, who were to become "agents of change, rather than targets of aid" (Cutshall et al. 2009, 8). It spoke of democratic institution building and of the need for participatory programs that, based on "self-development," were sustainable over the long term (Hellinger, O'Regan, and Hellinger 2009, 15).

These hardly new calls for self-help could serve as a convenient pretext for public disengagement from development. Grants-in-aid, argued neoconservative development practitioners, created abject dependencies and sapped individual initiative in aid no less than in social welfare recipients; multilateral lending institutions and private sector investment in labor-generating industry alone could keep poor countries dependent on their benefactors. Something very much like it was the Nixon Doctrine, whose "central thesis" gave so much impetus to NDL (Ruttan 1996, 96). While announcing at the beginning of the decade that "We have no intention of withdrawing from the world," Nixon made it clear that "Americans cannot—and will not—conceive all the plans, design all the programs, execute all the decisions and undertake all the defense of the nations of the free world. We will help where it makes a real difference and is considered in our national interest" (98, 102). Such statements have been interpreted as efforts to phase down America's foreign aid program. Title IX of the Foreign Assistance Act as amended in 1967, an important forerunner of NDL, enshrined the doctrines of self-help and participatory governance. But if under New Directions the United States was to leave the task of economic development primarily to host governments, empowering "the poorest of their people" to participate, then *more* US involvement would be needed to assure that these responsibilities were met, in such a way, as Nixon had said, that they serve the US national interest (Mickelwait, Sweet, and Morss 1979, 2). NDL foundered on these internal contradictions. USAID was never able to resolve the fundamental question of how to involve the rural poor in their own uplift. An "adequate planning requirement" written into the Foreign Assistance Act of 1961 paralyzed the approval process in endless paperwork—feasibility studies, engineering and financial plans, budget estimates, time-on-project predictions, cost-benefit analyses, worst-case scenarios, all prosecuted in an office in Washington thousands of miles away from the problem—before any funds could be obligated (Mickelwait, Sweet, and Morss 1979, 60). Development Alternatives, Inc., subcontracted by USAID in 1975 to aid in the design of 12 rural development projects in ten countries with a budget of $65 million, complained in a preliminary report that this "blueprint approach" (cited in Mickelwait, Sweet, and Morss 1979, xv, 5) militated against good, field-tested practices. The rush to design projects that would pass muster with Foreign Assistance Act General Counsel caused USAID to create "unrealistic implementation schedules and plans" and to ignore the "local environment, culture, and economy" (cited in Mickelwait, Sweet, and Morss 1979, 6). The contradictions of an aid program designed to help the poorest members of society to help themselves, a congressional gauntlet conducive to institutionalized gridlock, and brewing conflicts in the Middle East and Central America were enough to doom the war on poverty implicit in the New Directions mandate. "Lost Directions" is the name Ruttan (1996, 116) has given to the period that followed, when development assistance dropped from 40 percent to 24 percent of the aid budget and the ESF share of the budget increased from 26 percent to 42 percent. A 1988 Congressional Quarterly Research report describes ESF as "government-to-government cash transfers that are not earmarked for any particular program and require no U.S. accounting for how the money is spent" (Gimlin 1988, 6). To gain greater latitude in determining aid priorities than the more narrowly circumscribed authorizations of the Foreign Assistance Act of 1961, Congress and the executive branch after 1986 resorted to the so-called continuing resolution, a stop-gap appropriations measure that only further politicizes the review and approval of foreign aid. Budget analysts argue that CRs "restrict program implementation, and present operating component managers with the need to delay work and re-plan and reprogram repeatedly...CRs...reduce the efficiency of work processes" (Bockh and Blakeslee, n.d.). An ancillary effect of bureaucratized appropriations process is to channel a growing share of foreign aid from economic assistance to security programs in high-profile strategic areas such as the Middle East, Central America and the Caribbean, and South and Central Asia (see chapter 10).

In the mid-1980s, FOA appropriations began their decline from a peak of $20.7 billion in FY 1985 to a low of $12.3 billion in FY 1997 followed by an increase in FY 2000 to $16.5 billion, which included, however, a one-time allocation of $1.8 billion for Israel, Jordan, and the Palestinians in support of the November 1998 Middle East peace accord and $1.9 billion largely to fund Colombia's antinarcotics initiative (Tarnoff and Nowels 2004). In the aftermath of 9/11, these numbers have increased substantially. Not including US contributions to multilateral institutions or export aid, appropriations for economic and military assistance doubled to $29.4 billion in FY 2011 (Lawson, Epstein, and Resler 2011, 23–24). USAID must cope with an allocation system extending across a multiplicity of competing accounts (Cutshall et al. 2009, ii). Development assistance is but one of those accounts, in FY 2011 accounting for only 12 percent of the total amount of bilateral economic assistance. State Department/USAID officials now seek funding for 19 separate accounts under the five broad categories listed at the beginning of this chapter. To do so, each year they need to navigate an administrative boondoggle consisting of up to 92 General Provisions and another 30 historical objectives embedded in the Foreign Assistance Act of 1961 and its amendments (Rennack, Mages, and Chesser 2011, 3). This helps explain why, with the exception of military assistance and ESF aid, which vary with security threat levels, there is so little relative variation among functional accounts that are nevertheless the most politically controversial, the most likely to make a difference, and traditionally the most underfunded.[15]

There was plenty of precedence in the Foreign Assistance Act and its amendments for the self-help, self-governance, population planning, and basic needs doctrine of the New Directions mandate. By contrast, the Consolidated Appropriations Act of 2009–2010, a series of earmarks and CRs, subsumed a narrow doctrine of restrictions (Concerning the Palestinian Authority), limitations (on Assistance to Countries in Default, and on Availability of Funds for International Organizations and Programs), prohibitions (on Funding for Abortions and Involuntary Sterilization), sanctions (against Iran), protections (from Impact on Jobs in the United States), and new, more onerous eligibility requirements for Assistance generally (Rennack Mages, and Chesser 2011). Country earmarks read like a who's who of the post-Cold War period: Saudi Arabia, Colombia, Afghanistan, Sri Lanka, etc. Assistance to the former Soviet Union nations was, on the other hand, too important to leave to a General Provision; Congress established free-standing aid programs under the Support for Eastern European Democracy Act of 1989 and the Freedom for Russia and Emerging Eurasian Democracies and Open Markets Support Act of 1992 (Tarnoff and Nowels 2004, 13–14). Reconstruction and security assistance to Iraq, Afghanistan, and Pakistan were also covered but under supplemental emergency measures that in FY 2010 alone amounted to $4.4 billion, not including Defense Department funding.[16]

Thus, several trends mark the evolution of development theory and doctrine over the last decade. The 9/11 attacks gave new impetus to these trends, but their roots lie as much in end-of-the-century events and concerns. A more self-conscious commitment to bilateralism has emerged in recent years in part as a legacy of the Reagan administration's criticism of the multilateral development banks. In the conservative atmosphere of the 1980s, World Bank President Robert McNamara's promotion of basic human needs was viewed by some as socialism.[17] In his first term in office, Reagan slashed funding to the banks by over 30 percent (Wills 1987; Ruttan 1996, 130). Bill Clinton's proposed Peace, Prosperity, and Democracy Act, designed to balance a neoliberal aid package of tied trade and investment guarantees with humanitarian assistance, including democracy building for "growth with equity," fell victim to budget cuts and to Clinton's commitment to administrative downsizing.[18] Between 1994 and 1997, under Vice President Albert Gore's initiative to "reinvent government," USAID closed more than 44 overseas missions and cut its overall staff by close to a third.

The new bilateralism finds perhaps its most sophisticated recent expression in the Bush administration's 2002 National Security Strategy. Though affirming support for cooperative agreements among the United States, its allies, and the multilateral banks, it declared the intention "to act apart when our interests and unique responsibilities require" (White House 2002, 31). Questioning the ability of development assistance to stimulate sustained economic growth and reduce poverty "without the right national policies," it proposed a 50 percent increase in aid to countries that could demonstrate that their governments "rule justly, invest in their people, and encourage economic freedom" (21–22). Bush advocated and Congress enacted the legislation to create a new Millennium Challenge Corporation (MCC), a performance-based assistance program based on these criteria. MCC received an initial appropriation in FY 2004 of $994 million, up to $1.5 billion in FY 2005, but declined thereafter.[19] In the end, compared to the total State Department and USAID budgets of over $33 billion, MCC appropriations did not amount to much.

Thus, despite the challenges that USAID has faced, it clearly held the day with three-fourths of all foreign operations appropriations. Six percent of these appropriations go to support multilateral assistance, principally for the World Bank Group's International Development Association for concessional loans to the poorest countries. The remaining appropriations go largely to military/security assistance, with great year-to-year variations reflecting perceived assistance needs of security hotspots such as Iraq, Afghanistan, and Colombia (see chapter 10 on the development assistance-military assistance nexus). New accounts such as Transition Initiatives, the Freedom Support Act, and the Support for Eastern Democracy are intended to aid the transition of former Soviet Bloc countries and postconflict countries to liberal, free-market democracies. The President's Emergency Plan for AIDS Relief, a George W. Bush initiative, also has a security dimension with built-in performance indicators. The control of infectious disease risk that "directly affects public health in the United States" gave rise to Child Survival and Health Programs. Finally, a set-aside for Nonproliferation, Anti-Terrorism, Demining and Related Programs added to the expansion of bilateral aid. The cumulative effect of these supplemental accounts has been to shift for geopolitical strategic purposes the original priorities of development assistance, never wholly altruistic during the Cold War era, development per se was at least an equal partner with security concerns. What makes those concerns so different today is the problem of terrorism, a transnational, nonstate phenomenon with no clear solution except long-term socioeconomic development, a task that no country, however powerful, can achieve.

Summing Up

As defined under the Foreign Assistance Act of 1961, American aid was originally intended to encompass efforts to alleviate poverty, promote sustainable growth with equity, ensure civil and economic rights in the development process, integrate developing countries into the global economy, and promote norms of good governance. It still serves these functions but shares them with a growing class of new, ad hoc accounts under the general heading of Bilateral Economic Assistance, designed to address sources of local and regional instability. Inconsistent priorities degrade the content and delivery of economic assistance.

The inconsistency of objectives is certainly due to changes in the external environment—the rise or fall of competing aid providers and shifting geopolitical and security challenges. But the problems of inconsistent objectives and poor coordination are also rooted in the absence of an effort to clarify goals on a broad strategic level to maintain coherence in the face of changing circumstances and a realistic assessment of the resources required to achieve the combination of goals. Without such effort, US assistance effort will continue to fall prey to the goal displacement described in our introductory chapter and the unproductive posturing that has substituted for doing it right.

Development scholars and practitioners have learned that "there is no apparent simple relationship between aid and growth. Some countries that have received large amounts of aid have recorded rapid growth, while others have recorded slow or even negative growth. At the same time, some countries that have received little aid have grown, while others have not" (Radelet 2006, 7–8). Yet this is not a justification for doing nothing. Nor has the United States reaped fully the benefits for its generosity: greater investment opportunities for American business and a more peaceful, cooperative global order. That these goals and the institutional mechanisms for achieving them have often been in conflict is a large part of the story of the evolution of development assistance and its continuing challenge.

CHAPTER 9

Evolving Roles of the Military in Developing Countries

Introduction

Military roles in developing countries are numerous, complex, and highly variable—both in specific details and how these details change through time. Any overall survey of these matters is a challenge. Fitting the military as a distinct institutional entity into our organizing framework of theory, doctrine, and practice proves helpful.

Being clear about the core role of the military in general is also useful, especially as we explore many different theories and practical variants linked to them. *Armed forces exist to deter external threats to a nation and to deal with these threats if deterrence fails.* Adding missions and responsibilities to a nation's armed forces that extend or deviate from this core role must be considered carefully. In developing settings, such considerations are often hard to exercise, especially when alternative and countervailing institutions are weak or nonexistent.

Civil-military relations in development are important. The standpoint or perspective one takes matters. Should one focus on a single nation's military, or will a broader perspective of regional or global military powers work better? Where possible, we focus on the national domestic level, but still recognize influences of external powers.

Even at the national level, the military does not operate in isolation. Nongovernmental organizations (domestic, regional, and global), private sector activities (home-grown as well as those linked to increased globalization), foreign direct investors, the World Bank, IMF, other lenders, the United Nations, and so forth, all may have impacts on military forces.

Military doctrine is not static. For example, wars of national liberation relate directly to global conditions and events, ranging from the legacy of the Cold War to wars on terrorism. Adversaries and alliances shift; so do doctrines to accomplish changing military ends. Shifts may be negligible or slow to happen, as when forces are balanced or they may be rapid during times of crises, especially those threatening a regime or territorial integrity of a country.

The theories, doctrines, and practices in play for specific national military forces are best understood by attending to local details, not grand theories or strategies. Nevertheless, the interplay of national realities with regional and global realities in different time periods or eras should not be overlooked, and in making these connections, larger theoretical and strategic constructs may be helpful.

Eras of the Analysis

Different cutting points distinguish different eras in civil-military development, and we hope our choices clarify some notable trends and provide a guide to civil-military relations since the end of WWII (table 9.1).

As with NGOs, Era 1 covers the post-WWII period of humanitarian relief, relocation, reconstruction, the Marshall Plan, and the settling in of the Cold War antagonisms around the world. International civil relief organizations and systems were created from scratch. The NGO sector hardly existed prior to WWII. The League of Nations' humanitarian efforts prewar were limited; they no longer existed postwar. The United Nations was barely underway and offered little. The military was the default institution available to confront postwar development challenges in nations around the world.

For the Soviet Union and other Eastern Bloc countries, supporting wars of national liberation was regarded as separate from whatever progress of détente with

Table 9.1 Eras of military roles and civil-military relations

	Era 1 (1945–1960)	*Era 2 (1960–1990)*	*Era 3 (1990–2001)*	*Era 4 (2001–2015)*	*Era 5 (2015–)*
	Relief/expand/ civil-military affairs	*Community development/ wars of liberation/ counterinsurgency*	*Fall of Iron Curtain/ globalization*	*War on terror/China-Islam*	*Natural events/ resources/ complexity*
Focus	Specific	Local/specific	Regional/ national	Regional/ global	Ecosystems/ mega-cities/ disaster relief
Time frame	Immediate/ finite	Project life/war's end	20–50 years	Open-ended	Variable
Scope	Individual	Local/nation	Region/nation	Nation/ region	Ecosystems/ natural resources
Participants	Militaries	Development agencies/military/ NGOs	"Everyone"	Militaries	Super-networks
Military role	Primary/ central	Central/ coordinate	Various	Active/direct involved	Technical/ logistical/ security support

the Western powers was being made. The Soviet doctrine was that "while concluding agreements on the normalization of relations with the United States, the Soviet Union nonetheless would continue to support the struggle of the peoples for their social and national liberation" (Trofimenko 1981, 1087). Soviet training, indoctrination, and provision of weapons mirrored that of the United States, to the insurrectionist groups and the government these groups established. In border countries such as Afghanistan and Mongolia, Soviet military support was also provided, and Soviet troops were deployed in the failed attempt to maintain the Afghan government. Soviet foreign assistance to Cuba was at least implicitly tied to Cuban military activities in Sub-Saharan Africa.

Somewhere around 1960, explicit modernization efforts, aided by government development programs and the emergence of effective international NGOs, took form (Pye 1961).[1]

Often civil development programs were combined with localized military ones to mount counterinsurgency actions against indigenous forces engaging in wars of national liberation. The merging of economic assistance and military objectives took many forms, including infrastructure projects, pacification and land reform programs, and training and arming of national military forces to combat revolutionary guerilla forces (Lansdale 1962; Walterhouse 1962). US professional military training programs educated and shaped the perspectives of thousands of officers from the developing world during this period. Fort Benning, the Command and General Staff College, the Naval War College, and even the elite service academies at West Point, Annapolis, and Colorado Springs took in and indoctrinated elites from developing countries around the world (Pye 1964). Hand-me-down weapons (at first) and then increasingly sophisticated ones were also distributed, along with the associated traditions, doctrines, and mores. The USSR was similarly involved in training, indoctrinating, acculturating, and transferring weapons to its various client states in the developing world.[2]

The fall of the Iron Curtain and the collapse of the USSR opened Era 3, a time of emerging American hegemony and high hopes for reduced military expenditures and improved economic and social development around the world. This era is marked by longer term development initiatives of 20–50 years in the form of massive electrification (especially hydropower), transportation, education, and public health projects funded by national, regional, and international sources. A flourishing array of NGOs came on the scene, as we have noted elsewhere. A confident and integrated Europe took on development projects around the globe—collectively as the European Union came together and nationally in the form of creative and often generous directed aid and assistance programs. True development progress occurred in this period.

The jarring experience of the September 11 attack marked a sharp change of state for America and the rest of the world. Era 4 is the time of various "Wars On...," especially on terror and drugs. It is also a period when natural disasters seemed to happen more frequently and devastatingly. Counterinsurgency and special operations military forces teamed up with national police and border control paramilitary forces to suppress drug trafficking, fight insurgents, build and support communities, and provide humanitarian assistance during natural disasters. Demands on national military forces multiplied to do more than simply protect a nation's sovereignty. Questions surfaced: What is the purpose of a nation's military? What is its core competence? How effectively and efficiently is it delivering this? What other roles can and should a nation's military take on and still not compromise its ability to do its fundamental job? What happens when its fundamental job is no longer needed?

As Era 4 progressed, around 2010 is a good reference point, China and Russia began to assert themselves in two different ways, each having implications for regional and national development prospects.

Resource-seeking behavior on China's part grew apace with its remarkable economic growth. China's reach around the world for natural resources to guarantee its own future economic success became clear. Development assistance to countries providing resources to China began about this time and continued to grow. The use of Chinese military assistance to protect resource bases also became evident. Energy-providing behavior characterized Russia's reemergence as a significant regional actor around the midpoint of Era 4. The protection of sea lanes, pipelines, and transportation routes for strategic reasons created opportunities to increase the size and capabilities of related military forces. The regional focus in the early stages of Era 4 becomes increasingly global as resource-seeking and energy-providing demands began to grow in 2014 and beyond.

Theories and Doctrines

The following is a general guide to understand how militaries operate and contribute in developing countries. Specific conditions and details always matter as they shape future outcomes. They also do not apply across the board to all countries; they may change through time even within one specific country.

Theory of Weak Civilian Capacity

As WWII ended, leaders around the world realized that their armed forces were among the few, often most dominant, organized institutions they had. Nonmilitary institutions were by comparison chaotic, if they existed at all. Relying on the military to take on economic and social development seemed obvious under these circumstances. The military thus should take the development lead. And it did in some, but not all, cases.

The theoretical premises related to the vacuum of other, countervailing civilian institutions are key. Deficiencies in nonmilitary institutions urged the imposition of order—coercion even—that the military could provide. The underlying and usually poorly articulated assumption was that the military as specialists in violence would go back to their barracks once civilian institutions came on the scene. This assumption often did not work out well. The fundamental responsibility of a military is to protect the nation from *external* enemies. This is clear and hardly controversial. However, protecting a nation from *internal* disruptions is also often left to the military. When other security institutions such as the local police, Guardia Civil, Carabinero/Carabiniere, and the national guard are limited or unable to control internal disruptions, the military is called on to perform. This can be problematic.[3]

A third kind of responsibility, termed *management* with respect to the rebuilding of war-torn areas after WWII, often found the military constructing ravaged infrastructure, housing and feeding refugees and the displaced, and training and employing citizens on an interim basis until civilian authorities could assume these responsibilities and tasks. The Theory of Weak Civilian Capacity thus has at least three different outcomes, only one of which—*defend the nation from external threats*—is clear, fundamental, and uncontroversial. Problems occur when the military is *responsible for internal matters*—protecting the nation from internal threats because of limited police or paramilitary institutions, or fulfilling national management needs because of gaps and vacuums in civilian institutions. We next sketch out some of these problems beginning with the post-WWII era.

Where imperialism was challenged or shattered, *indigenous* military remnants of British, French, Dutch, and other former colonial powers emerged to fend off external threats posed by former colonial masters. New indigenous militaries also filled gaps and vacuums in civilian institutions. Military skills indigenous groups possessed often became the base of independence and resistance movements. These skills also enabled opposing groups in numerous civil wars. China's civil war between Communists and the Kuomintang, long simmering during WWII, took full form this way (Westad 2003).

In areas savaged by war-fighting, constructive military actions helped rebuild war-torn infrastructure. Training and employment also followed and helped restore nations. Whatever the specific national experience, the post-WW II era generally expanded the concept of civil-military affairs.

> Post-World War II Europe witnessed what was perhaps the most extensive use of the military in civil affairs. It is important to recognize the influence this had both on military doctrines of civil involvement and on development of the international relief system and the approaches that relief agencies have used since that time... The role of the military was expanded as never before.
>
> From the beginning, the objective was to establish martial law in the occupied territories, then quickly rebuild indigenous capacity to manage cities, the provinces and, ultimately, the national governments. In Germany, the process took longer but the goal was the same—the military role was to shift from *security and management* to strictly security as quickly as possible. (Cuny 1989, 2; emphasis added)

Shifting occurred, although not always quickly or without long-lasting consequences. Cold War and postcolonial maneuverings of the victorious major powers worked to sustain and entrench the military for years in different places around the world.[4]

A countervailing doctrine to that of expanding professional scope essentially says that the military should stay out of everything except responsibilities that are clearly and unambiguously *military*. Classic treatments of civil-military relations, notably Huntington's *The Soldier and the State: The Theory and Politics of Civil-Military Relations* (1957), make a compelling case here. Practical, empirical treatments of this argument and doctrine provide evidence (Stepan 1971, 1988; Trinkunas 2005). The main question is whether eventual civilian control of the military is compromised when the military leads and then dominates in matters of internal security and economic development (Goodman 1996).[5]

Theory of Professionalism

The hierarchical structure, the capacity to impose discipline, and the capacity to train effectively often make the armed forces the most professional organization in a specific developing country. The professionalism label is tricky, however. For many, being a military professional means doing your military job as well as possible. Getting involved in political, policy, and development issues is not in the job description. Involvement sullies one's military professional reputation.[6] The theory of professionalism thus has two different interpretations, each having different implications for the role and activities of a nation's armed forces.

The relationships between this theory and the previous one are close. A shared premise with doctrinal implications is that since the military is the most competent institution in a country, it has a professional duty to take on many more national responsibilities (Daalder 1962; Johnson 1962; Jones 1965).[7] Paradoxically, military competence enables and encourages them to go beyond what they know how to do.

To maintain a good military requires that the ports, airports, air traffic control, communications, and roads are in good working order and operate well. In Brazil, for instance, airports and air traffic control remain under military control, albeit not without issues and controversies (ABAG 2013).[8] Infrastructure is a military responsibility—in theory, at least initially as a country develops. Again, as with airports in Brazil, responsibilities outside those only required to fight wars sometimes lead to problems. Building roads, bridges, sewers, and communications systems is certainly within the scope of military engineers and engineering (Johnson 1962). However, if these activities involve complex technological demands beyond the rudimentary ones that soldiers need to fight and defend a nation, problems of quality and effectiveness are inevitable. Beyond competence overreach, huge temptations for corruption and personal gain cannot be overlooked. Overreach also occurs when military core competencies extend into entirely new kinds of economic activities. Building hotels, resorts, ecotourism, and other businesses are illustrative.[9]

Keeping the military focused on its basic mission is a fundamental political responsibility. Defining fundamental survival threats to the nation and ensuring that the military does not stray unnecessarily away from these is essential.[10] This is the basic motivation and glue that shapes and guides professional militaries everywhere. One version of this is an external threat to national sovereignty by foreign invaders. For global powers, especially in the Cold War era of Mutual Assured Destruction, a threat to survival was literally true. For many developing countries, focusing on threats is not so simple. It might be a border war with a neighbor: Ecuador, Colombia, Venezuela, and Peru, and perhaps several other neighbors each fit this pattern. It might be an internal revolutionary or guerilla threat. There are too many examples around the world. Often what passes for a mortal threat by one group or another may not pass muster with the military itself. Doing police work to combat drug trafficking and gain generous US drug suppression money has been less appealing to the Colombian military than going after the Revolutionary Armed Forces of Colombia (FARC) or National Liberation Army (ELN), where the threat to the nation was generally accepted and taken as a given (Jaskoski 2013, Chapter 7). Doing police work today turns out to be a different matter in Honduras, El Salvador, Guatemala,

Nicaragua, and Mexico, because the scale and violence of drug trafficking pose basic threats to these nations.

Sustaining a focus on the basic professional mission for a nation's military is complicated. Deviations and additions are referred to as "mission creep" (Wilson 2011). Mission creep is a theoretical and doctrinal quality shared by the military and NGOs (Siegel 1998).

Theory of Beneficent Charity and Humanitarian Necessity

Natural and human-caused disasters often exceed the capabilities of civil authorities. The manpower, professionalism, and organizational strength found in military organizations open opportunities for help. Other national institutions cannot compete, especially if disasters and catastrophes increase in number, scope, and scale. Should climate change figure increasingly in natural disasters—as scientific consensus warns—expect increased military as a matter of human necessity (Scheffran et al. 2012; IPCC 2014). Beneficent charity is not subtle: do whatever is required with whatever resources are available to alleviate human suffering. The Ebola epidemic in Western Africa in 2014 is so severe that national and international military forces are being called into service in the absence of effective responses by national, regional, and international health organizations (Walt 2014, 32–37).

This theory and derived doctrines have mostly been successful in the post-WWII era. However, the dire circumstances in the immediate postwar are not the same as contemporary civil wars, famines, and other disasters, both natural and medical, in the developing world (Rana 2004). A main issue now is to gauge whether, how, and when to use military means, particularly to solve fundamental, long-term development needs.

> Provision of tents to victims of an earthquake or hurricane often delayed reconstruction and failed to address critical land issues. Construction of refugee camps for famine victims drew people away from their land, making agricultural recovery nearly impossible and creating an even larger relief requirement…
>
> The military forces committed to these operations also continued to use the same modes and doctrines. Planes are used in ever-increasing instances to deliver food and supplies; engineers are still committed to building refugee camps. Yet there is increasing concern that these uses are not without costs. (Cuny 1989, 3)

Flying large amounts of food and medical supplies into airports capable of receiving transport aircrafts is a positive, but only if there are sufficient means to distribute the goods into more remote areas where they are needed. When there are no roads, it usually means finding expensive helicopters to do the job. Military aircrafts are not cheap and are often in short supply. Indigenous military forces in general are not all that well financed, with the result that paying the bills for military operations in humanitarian situations comes out of existing domestic, development, and humanitarian budgets (SIPRI 2008). This would not be as consequential except that indigenous military forces are more involved in humanitarian relief efforts than ever before.[11]

Unfortunately, they are often ill-prepared for these responsibilities; the military doctrines they relied on are ill-suited to these nonmilitary emergency circumstances; and important details defining the disaster or emergency are seldom considered until boots hit the ground and operations become a trial-and-error, learn-by-doing experiment.

Theory of Universal Conscription

The core idea is that the military has a capability to educate and so to convey information, skills, and a sense of political identity and acceptable political behavior. Universal conscription or other means to direct or induce service in the armed forces creates a national armed force and promotes national loyalty.[12] The military has comparative advantage over other institutions in educating, socializing, and controlling key demographic groups. Common means are training for skills such as reading, preparing for jobs in manufacturing and increasingly in services such as information technology (IT), and doing all this by emphasizing order, discipline, and professionalism. Developing nations lack advanced educational opportunities, so this theory and associated doctrines are rational and often effective. Loyalty and a sense of national identity are real, practical outcomes. For other nations, especially those relying on highly skilled and technically adept individuals to operate high-tech weapons and systems, more than raw recruits are required. The rise of highly skilled and compensated professional armies and proliferation of private military contractors are illustrative (Schaub and Franke 2009). Some now realize that these trends run counter to the national loyalty goal embedded in the theory of conscription. Mercenaries are loyal to whoever pays them (Scahill 2007; Prince 2013).[13]

The crucial importance of context is emphasized for universal conscription in Israel, which has evolved and adapted to the needs of a high-tech, information-based economy.

> Each year, Israel's military puts thousands of teenagers through technical courses, melds them into ready-made teams, and then graduates them into a country that *attracts more venture capital investment per person than any in the world.* Military service in Israel is generally compulsory, lasting two or more years. Many would-be entrepreneurs apply to the IDF's [Israeli Defense Force] computer training academy, known as Mamram [which] acts like a school for startups, teaching programming and project management to cadets in olive-green uniforms. Young hackers with proven skills get recruited by specialized intelligence units such as Matzov, the army's cybersecurity division, or units involved in signals intelligence and eavesdropping. (Kalman 2013; emphasis added)

Israel now has one of the largest concentrations of big-data engineers and analysts in the world and is creating and spinning off high-tech start-ups at a phenomenal rate. Nevertheless, Israel's lauded national service has also run afoul of a growing number of ultra-conservative religious groups that resist military service entirely. The socialization goal of the theory of conscription has evolved in the case of IT and high-tech and created domestic conflict in the case of the ultra-orthodox.

The specter of mission creep emerges as unbounded lengths of service create demands for those in uniform to be involved in different activities than just being a soldier and defending the country. The People's Liberation Army (PLA) in China is a monumental example of this problem.

Structuring means to end military service exist, but do not always succeed. For instance, the US military and others modeled on it have clear rules for promotion that feature up-or-out options. Steep pyramids to attain top leadership positions are coupled with exit and retirement expectations at the end of mandatory service for constripted soldiers (one to two years), or for an enlistment (typically three to five years) in the first instance and in the second, retirement case, for a fixed period of time, typically 20 or 30 years.

Theory of Corporatism

This theory divides and organizes different functional groups and categories so that they are made subordinate to the state. Groups could include business, labor, agriculture, military, intellectual/scientific, and other defined interests. Corporate in this theory does not solely refer to business corporations; rather it refers to the society as a whole body composed of many different, connected, and essential parts. In this version, the role of the state is to create conditions that ensure the success of the body by maintaining a healthy balance among the essential parts (*Encyclopedia Britannica* 2014). Another source, related directly to our topic of military and development (especially in Latin America), summarizes corporatism in terms of structures, subsidies, and controls that a state provides to each constituent group:

> A system of interest representation is defined as corporative to the extent that it is characterized by a pattern of state *structuring* of representation that produces a system of officially sanctioned, noncompetitive interest associations which are organized into legally prescribed functional groupings; to the extent that these associations are *subsidized* by the state; and to the extent that there is explicit state *control* over the leadership, demand-making, and internal governance of these associations. (Collier and Collier 1977, 493)

The theory has had many different manifestations over the years, and, unsurprisingly, it has also had varied rationales, interpretations, and practical outcomes. Italy under the Fascist Mussolini is commonly noted as a modern illustration of corporatism—one that obviously did not turn out well. Other European examples post-WWII are sometimes identified as Austria, Norway, and Sweden, all of which took aspects of the theory to structure, subsidize, and ultimately control conflicts between business and labor in the interest of economic development. These experiences turned out rather well.

In the developing world, corporatism's notable practical examples are primarily found in Latin America, but here too the lessons to be learned about the utility of the theory with respect to military involvement in development are at best mixed and at worst fraught.[14] In the absence of stable state institutions, the maintenance of control required by corporatism resulted in extreme volatility, with a long list of military takeovers, notably in the 1964 through 1973, for example, Argentina, Peru, Panama, Ecuador, Bolivia, Uruguay, and Chile (Chalmers 1977), but continuing into even recent times.[15]

During the Cold War, a National Security Doctrine provided a rationale to consider the military as *the* central state institution. Internal threats to national security were often tied to problems of underdevelopment and so enabled the military to force whatever government was in power to provide conditions for development. This simplistic idea typically justified repression of internal enemies and simultaneously revealed military ineptitude along a broad development front.[16]

While each nation's experiences differ in detail, a common practical outcome of authoritarian military involvement has often been a slowing or constraining of overall potential development performance. Political instability does not readily attract investments for the long term, especially from foreign sources. Military control of state instruments of power and productivity likewise has a poor record of achieving economic gains over meaningful periods of time. Soldiers are not the same as business people, something we take up in more detail next.

As relevant as total economic performance or its lack, equitable distribution of a nation's economic bounty is also problematic and challenging, no matter who is in overall control. Again, the Latin American experiences with authoritarian military leadership are not encouraging. When the military controls the state and when political institutions are nonexistent, volatile, or unreliable, the theory of corporatism and its sundry doctrines and practices related to development have been disappointing at best.

Theory of Enterprise

Considering rationales for military involvement, it is no surprise that militaries in developing countries are often active in economic enterprises, too.

Different motivations and opportunities justify and explain why the military gets into business. Sometimes it is the most competent institution in a country and so does business almost by default. The military may find itself in charge of or owning valuable lands, precious minerals, oil and natural gas access and rights, or other profitable commodities that can be turned into businesses. Sometimes making money is rooted in fears that an impoverished nation will not be able to provide sufficient funds (from the perspective of the military) to support defense requirements adequately. Whatever the country specifics, the military's ability to defend the nation is often put at risk by avaricious military businessmen. Corruption, mismanagement, lessened military professionalism and capabilities, and other symptoms are evident, as we shall see in the following vignettes.[17]

Egypt

No one knows how much of Egypt's economy the army controls. Guesstimates range from 10 percent (*The Economist* 2011; Stier 2011) to more than 40 percent or even 60 percent (Tadros 2012; Hauslohner 2014). The diversity of businesses controlled by the military is quite stunning: cement, olive oil, household appliances, pest control, catering, child care, tourism, job placement for military retirees, flatscreen televisions, gas stations, technology projects, land on which they have built resorts, ports, and housing complexes, tax breaks and other privileges, including the use of conscript labor. Hauslohner (2014) reports that former President Morsi government was about to negotiate with "Qatar over development of the [Suez] canal zone in a way that didn't directly involve the military…Anti-brotherhood media quickly howled that Morsi was 'selling' the canal to a foreign country. Sisi portrayed it as a national security decision."

The special circumstances created by the multibillion-dollar annual aid payment the United States sends to Egypt for military equipment confound the corporate picture—for Egypt and also for US military arms producers such as Lockheed Martin, Boeing, and General Dynamics for aircraft, tanks, missiles, guns, and a full range of weapons (Valente 2011). Justification of the military's dominance in the economy was "because it helped ensure regime stability and security," at least until Tahrir Square when it no longer did (*The Economist* 2011).

Pakistan

Ayesha Siddiqa (2007) claims that the army controlled at least $15 billion of Pakistan's economy in 2007. Her book, *The Military, Inc.: Inside Pakistan's Military Economy*, was a sensational exposé whose implications reverberate to this day. Others have pursued her lead. Najad Abdullahi (2008) suggests that the total may have been $20 billion or more. Evidence and precise data are hard to come by, but nothing suggests that this dominance is any less in 2014. It is likely to be even higher (*The Economist* 2011).

The military is everywhere, including serving officers, retired ones, and those in charge of various foundations and pension funds that the military controls (Abdullahi 2008; Khan 2012). Claims and rationalizations of the military's superior managerial and administrative capabilities abound, based on its cultural and professional ties to the British military and colonial civil services. Hopeful arguments based on the indisputable economic successes of the PLA are also deployed.

Actual lackluster performance of the Pakistani military in a wide array of true business activities calls this myth into question. The early hope for military control of the economy was grounded on its good performance meeting IMF structural requirements in 1999–2000 and 2005–2006 periods. Satisfying the IMF and running a successful business are not the same thing, as the subsequent penetration of the military into so many parts of Pakistan's economy reveals (Khan 2007, 2012).

Many military businesses exist and depend heavily on hefty public subsidies—often sourced out of legitimate development line items and public sector budgets. "[P]rivate sector activities are more complex and require more than military training which could be accounting for the high failure rate of the military's venture into economic activities. There is no compelling case to support the military's venture in private sector activity" (Khan 2012, 6). As the military took on more political power, especially after 1990, it controlled more and more of civil (especially economic) life in Pakistan. At some point, the common, public interest was overtaken by the narrower and personal one of sustaining and enhancing the military for that institution's own sake (Siddiqa 2007, Chapters 1, 4; Khan 2012). Civilian control of the military has fared poorly over the years—most recently in the case of current Prime Minister Sharif, despite his strong election victory in 2013 (*The Economist* 2014a).

The greatest, consistent problem according to two recent critiques "is an army (and its spies) with too much power and no accountability" (*The Economist* 2014c, 78, reviewing Jalal 2014 and Fair 2014).

China

What is sometimes called the "Beijing Consensus" summarizes in a simple label what has been in fact a monumental, complex, and long-running process for economic development. The process is founded on a strong military basis and has used, even depended on, the military to operate the key economic elements and drivers (Cheung 2001). Almost as remarkable as the story of economic success is the subsequent tale of political assertion and restoration of control over the military—primarily because of widespread corruption, fears about compromised force readiness, and concerns that an independent and self-funded PLA could lead to decreased loyalty to the Communist Party (Goodman 1996; Kwong 1997; Busch 2008).

The full story of how the PLA got into business is fairly complicated, but the basic impetus can be summarized.

> Between 1978 and 1998, the Chinese People's Liberation Army (PLA) underwent a dramatic transformation. Faced with the contradictory forces of a declining military budget and pressures to modernize at the end of the Mao era, the army reluctantly agreed to join the Chinese economic reform drive. It converted and expanded its existing military economy to market-oriented civilian production, in the hope that the resulting profits could replace lost expenditures and help to finance the long-needed modernization of weaponry and forces. (Mulvenon 1999, 2)

The Chinese military's efforts to generate cash to sustain and improve the military post Mao (1976–1980) resulted in numerous, small, out-of-the-way, and out-of-sight businesses making military equipment for the PLA's own use (Goodman 1994; Goodman, Segal, and Yang 1996). The military soon realized that basic military equipment was a marketable commodity around the world and so started exporting it and making money (Bickford 1994, 1999; Welker 1997, 3).[18] Profits from these exports financed many more and different businesses. No one planned this, and no one could have imagined how everything would eventually shake out in the next decade or two. Simply stated: PLA businesses generated a transformative industrial revolution and succeeded in lifting millions of Chinese out of poverty.

The mysterious nature and array of these businesses did not get noticed in the early years—certainly not by Western businesses, markets, or governments and probably not even by the central Chinese leadership. At some point the political leaders did notice and were not entirely pleased by what they saw (Solomone 1995). The central political leadership did support various small successes. It also ordered the creation of joint companies, conglomerates meant to consolidate and rationalize management and supply chains and so improve control and overall efficiency (Mulvenon 1999, 9–11).[19] Many of the companies employed retired PLA officers in top corporate positions. The political leaders also regionalized and then globalized these growing enterprises with listings on exchanges—primarily in Hong Kong where transparency and disclosure rules are slack; joint ventures and partnerships; and the casting off of subsidiaries around the world. Securing financial access in the markets was supplemented by the acquisition of holding companies in strategic international tax havens (Welker 1997).[20] The businesses insinuate themselves into different markets in different ways, and keeping score here has taken on huge importance for intelligence agencies everywhere.[21]

The penetration of Chinese firms that control ports is particularly impressive. Controlling ports worldwide is the business of Hutchison-Whampoa and Hutchison Port Holdings (HPH), which is itself controlled by Cheung Kong Holdings in Hong Kong. It seemed innocent enough in the 1980s when operations were mainly confined to Hong Kong and several other ports in the region. But now HPH controls more than 136 important ports around the world, which makes it the world's largest seaport operator. Significant among all its port operations is the Panama Canal, over which it has an exclusive contract (Busch 2008; Hutchison Port Holdings 2014).

The divestiture and reform of the PLA's businesses was complicated and challenging. A slow-motion competition for dominance occurred between military and civilian/political factions. It also took nearly a decade. There is evidence of sustaining attention to this day, especially the vigilant rooting out of military/business corruption (Mulvenon 1998; Mayama 2000; Mulvenon 2001a, 2001b; Buckley 2014). It is an amazing accomplishment that may not be replicable—especially in countries where countervailing political power and authority does not exist to challenge khaki-clad capitalists. This is a cautionary note for countries seeking development by embracing the Beijing Consensus.

The military's role in China's development could not have been forecast, much less imagined beforehand. It happened as the military struggled to keep itself on an even financial keel in an era of economic upheaval. Domestic economic successes and benefits of the military's involvement in the economy were not planned, although the results are appreciated and have been leveraged for other purposes. China and the rest of the world must now learn to live with the consequences.

Indonesia

In a world accustomed to the economic promise of the BRICS (Brazil, Russia, India, China, South Africa), it is unlikely to highlight Indonesia as being at least as interesting. For one reason or another, Brazil, India, and South Africa have faltered since 2001 when Goldman Sachs coined the term BRICS.[22] The last four or five years have been especially problematic. Each of these countries promises to be "the development miracle of the future" for what now seems like a long time. Russia is a different case and depends on oil and gas and a declining and aging population, neither of which can possibly last—not for long and certainly not forever. The military's role in all of the BRICS, save China, is negligible to modest—even in Russia despite Vladimir Putin's favoring of it in recent years. Indonesia is a surprisingly different case and proposition, as summarized by Abdul-Latif Halimi (2014):

> Indonesia has achieved a 5.4 percent average annual growth since 2001 and sustained only a slight impact after the 2008/09 global recession. Its economy ranked 27th in the world in 2001; today it is 16th. The Asian Development Bank forecast for 2030 places Indonesia at 7th largest, which is larger than both Germany and the United Kingdom. Citibank estimates that Indonesia could rank 4th in the world as soon as 2040. Only the U.S., China, and India will be larger. Government debt has fallen from 95 percent of GDP in 2000 to just 26 percent in 2014. Fitch and Moody's granted Indonesia investment grade status in 2013. Indonesia's strategic importance in Southeast Asia is growing, especially with the ascendance of China and the U.S. "rebalance" to Asia and Southeast Asia in particular.

The role of the military has been central since the end of WWII, and not always beneficial for economic development. The long-standing involvement of the military in nearly all sectors of the economy began under the reign of Suharto with investiture in government and state-owned corporations. It persists there and continues to show up in various military-controlled foundations "which operate businesses in the financial sector, travel industry, manufacturing and resource extraction. These foundations fund, in part, the underfunded military" (Beeson 2008). Military involvement in Indonesia's vast natural resource sector has been extensive and challenging for many years (Ascher 1998; Mishra 2014). Problems with corruption persist well beyond the natural resource sector. A growing trade deficit, unsustainable fuel subsidies, inadequate infrastructure, poverty, and widening inequity between haves and have-nots all demand attention and solution (Cochrane 2014; *Christian Science Monitor* 2014). Despite all the rosy economic forecasts, resolving and coming to terms with a fundamental change in national leadership in the form of populist President Joko Widodo—at the same time as a lingering list of basic problems beg for attention and solution—will test mightily the balance between civil and military sectors in Indonesia (*Christian Science Monitor* 2014).

Theory of Economic Stimulus

The influence of the military may be a positive economic factor if government expenditures increase demand for military goods and services and create a multiplier effect for consumer spending. Employing armed forces removes pressure on the civilian labor market—especially that from the young, poorly educated, and unskilled—while providing education and basic skills for subsequent deployment in civilian life.[23] Military expenditures for research and development sometimes benefit the civilian economy, for example, the internet, GPS, radar, nuclear power. Government expenditures contribute to a nation's economy, but how *military expenditures* specifically contribute, particularly in different settings and circumstances, is often hard to figure out.

Government expenditures are frequently assumed equal in their effects. Actually they take many different forms, and few of these are meant to maximize simple financial returns on investment. There is the unresolved question of what multiplier to assign to different expenditures (de Rugy 2012). Most assume that weapons are less useful for development than many other public investments are.

The theory of economic stimulation was formulated on 1930s' Depression-era experiences of developed countries, not of countries desperately seeking a way out of poverty (Benoit 1968, 1973). The perspective of five great powers from 1870 to 1939 is not clear about what military expenditures contributed to economic growth. Germany, France, Russia, Japan, and the United States invested in their militaries, but the longer term economic effects for each were not the same. Often military spending is fear-motivated and forecloses other growth opportunities (Castillo et al. 2001).

Likewise, military spending in advanced countries features state-of-the-art technologies and manufacturing processes and so creates outcomes for an economy different than arms transfers to poor countries might (Looney 1989; Graham 1994). Even among somewhat similar developing countries (e.g., Egypt, Israel, Syria), evidence relating military investment to economic growth is inconsistent (Abu-Bader and Abu-Qarn 2003). Military spending has opportunity costs, especially when basic health, education, and social needs are neglected to favor military priorities (UNICEF 1996). Military expenditures expand opportunities for corruption. Reservoirs of cash pose temptations, whether these are directed to military projects or toward businesses that the military controls (Mauro 1995; Gupta, De Mello, and Sharan 2001).

External versus Internal Theories

The seven theories roughly fall into two distinctive categories and ways of thinking about and using the military in development. The division emphasizes different perspectives, either external or internal, and reliance on force, either coercive or noncoercive. *External/Coercive* perspectives include theories of weak civilian capacity and corporatism. *Internal/Noncoercive* perspectives include theories of beneficent charity/humanitarian necessity, universal conscription, and enterprise. Professionalism, for reasons earlier mentioned, seems to straddle the basic divide between External/Internal and Coercive/Noncoercive.

Key Considerations and Conditioning Factors

Considerations and conditioning factors focus attention on activities and processes within the context of a problem that will likely influence subsequent outcomes and events. Which of these will matter and by how much is not predictable. Military involvement in the development of a society is dynamic and at times volatile. Historical trends in civil-military relations are seldom purely good or clearly harmful to the larger objectives of a society's development. There is no best military case or pathway to development success. Context matters, as any close examination of specific cases since WWII will attest. Nevertheless, several key considerations, conditioning factors, and questions are identifiable and may provide guidance to future evolutionary pathways.

Down-Shifting and Right-Sizing

Getting the military back in the barracks and out of politics and business has, as illustrated previously in this chapter, long been essential for development to occur.

Reducing a nation's military after guerrilla, civil, or border wars, when full-scale mobilization is no longer necessary, is a related but different objective. Demobilized soldiers (and their former guerilla enemies) need opportunities rare in the developing world. Wealthy countries provide ideas for what might work however: grants for job retraining, low-cost loans to start businesses, aid for education, and subsidized home loans and medical insurance are indicative.

Down-shifting from active military status into the national guard, to the reserves, or into retirement is not simple or attractive (De la Rosa 2007). Assurances about continuing credit toward retirement, pay for refresher training and other days on duty, promotions while in guard or reserve status, and several other practices should be considered to ease these transitions and maintain political control while providing human capital and resources for development. Civilian jobs taking advantage of military skills such as logistics and civil engineering (for shipping, transport, airlines, and construction infrastructure) could provide preferential treatment to outgoing service personnel.

Conscripted forces differ from more professional and long-serving forces. Conscripts provide a labor force buffer in times of economic downturn. Keeping them in the military employs them. It also offers educational and training opportunities before they return to civilian life. Conditions therefore occur when holding military members in service may help basic economic conditions improve.

The officer corps in many developing countries also needs incentives to relinquish power to political and civilian authorities. Many countries exhibit long-standing cultural and class identifications with the military as a profession. Often middle-class officers and their offspring are directed to military academies and subsequent military leadership positions. Legacies routinely persist over generations. Experiences and exposures are narrow and mostly limited to military and professional subjects and skills. Many developing countries should explore chances to broaden experiences beyond conventional military experiences and so create careers in the national guard, reserves, and retirement or second careers in civilian life. The Internal and Noncoercive theories previously characterized contain constructive suggestions for means to succeed here. At a minimum, military leaders need to know they will continue to receive pay and benefits and that the military will not be emasculated in a country's transition to a right size for its national forces. This point is demonstrated by the contrast between Chile post-Pinochet and the former Soviet Union post-1989. The Chilean armed forces transitioned gradually and continued to receive respect and resources. The Soviet collapse was humiliating to its military and helps one understand its current support for Putin's efforts on its behalf.

Professionalization

Educational opportunities for servicemen and women matter for current levels of professionalization and to shore up future development capacities. Are there recruitment and educational opportunities for a domestic officer corps? Does the officer corps benefit from advanced professional education at home? To what extent is military education provided outside the country and in various professional subjects that contribute to development? American, Russian, Chinese, and different European countries have provided such opportunities since WWII, so the issue is whether a country takes advantage. But does it?

Prevalent military careers and expectations are important. Lacking regular promotional pathways, with objective and fair criteria for evaluation, military organizations often become ends and means in themselves rather than ways and means to defend a nation or protect its sovereignty.

Mission Creep

Mission creep occurs when basic military objectives open up and give way to civilian, humanitarian, nation-building objectives (Siegel 1998). The melding and blurring of military with other different missions have long been with us, as Huntington (1957) pointed out more than a half century ago distinguishing between core security and peripheral management goals and objectives. Mission creep is traditionally regarded as an unwanted and harmful diversion from the military's fundamental purpose of national defense (Janowitz 1964). However, it may now be time to reconsider what constitutes defending the nation for many developing countries and regions around the world.

If climate change continues on tracks the scientific community foresees, genuine security threats to nations will become common. Defending the nation in the future is likely to mean protecting it and its citizens from environmental challenges of enormous potential and shocking consequences. Similarly, recent experiences dealing with the Ebola virus in West Africa have so overpowered existing health institutions that military forces from several nations have been employed and justified with threats to national and regional security (Gettleman 2014).

If population growth and wholesale displacements of populations increase, then the already large number of 50 million refugees worldwide will grow larger (Kimmelman 2014, A-8).[24] Some of these displacements will be voluntary, as for vast movements of rural populations for economic reasons to existing and newly made urban areas.[25] Other displacements will not be voluntary, but will result from wars and other conflicts. Syria provides awful evidence of this sort, and the US border with Mexico illustrates the former.[26] Whichever stimulant, protecting and supervising large numbers of displaced people will require traditional military skills and capabilities. Also required are creative and innovative thoughts and actions well beyond guns and tanks to reduce human suffering on this grand scale. Mission creep under the circumstances becomes essential.

Political Institutions and Public Administration

The civil-military equation indicates who controls decision-making power, but also how this power is deployed. Civil service and administration, when professionalized and competent, serves as a countervailing force to the military—especially as the latter intrudes into economic or other management roles and functions. The existence of national police and border protection forces as parts of a comprehensive civil service could be decisive, as the experience of Costa Rica demonstrates (Barash 2013).[27]

Classic public administration theories, doctrines, and practices emphasize the importance, even centrality of bureaucracy as a means to achieve development.

> [E]conomic, social, educational, political changes emerge in large measure as the result of direct governmental intervention. Whereas much of the Western world developed with relatively little direct intervention by the "public sector," this history clearly will not repeat itself. For reasons that range from economic necessity to ideological rigidity, the developing nations insist that government—particularly the bureaucracy—should play a major, even exclusive, role in effecting the changes that are sought. (LaPalombara 1967, ix)

The development role of the military during the 1945–1990 period was commonly conceived, based largely on its professionalism, as just another bureaucracy (LaPalombara 1967, 502–503).

The competence, professionalism, and defender of democracy themes noted earlier were generally presumed to guide and limit military excesses and self-dealing (Janowitz 1960; Johnson 1962). These presumptions routinely proved to be misguided, as did those that treated the military as "just another kind of bureaucracy." It is not.

Control and Ownership of Business

A great deal has happened soon after WWII. Myriad patterns and pathways have developed and evolved with respect to military activities in different national economies. Asking what these patterns are and how they evolved is essential. Dominance of the military illuminates the total development experience in Egypt, Burma, China, and Indonesia in fundamentally different ways than it does where the control and ownership of business have been more widely distributed and shared among private and public entities. Transitions from predominantly military control of business to a more balanced array, as in Chile, are special cases to probe and understand. What prompted relinquishment from military to civil control? Can parallel conditions be constructed in other countries to accomplish these ends?

Chile has since 1976 dedicated 10 percent of the gross revenue from the copper industry to its military as a means to ensure steady financial support to the armed forces while keeping them from having direct, hands-on control of the industry. This arrangement has been challenged, unsuccessfully, by Chilean presidents since the first Bachelet term in office. It was being challenged more forcefully in 2014.[78] The Chinese experience provides a different story, but one that also features a strong military presence in business and the economy being challenged effectively by powerful political interests (Marquand 2014; *The Economist* 2014b). The Egyptian experience of seeking a balance between military and civilian control of business is yet another story altogether with an unforeseeable ending.

These considerations and conditioning factors are selective of the numerous possibilities. Nonetheless, each of them—alone and in different combinations—will likely help shape and direct events in the coming era.

Exploring the Next Era of the Military's Involvement in Development

The military's development roles in coming years will continue on many traditional pathways. Business as Usual is likely to dominate. Great variability in performance will also persist from country to country and between different regions around the world, as has been the case since WWII. However, several large and significant factors in the coming years set the stage for and necessitate major changes. *Natural events* related to climate change and public health emergencies, *natural resource competitions*, especially those having serious ecological consequences, and *burgeoning populations*, especially in cities, will combine and challenge in coming decades. Militaries around the world are beginning to respond, but only tentatively and with less effectiveness than what will be demanded of them.

International Rapid Response/Humanitarian Disasters

Possibilities following abrupt climate change have captured scientific attention since at least 2002 (National Research Council 2002). More recent studies refine the underlying science and focus usefully on implications flowing from it (Gillis 2013; Kerr 2013; Parker 2015). At the moment, general catastrophic, doomsday predictions are not supported by scientific consensus. Specific concern for Most Vulnerable Countries, usually specified as those with low-lying coasts, is a present and growing scientific concern (Barbier 2014; IPCC 2014). Military involvement in disaster recovery, such as Haiyan in the Philippines, Sandy in the greater New York/New Jersey area, and Katrina, Mitch, and Joan in the Gulf of Mexico, all provide important lessons about humanitarian responses involving the military and many other organizations. Lessons being learned in a cascade of natural disasters in recent years will eventually refine a bill of particular responsibilities.[29]

High on this list is something called "a global emergency task force" that can supply communications, control, and logistical support. It "must be the 'commandos' of any international relief operation" (Scheffran et al. 2012; Barbier 2014, 1250). The military reference is telling and squares well with our own assessment in the NGO chapter for Rapid Response. In both, drawing the military more directly and constructively into humanitarian and disaster relief activities is urged. A useful turn to a regional focus for dedicated task forces, for example, Latin America, Asia, or Africa, helps align recovery efforts with specific geographic and social cultural realities. It also links recovery efforts better with existing development banks and related funding sources

Reimagining and then recasting military forces in developing countries for larger and more sophisticated roles potentially benefit them, especially when budget cuts and reduced strategic threats force downsizing. Assisting disaster relief can improve the armed services' image if they are properly trained and behave appropriately. It also provides training opportunities with respect to infrastructure repairs, medical operations, logistical support, and command and control. Achieving these benefits depends on careful planning and respectful integration of the military with other civilian organizations (Hofmann and Hudson 2009; Svoboda 2014).

The number, scale, and devastation of natural disasters call for extraordinary measures in response. War-like images abound in the aftermath of typhoons, hurricanes, floods, Tsunamis, forest fires, droughts, earthquakes, volcanoes, and other disasters. Increasing concerns about human health threats on a large scale present similarly huge challenges to conventional organizations (Walt 2014). The war-like images are reinforced by the presence of regular military forces bringing their own talents and resources to bear for humanitarian objectives. Good and humane intentions aside, the blending of military and humanitarian cultures, people, and organizations is seldom easily done.

Natural Resource Competitions

The resource-seeking and energy-providing activities identified earlier as coming to prominence at the end of Era 4 will continue and increase around the world. Add to these a third element of food security and production and a full array of natural resource possibilities emerges to engage and require military, paramilitary, and other security forces.

Inventories of national resource assets to establish baseline inventories will create demands for satellite monitoring, remote sensing, and other large-scale data-collecting and handling activities within the scope of many existing militaries' capabilities. Real-time monitoring of national resource assets provides important intelligence on resource use, both legal and illegal. Protecting and managing forests, wilderness areas, and offshore oil, gas, and minerals create opportunities for military involvement, preferably by designated and properly oriented units such as coast guards, engineering battalions, drug enforcement and border patrols. For countries providing energy for world markets, the protection of development and production facilities as well as pipelines, ports, and other transport means will likely increase in coming years as many of these resources become more valuable and in demand. The logistical strengths of services along with their C^3I (communications, command, control, and intelligence) abilities all have potential uses in managing resource competitions.

Food production and security means protecting lands and crops as well as feeding people. As world population increases, disproportionately so in the developing countries, feeding populations will probably increase in importance—translating into more demand for monitoring and protective services, including protections for intact ecosystems to sustain them.

Each natural resource competition offers many opportunities to link military training, skills, and experiences to appropriate nonmilitary needs and careers having distinct security implications. Down-shifting and right-sizing the military to transition away from destructive and into constructive roles comes into play here. Considerable creativity is called for to identify, align, and link these opportunities—starting with education and training within the armed forces themselves.

Population—A City Focus

Cities are increasing in number and size around the world (as discussed in chapter 6). Vast deficiencies in providing urban infrastructure exist and are growing worse. Roads, bridges, transport, water, sewers, electricity, information (fiber optic, internet), public health, education, and many other kinds of infrastructure are inadequate.

The sheer size of mega-cities creates specific needs for crowd management and control that outstrip conventional means, for example, traffic congestion of epic proportions. Size combined with modern social media presents a different management and control imperative to political leaders, as demonstrated in Cairo, Hong Kong, and even New York (Occupy Wall Street).

Down-shifting and right-sizing come into play again in seeking links between specific needs of urban populations with existing and potential training, skills, and experiences found in the military. Specific attention could be directed to veteran affairs organizations to develop adaptive retraining and education programs to enhance these links. The potential of large-scale, even super-networks to connect retraining and education to different civilian jobs is intriguing, as is the prospect of plugging into internet-based training and education opportunities anywhere in the world.

Mitigating any of these deficiencies goes well beyond the capacity of any one institution, even a robust military. Many institutions and many more resources will be needed to improve matters to open more prospects for development. Securing and sustaining a balance between political, economic, and military interests and capabilities becomes more important as the complexity and stakes in development all increase.

CHAPTER 10

Complementarity of Security and Development Doctrines: Historical Cases and Aftermaths

As seen in chapter 9, the roles of the armed forces within developing countries are often heavily influenced by their interactions with governments and with the armed forces of other countries. Chapters 7 and 8 demonstrate that these relationships are even more complicated when foreign assistance is entwined with security initiatives. This chapter examines these complications through US assistance in three of its most difficult development and security challenges: Korea, Vietnam, and Chile. We also explore the aftermath of these and the more nuanced security development balance applied to today's Venezuela as the closest Latin American parallel to Chile in the 1960s and 1970s. Thus, combined with the governance progress made in Korea and Chile, in part attributable to development policy, this chapter demonstrates that the tensions between security and development imperatives can be mitigated in the current, less ideological context.

When USAID was created in 1961, in the throes of the Cold War that would continue for 30 years, the overriding doctrine guiding US bilateral aid was simply stability and security.[1] This objective turned out to be elusive. The prevailing modernization theory (see chapters 3 and 5) appeared sound enough. Rostow's 1960 noncommunist manifesto was a refutation of revolutionary socialism and an account of development as a guided series of steps leading gradually to prosperity and therefore to democracy, as long as certain decisions and leadership capacity prevailed. In this view, the first and most important task for donor countries like the United States—before economic growth, literacy, or mass communications—was to stabilize government and make the state secure. If this could be accomplished, at the "end of the road" people would be "well fed, well educated, and well provided for with consumer goods, medical care, and social security" (Black 1967, 148). The intended outcomes were straightforward: economic aid would increase agricultural and industrial productivity, build desperately needed physical infrastructure, improve health and educational services, and produce a new managerial and civil service elite ideally educated in the United States. Large capital investments across all sectors of society would create the necessary balance between economic development and cultural and political stability, an important precondition for peaceful, functioning democracies. Stable government was essential to economic growth, but so too was a stable economy to the development of good governance, itself a vital ingredient to successful aid programs. Economic growth and political stability would create the conditions for a new prosperity by removing the causes of radical social revolution—a discontented working class, systemic inequalities, and corrupt, unresponsive politicians—and clearing the path to greater foreign private investment, the traditional barometer of successful development (Rostow 1956, 1960). A flexible balance of good government, foreign aid, and either direct or indirect military assistance would need to prevail, however, in countries, such as Chile and Vietnam, where conditions fueling civil unrest and encouraging local insurgence seemed intractable. Economic aid linked to social and political reform and backed by the use or threat of military force, whether overt as in Vietnam or covert as in Chile, was, according to Cold War policymakers, the best antidote to violent revolution.

By the end of the 1960s, however, many writers on the right *and* the left began to question that peace and development were so entwined. Economic development might be an "inherently self-destroying process for the large group of countries recently emerged from colonial status" (Muscat 2002, 106, 110). Another argument, that communist-backed "wars of national liberation" would envelop the developing world and destroy free institutions,[2] was also prominent, although free institutions were hardly prevalent. Their notable absence was the primary reason for stressing the political development of countries scheduled to receive US aid.[3] But political change did not necessarily equate to democratization. Where efficient and effective participatory democracy was absent, historically untenable, or too weak to make a difference, political change often aimed at establishing security and stability *at the expense of* democracy. In Latin America the nightmare scenario for US policymakers was not military government, which had often been turned to US objectives, but "intervention from outside the hemisphere by the international communist conspiracy" (Patterson 1989, 115–116). However much to be desired, democracy and constitutionalism were pipedreams where democratic governments could not stand up against communist insurgents.

As Pye observed in an early memorandum to USAID, "If we go beyond the notion of democracy in searching for the basis of political development, we are likely to note that development is often associated with the concept of stability…stable political systems are assumed to be more developed than unstable ones" (1960, 6–7). The need for capacity building in countries having better law and order and the ability to manage a crisis was regarded as of greater practical importance than unrealistic efforts to install democratic regimes. Going back to Truman's Point Four program, the goal of US foreign policy, for Pye and others, was the achievement of "orderly social processes," democratic or otherwise (Kaplan 1950, 55). The political realism behind the containment doctrine dictated prudence in the face of "the actual facts and their consequences," a preference for limited action that would minimize risk and maximize benefit and achieve "interest defined as power" (Morgenthau 1978, 4–5).

While Pye questioned whether it would be possible to know how to change social structures or value systems, USAID leaders and Kennedy's foreign policy circle were not so dubious, especially when fearing that US national security was at stake. Nor did Pye's appeal for "developmental diplomacy" (1960, 10–11, 13)—a rare sensitivity to the needs and desires of aid recipients—prevent the passage of the Foreign Assistance Act of 1966, urging "the building of democratic and public institutions *on all levels—local, state, and national.*" In 1968 the clause—"In particular, emphasis should be given to research designed to increase understanding of the ways in which development assistance can support democratic social and political trends in recipient countries"—was added (Committee on International Relations and Committee on Foreign Relations 2003, 143). This was a tocsin call for research relevant to American national self-interest.

Implicit in the democratization rhetoric was the reality of American security needs—the need for the proper mix of military assistance and economic aid. The Eisenhower administration set the tone. The 1959 President's Committee to Study the United States Military Assistance Program (MAP), chaired by former undersecretary William Draper, deemed a purely defensive approach to Soviet-Chinese expansionism as inadequate and that "our best counteraction" was "a strong and growing economy throughout the free world. Our aid program is an important element in the achievement of this end." Economic development and the resulting political stability of countries exposed to "the threat of communism" were regarded as the best deterrence against aggression (USAID, 1959, 6, 8, 10). Kennedy himself, arguing for his comprehensive new foreign assistance program, cautioned that "This is not an effort... [to benefit] those who are less fortunate[;] this is a program that involves very importantly the security of the United States." The American people needed to understand "the real nature of the struggle in which we're engaged" (Kennedy 1961a, 1). In January 1962, Kennedy instructed the Joint Chiefs of Staff to "plan in terms of which mix of military and other forms of aid will best serve our overall national security aims." MAP and USAID programs were complementary, not competitive. Not only should they be "effectively coordinated" but efforts should also be made to "mesh" their planning cycles to achieve optimal complementarity (Gerakas et al. 1995, 32). USAID was to take the lead, not the State Department or the Defense Department, in ensuring that military assistance and economic aid did not work at cross-purposes by, for example, competing for appropriations or juggling between accounts(Committee on International Relations and Committee on Foreign Relations 2003, 290). In response, in May 1962, outgoing USAID administrator Fowler Hamilton outlined recommendations for more effective coordination between economic and military aid programs or mix initiatives. He proposed incorporating mix studies into all aspects of the planning process in target countries on the Sino-Soviet periphery. These studies would identify variations in budgetary needs, depending upon each country's unique circumstances (e.g., its ability to self-finance some defense expenditures), and highlight direct and indirect "side effects" of greater "complementarities" (Hamilton 1962, 1–4).

Two months later, Secretary of State Dean Rusk issued a secret memorandum to all foreign mission heads, the Military Assistance Advisory Group, Foreign Policy Advisors, and command headquarters. Rusk called for "explicit consideration" of

MAP/AID joint operations—being "a regular part of the planning process" and announced the Department's intention to formulate "a single long-term foreign assistance strategy, military and economic" (1962, 1).

Yet a year earlier, Kennedy had outlined an approach that clearly separated military assistance from social and economic development. The latter, he asserted, "must be seen on its own merits, and judged in the light of its vital and distinctive contribution to our basic security needs." Was Kennedy being disingenuous, or had the lessons of Berlin, Cuba, and Korea, among others, taught him the futility of exclusively military or narrow economic solutions to address insurgencies? How could he loudly declaim "our economic obligations as the wealthiest people in a world of largely poor people" (Kennedy 1961b) while presiding over one of the largest military build-ups in recent US history? This was a "development race," in Kennedy's words, to demonstrate "that the United States, not the communist nations, represented the wave of the future" (Rabe 2010, 143). The Cold War logic dictated repressive policies abroad (e.g., the use of counterinsurgency forces to suppress popular dissent) at the cost of freedom at home. The pursuit of national self-interest was the rationale for American altruism through development aid, especially in countries under communist challenge. Kennedy, like so many others of his generation, viewed the United States as a sanctuary of freedom and democracy that required neglecting democracy in certain developing countries.

With a view to improving our understanding of the doctrinal relations of security and development, this chapter examines USAID policies in Chile, Korea, and Vietnam between 1961 and 1967 and the aftermath of each. An assessment of military, political, and economic considerations illuminates how doctrines shape how foreign aid has been utilized and how the United States is perceived—donors and recipients share a symbiotic relationship that is impossible to disentangle.

Chile

The most progressive Latin American country in the 1960s—Chile—was also the most problematic for American policymakers. Political forces ranged from radical Marxist and socialist forces to an increasingly beleaguered Right wing willing to take drastic measures, all operating within a liberal Christian democratic framework of constitutional reformism. Chile—considered by administrative officials as "the showcase" for Kennedy's Alliance for Progress, the president's response to Castroism—thus aroused their gravest fears: a communist regime emerging not through violent revolution but through legitimate democratic processes. Unlike its neighbors, Chile enjoyed an uninterrupted record of democratic government since the early 1930s. Despite many of the same economic problems of other Latin American countries, Chile's constitutional order was the envy of the region's liberal democrats. Top US officials feared that the coalition of Socialist, Communist, and other Leftist parties, centering on a young Salvador Allende, threatened all that (US Senate Select Committee 1975, 4).

The economic problems were universally regarded as severe: the social problem of poverty and inequality, dependence on volatile revenues of mineral exports, and low productivity in both agriculture and industry. Chilean officials and US development experts viewed land reform, although an explosive issue dividing Right and Left, as nevertheless the key to transforming the social and economic structure inherited from colonialism. With 2.5 million peasants, largely living and working on large, poorly managed private estates, and such low productivity, despite having some of the most fertile lands in Latin America, Chile by 1960 was a net food importer (Winn and Kay 1974, 135–136).

While disenfranchised peasants flooded into the cities looking for work, urban jobs were scarce. CORFO-led[4] import substitution policies such as tariff protection and tax credits failed to stimulate sustainable industrial growth and worsened the urban-rural divide, as employed urban residents earned nearly triple the wages of rural workers (Collier and Sater 1999, 272). Reform (though meaning different things to different people) suddenly became an extremely prominent issue.[5] Conservatives saw land reform as a hedge against social revolution; Marxists as a means to break the backs of the capitalist, landowning classes as well as achieve redistributive justice for the peasant masses; Christian democrats as creating a "Revolution of Liberty," a communitarian society of equals (Loveman 1976, 293, 187, 199, 225). Kennedy administration officials were unsure about land reform, even if carried out within the legal framework of Law 15.020, which in 1962 declared that "The exercise of the right of property in rural land is subject to the limitations required for the maintenance and progress of the social order." State Department officials were more interested in "the existence of favorable investment conditions [that] included low levels of inflation, balanced government budgets, and open markets" than in structural reforms (land reform being chief among them) and feared that they would only turn Chile into another Cuba (Michaels 1976; Taffet 2007, 73, 86).

Therefore, USAID was instructed to hold up much needed grants and loans to Chile, pending a demonstration of the government's willingness to pursue stabilization policies. Approving program and project loans to Chile on the basis of curbing inflation was, in the words of US Ambassador Edward Korry, "primordial." However, one interpretation of the ultimate consequences is that requiring inflation control policies and reducing balance-of-payments problems harmed American interests in the region and reduced the prospects for peace (Taffet 2007, 91). By tying aid to government cutbacks, US policies antagonized the center-left social democrats like Chile's President Eduardo Frei, pushing them further to the left. By discouraging public spending, they eroded popular support for the moderate reform agenda of the Alliance for Progress. Aid conditionality expanded the base of support for the Marxist Left. The State Department and USAID were pushing antistatist, neoliberal doctrines—free market competition, privatization, free trade and export expansion—20 years before the neoclassical counterrevolution of the 1980s.

Bound to, yet wary of, the reformist agenda of the Alliance for Progress, USAID walked a tightrope between demands for land reform, which raised the specter of expropriation and nationalization, and more conservative agricultural adjustments like livestock improvement and credits to small farmers. However, Christian democrats and Marxists were not immediately at loggerheads over solutions to the land problem. Leftists organized into the Popular Action Front and worked with ruling Christian democrats to push through a land reform bill that eschewed the wholesale confiscation and redistribution of large estates, preferring a "mixed form of individual holdings, cooperatives, and state farms" that would preserve and extend "gains already won within the framework of the established system," gains like the constitutional right to unionize and the right to strike (Kaufman 1972, 197). Yet the constitutional amendment that permitted land appropriation without immediate full compensation was a red flag to the Right, and upon the end of Frei's administration, the Left launched illegal land takeovers. There was much in the Christian Democratic Party program of noncapitalist development to which the Left could subscribe. But US officials resisted this interpretation, insisting instead that political stabilization dictated a policy of counterinsurgency.

The objective was to provide technical assistance, training, and riot control equipment to the national police force (Carabineros de Chile) to "more effectively prevent and control public disorders and to counter acts of subversion, terrorism, and other threats against the internal security of the country" (USAID Office of Public Safety 1964a, 1). Despite little evidence of real terror in Chile, for the Cold Warriors in USAID's new Office of Public Safety there was always the imminent threat of it, especially following a contentious presidential race pitting Marxist-Socialists against liberal democratic moderates. With an overriding US aid policy of stability and security, the lack of police preparation became the real enemy.[6]

In collaboration with MAP/Chile and a special US Military Assistance Team from a police training school in Panama, USAID and the Embassy in Santiago developed a Special Emergency Equipment and Training Program that involved the provision of basic riot control equipment to the Carabineros and in-service training for new recruits in seven regional training centers and cities where the potential for civil disturbance was believed to be highest. Undersecretary of State George Ball himself weighed in on the looming problem of Chile's internal security and its maintenance both by military and police forces. "In apportioning aid to military forces in riot control," he directed in a secret memo, "the role of the Carabineros should be considered paramount, and US aid to military forces for this purpose should not become disproportionate to that being furnished Carabineros under the AID program" (Ball 1964). The government in cooperation with USAID stationed over 600 noncommissioned police officers in Santiago alone. These new centers of police protection were now at good capacity, bristling with "the supplies and weapons for handling public disorder" (USAID Office of Public Safety 1964b, 2). The Carabineros, after the 1973 coup against Allende, were among the most brutal repressors of suspected Leftists. US assistance continued after Allende took office, but only for assistance to the Chilean military, which received $45.5 million in military assistance in 1970–1974 (Farnsworth 1974, 139).

Did the $1.7 million in police assistance between 1963 and 1970 and MAP loans and grants of $100 million compromise hard-won gains in the country's socioeconomic development, many of those gains paid for by US loans and grants totaling $279.7 million between 1963 and 1968?[7] An angry Eduardo Frei, Chile's new president, wrote in *Foreign Affairs* in 1967 that the militarization not only of Chile but of Latin America generally had divided the region, exacerbating the threat of internal subversion and violent revolution. Militarization, argued Frei, was precipitating an arms race that fueled nationalization and dashing hopes for the regional integration annunciated by the Alliance for Progress. The focus on security, combined with a string of structural adjustments, distracted attention from the very reforms that

Kennedy had originally called for as required for peaceful revolution (Frei Montalva 1967, 437–438).

Between 1962 and 1969, Chile received over a billion dollars in direct loans and grants from US government and international agencies, making it the largest per capita aid recipient in Latin America. In 1975 a Senate committee estimated that between 1963 and 1973 Chile received an additional $13.4 million in covert aid, nearly $4 million of that between 1962 and 1964 alone (USSC 1975, 7; Patterson 1989, 116). These monies were approved by high-level sub-Cabinet committees like the Special Group set up by presidential order to monitor and fund counterinsurgency activities throughout the developing world. In Chile, in the run-up to the 1964 presidential election, the Special Group authorized funding for some 15 covert action projects in four categories: propaganda, including black propaganda designed to create internal divisions within communist ranks and between communists and their labor movement allies; bribery and manipulation of the media; trading of influence within and between public and private sector institutions; and direct subsidies as well as intelligence sharing with the military to support a potential coup (USSC 1975, 14, 7–11).

Set up by Kennedy in 1962, the Special Group included the Attorney General, the Deputy Secretary of Defense, the Chairman of the Joint Chiefs of Staff, the Director of the Central Intelligence Agency, the Administrator of USAID, and the Director of the US Information Agency. It was to ensure that the importance of counterinsurgency was reflected in the "political, economic, intelligence, military aid and informational programs conducted by State, Defense, AID, USIA and CIA." It was to create and coordinate interdepartmental programs, ensuring free and unobstructed flow of resources and information within and across those programs (Gravel 1971, 660–661). The Special Group reinforced the growing linkage at the highest levels between military assistance and economic aid.

Where nation building and modernization programs fell short in removing the teeth from rural and urban discontent and where organized insurgents threatened to seize the upper hand from unresponsive government, American planners turned increasingly to a counterinsurgency doctrine emphasizing military and police training and development, through special warfare training, direct provision of equipment and supplies, intelligence gathering, demolition, and communications. The counterinsurgency doctrine not only entailed efforts to seek out and destroy insurgent units but also to win the support of civilian populations by investing in education and health care, decent housing and labor conditions, and so forth.

The US government was indeed divided on the magnitude of threat posed by the Chilean Left and the 1970 election of Allende. Only a month after the election an Interdepartmental Group for Inter-American Affairs, with representatives of the CIA, the State Department, the Defense Department, and the White House, concluded that Chile was no longer a US security threat. At around the same time the Church Committee suddenly announced that "the United States had no vital interests within Chile" (USSC 1975, 48). This was startling, in light of the Allende government's expropriation of major holdings of US copper companies, yet reinforced by National Intelligence Estimates downplaying Allende's revolutionary intentions both in Chile and in Latin America generally. But whatever Allende's preference for consolidating the

Left's gains through more gradual changes, the more radical wing of the Left provoked land and factory takeovers, while the Allende administration's policies exacerbated the economic crumbling caused by capital flight and US-backed efforts to undermine the economy. Between 1970 and Allende's overthrow in 1973, an additional $8.8 million was authorized for covert actions in Chile (USSC 1975, 46–48).

Chile's Aftermath—and Is Venezuela the New Chile?

Although the US government encouraged the 1973 coup against Allende, by the mid-1980s the US government supported the transition back to democracy in several ways. Initially, the United States provided development assistance to the Pinochet government, in part because the Chilean economy was in shambles following capital flight, worker takeovers of factories, and Allende's reckless economic policies, and the fears that the Left would come back if economic collapse persisted.[8] Yet when CIA destabilization efforts against the Allende administration were revealed, and in 1976 the ex-Chilean ambassador to the United States under Allende was assassinated in Washington, DC, US assistance to the Pinochet regime was abruptly cut.

By the mid-1980s, with clear signs of political moderation from both the Chilean Right and Left in the protracted negotiations with the government and armed forces to restore civilian rule, the US government supported civil society groups financially, but also by signaling future political support. Thus, when Pinochet lost the 1988 plebiscite to extend his rule indefinitely, the US government strongly signaled its opposition to annulling the results. Constable and Valenzuela (1989, 184) note that "Chile's transition to democracy can be viewed as a success for U.S. policy, which has given strong support to democratic forces since 1985 and played an important role at several key moments in discouraging reversals in the political liberalization." They add that US officials "set the proper, low-profile tone in encouraging the transition," and they surmised that "by 1985 U.S. policymakers [concluded] that prolonged military rule was only strengthening communist groups... U.S. officials helped ensure a fair vote in the plebiscite by financing the parallel vote count and voter education projects, and by warning the regime against trying to doctor or abort the results" (184–185). Official US support was reinforced by the efforts of such US-based entities as the National Endowment for Democracy, Freedom House, and the Washington Office on Latin America (Sankey 2014). It is noteworthy that while all three are legally NGOs, the National Endowment for Democracy was founded by the US Congress in 1983 as part of an initiative "to promote the development and strengthening of democratic forces overseas" (2).

The US government has supported the consolidation of Chilean democracy through foreign assistance and a trade pact. The possibility of this pact rested on economic policy reforms pursued during the Pinochet years *and* the maintenance of these reforms under civilian governments despite strong pressures to stimulate the economy at the risk of rekindling inflation and trade imbalances. The democratic governments, the most mature among Latin American nations in economic policy as well as governance, undertook the market-oriented reforms required of trade pacts to harmonize economic policies of trade pact members. Chile's reforms were thorough enough to secure membership in the Organisation for Economic Cooperation and

Development in 2010—only one of three developing countries to be admitted to this "rich country" group.

What Has Changed in US Approaches to Radical Regimes in Latin America?

The closest contemporary Latin American parallel to Chile of the 1960s and early 1970s is Venezuela, which since 1998 has been ruled by Presidents Hugo Chávez (until his death from cancer in 2013) and Nicolás Maduro, under an increasingly populist radical Bolivarian Revolution—reflecting the aspiration to spread Venezuela's socialist approach throughout Latin America in the name of the pan-Latin American independence hero. Although the parallel is far from perfect, the US reactions to the increasingly Leftist and authoritarian Venezuelan administration illuminate both changes and continuities in foreign assistance and security strategy.

The significant differences include the fact that Venezuela did not face a truly far-Left threat in the 1990s; Venezuela's Communist and socialist parties and factions were weak, and Chavez himself, when he was first elected, was less rhetorically radical than he was later to become. The broader ideological landscape had changed dramatically as well. The Leftist guerrilla groups lost their inspiration following the collapse of the Soviet Union and suffered setbacks in societal rejection (e.g., community defense groups against Peru's Sendero Luminoso) and military defeats by armies and Right-wing paramilitaries. US foreign assistance supported some of these antiguerrilla efforts, but (as in Colombia) was typically framed as "the war on drugs," with the effect of justifying assistance in terms of addressing domestic US problems. The worry about the Red Tide in the 1960s through the 1970s has given way to tolerance of the Pink Tide. The political landscape of Latin America in the 1990s became divided between governments with fairly cordial relations with the United States, like that of Chile, and governments with populist, sometimes anti-American, rhetoric, but with no aspirations to launch a Communist tide across Latin America. Even as Leftist parties maintained their symbolism (Chile's Michelle Bachelet was inaugurated in 2006 with hammer-and-sickle flags fluttering behind her), most Left governments adopted surprisingly centrist economic policies (e.g., the preelection fears that rhetorically extreme "Lula" de Silva would radicalize Brazil politically and economically were unfounded).

Furthermore, the opportunities to win support for moderates like Chile's Eduardo Frei through foreign assistance were simply absent in oil-rich Venezuela. In fact, profligate spending of oil revenues prior to Chávez's first election was widely seen as corrupt favoritism bypassing the poor, a major factor triggering the 1989 Caracazo riots that undermined the support of the traditional parties.

Unlike the well-organized Chilean Left that brought Allende to power in 1970, Chavez's Fifth Republic Movement was ad hoc. His failed 1992 coup and two years of imprisonment did not give rise to stiff resistance; when he was elected in 1998, he appeared to be a garden-variety populist. Yet he orchestrated a constitutional referendum in 1999 that established a new constitutional assembly with the power to change the institutional structure, with subsequent changes that included a commission with the authority to dismiss the supreme court, the dissolution of the existing supreme court, and the appointment of new judges. Chavez was reelected in 2000 under the provisions of the new constitution, and again in 2006 and 2012.

In December 2006 Chavez consolidated the Leftist parties into the United Socialist Party of Venezuela. A constitutional referendum proposing increasing presidential powers, extending the presidential term to seven years, and eliminating presidential term limits was narrowly defeated in December 2007, but in 2009 another referendum abolished all term limits. Perhaps the most important contrast with the Allende administration was that Chavez purged the armed forces of officers likely to oppose the consolidation of his powers.

The US government's response fell far short of conspiring with the military, or trying to sabotage the Venezuelan economy. Despite the opportunity to invoke the erosion of political competition as a rationale for economic sanctions, it was not until 2014 that sanctions—of a quite selective nature—were imposed, following government crackdowns on protestors. According to a 2006 secret communication from the US Embassy in Caracas reporting on progress of the 2004 USAID/OTI Programmatic Support for Country Team 5 Point Strategy, the focus of US action, conducted through USAID's Office of Transition Initiatives (OTI), was directed at "1) Strengthening Democratic Institutions, 2) Penetrating Chavez' Political Base, 3) Dividing Chavismo, 4) Protecting Vital US business, and 5) Isolating Chavez internationally," justified on the grounds that Chavez had "systematically dismantled the institutions of democracy and governance."[9] The major thrust was to fund both US-based and Venezuelan NGOs. The cable noted that "OTI has supported over 300 Venezuelan civil society organizations with technical assistance, capacity building, connecting them with each other and international movements, and with financial support upwards of $15 million. Of these, 39 organizations focused on advocacy have been formed since the arrival of OTI; many of these organizations as a direct result of OTI programs and funding." The deployment of OTI funds to Venezuela is arguably beyond OTI's mandate, which is, according to OTI's self-description, focused on "countries transitioning from authoritarianism to democracy, from violence to peace, or following a fragile peace."[10] While OTI still functions around the world, its operations in Venezuela ended in 2010. US-based NGOs continued to support domestic NGOs;[11] many are increasingly critical of the growing infringements on political competition, and the Venezuelan government has been forcing foreign-based NGOs from operating in Venezuela or funding domestic NGOs. International NGOs are now largely criticizing the Venezuelan government and security forces from afar.[12]

Yet aside from the mild economic sanctions, the US government has apparently been waiting for the Bolivarian Revolution to collapse under the weight of its economic failure, thereby minimizing the risk of providing the Latin American Radical Left with additional rationales to mobilize anti-American sentiment and action.

Korea

For much of the 1960s, US foreign aid to South Korea was dictated by military necessity, the need to shore up the Republic of Korea (ROK) against the inroads of a communist regime in the North that seven years after the Armistice still remained aggressive, insistent, and antagonistic to the notion of a unified Korea under international auspices rather than the establishment of an all-communist state in the region. And it was not above exploiting sympathizers in South Korea, where unification remained a popular ideal. By the early 1960s officials within the Kennedy administration like Carl Kaysen, the president's Deputy Special Assistant for National Security Affairs, were convinced that "the principal external threat is not one of renewed aggression but of Communist exploitation of South Korea's weaknesses via the unification theme" (Kaysen n.d., 11) The defense of the ROK was no longer simply a matter of developing and maintaining the country's 600,000 man army (no match in any case, reasoned US officials, for an all-out invasion of North Korean and Chinese forces). In a return to the familiar doctrine of winning hearts and minds, what was most needed was a greater concerted effort to support South Korean economic and social development. Economic growth, Cold War logic dictated, would trickle down to the poorest elements of Korean society, lift them up, and in so doing shield them from the powerful allure of Communist ideology. Only by outperforming the communists could liberal developmentalists hope to prove the efficacy of their plans to the masses. As a case in point, US officials, who met the new military junta's Five-Year Economic Plan with enthusiasm, worried about its undercapitalization of areas crucial to the expansion of the private industrial sector—urban water systems, power transmission, coal mining, and cement manufacturing, for example—and so spoke of leveraging development funds to generate new sources of "power" (Kaysen n.d., 13–17).

This is not to suggest that American officials resisted the view popular among South Korea's military leaders that privatization would have to await the consolidating gains of economic development. In the minds of development officials, democratic reform, even the goal of reunification itself, was secondary to securing both economic growth and military security (USAID 1962, 1). In 1964, working with the South Korean government, USAID moved expeditiously to help South Korea combat what experts deemed was its greatest obstacle to growth: an unfavorable balance of trade exacerbated by a reputation on the international market for the poor quality of its consumer products. It helped set up an Export Promotion Sub-Committee of the Korean-American Economic Cooperation Council, securing a seat on the committee along with representatives from private industry and government. Over the next decade, USAID provided the moral and material support for a campaign to increase the amount of and improve the quality of Korean exports. It sent in experts to help institute new quality controls at the factory level, improve productivity, and establish new legal standards and guarantees, while using its own considerable influence with American businesses to bolster confidence in Korean-made goods. In just over a year, Korean products started appearing in large department chains such as Sears, JCPenney, Macy's, and Woolworth's (Butterfield 2004, 206–207).

Considered a "showcase of American effort and intention" like Chile, South Korea was nevertheless still under the thrall of a strong nationalist government that was often repressive and authoritarian in its methods (USAID 1962, 2). To compound this problem, its social and economic institutions were still largely rooted in traditional village life and in dynastic norms that, in effect, discouraged entrepreneurship and technological innovation. Rejecting the concept of private property dominant in liberal democratic societies and reinforcing instead ancient Confucian norms of status and prestige, traditional Korean society, according to development officials, placed the country at odds with the free-market doctrine and consumption patterns of the economically advanced countries. In the Confucian world of South Korea, "barely over the slave, came the businessman and trader… Business was despised and considered the legitimate expropriative prey of the officials who could not themselves engage in it; it reacted often in the resentful, clandestine role Korean society carved for it" (USAID 1962, 4). And yet if South Korea, as "one of the key outposts" of "free world forces," was to resist the advances of an aggressive Communist North, the growth, modernization, and sustainability of not only its armed forces but also its economy and society was essential (10). Although USAID officials recognized that the "breakdown of traditional society" would create "social and political instability," this was a temporary expedient to a deeper and more lasting stability based upon the values of "individualism, cosmopolitanism, liberalism, and individual initiative," all prerequisites for Korean prosperity (3, 4).

Just how much relative peace and conflict Korea's model of guided capitalism produced on balance is a difficult question, compounded, on the one hand, by a prior history of cultural and political authoritarianism and, on the other, by ambiguities among American leaders over the tradeoffs among security, democracy, and development (USAID 1962, 13). One answer, as in Chile, was in the peculiar mix of MAP and USAID funding and its militarizing, culminating in 1972 in President Park's declaration of martial law. At the outset of the Park regime in 1961, US officials were persuaded that they could have it both ways—a strong Korean government capable of making "the difficult decisions required for economic reform without excessive interference from the civilian population" and progress toward democratization that would open the country up to greater participation (Brazinsky 2007, 161). As it turned out, events and their own zeal to contain Communism proved them wrong. While looking to the Park government to combat internal corruption and factionalism, the United States also pushed, despite strong opposition in Korea, to enlist Korean armed forces in the Vietnam military build-up and to promote normalization with Japan, a distressing possibility for most Koreans (120). For its cooperation in these tasks, the ROK made fresh progress toward attaining a positive balance of trade, a perennial American economic statecraft instrument. South Korea earned $402 million between 1965 and 1968 in export sales to Vietnam and the US military, and made rapidly growing Japan its number one importer (140–141).

The Korean War was America's first proxy war and as such held a special place in the development thinking of counterinsurgency experts. While aid workers and government officials alike worried about the state of the Korean economy, especially under conditions of political repression, they were also sensitive to the need to strengthen local Korean police forces against the new and very real threat of communist infiltration and insurgency. In 1963, Bryan Engle, the director of Public Safety at USAID, outlined a MAP plan for the Korean National Police (KNP), the first line of defense against civil disturbances. Citing an absence of "standardization of weapons," "inadequate" mobility and communications capabilities, and "deficiencies" in police training and preparedness, Engle worried that at its current force level of 30,000 men, the KNP was ill-equipped to combat "continuous Communist efforts to infiltrate and then to undermine the internal security" of the country (1963, 1–2). With talk in the air of reducing the Korean army by more than half and sending the bulk of the men into the Korean countryside where, in the words of Secretary of State Dean Rusk, they might be "helpful to economic development," the problem of police preparedness was more urgent than ever. The military may be "essential to inhibiting insurgency," yet in countries like Korea, with little or no real democracy, the best way of achieving this mission was to cultivate friendly relations between the military and the populace. In practice this meant employing the former in civil works projects in public transportation and communication, health and sanitation, and only tangentially in counterinfiltration activities. But it was a double-edged sword. A large military presence in rural areas, no matter how helpful, might, US officials worried, only further alienate the local populace (Rusk 1962, 1–2). This only made the problem of creating an efficient and effective local police force all the more pressing. As in almost every case, the solution was a pragmatic doctrine of selective adaptation and adjustment (to changing circumstances on the ground) in the service of larger military objectives.

By 1968 the threat of communist infiltration seemed more credible than ever. At the beginning of the year a team of highly trained North Korean agents entered Seoul in a failed attempt to assassinate the South Korean president.[13] According to intelligence sources, an estimated 15,000 to 20,000 North Korean guerillas stood poised for action in the ROK (Engle 1968, 1). Engle was now requesting $5 million in FY 1968 (out of a proposed $100 million in military assistance for Korea) to build up the counterinsurgency capabilities of the KNP. These funds would go to USAID but be administered by the US Operations Mission (USOM) (Secret Seoul 4499 1968, 1). In a secret memo to the Department of State, the Department of Defense and the American Ambassador in Seoul spelled out the case for additional police assistance in language by now familiar:

> We have pursued this policy considering that an effective police force for this purpose would enhance the public image of a democratic civilian government, avoid degrading the ROK/US military posture to meet overt NK military aggression, and minimize the political and social strain within the country that could develop as a result of the use of military forces for counter-infiltration activities in their contacts with the civilian population. (1–2)

With $5 million in funding, USAID/USOM would be able to train and equip a 4,000 man Combat Police Force, provide for an elite Reserve Task Force of 9,281 police, and train and equip a Tactical Mobile Force of 407 men in Seoul itself (Engle 1968, 2). President Johnson gave his personal imprimatur to efforts to combat the growing threat of North Korean infiltration (Lee 2006, 55).

The eight-volume joint study of the Harvard Institute for International Development and the Korean Development Institute published in 1980 concluded "that, during a period of rapid economic growth, the existence of an authoritarian government accepting economic development as its first priority, able to maintain economic stability, and capable of making difficult economic policy decisions, and implementing these decisions, has been a positive factor in promoting growth" (Steinberg 1982, 45).[14] Yet in Korea, positive growth notwithstanding, the choice was not simply between political liberty and social and economic stability. In the context of Vietnam, increasing North Korean belligerence and growing democratic opposition to authoritarian government, US aid to South Korea was tied less to evidence of democratic reform (still a generation off) as to cooperation with US Cold War goals in the region. Success was mixed for both. Cold War containment policies made a virtue of militarized necessity, stalling the democratization process, yet the contentious nature of US-Korean relations was a great school of democracy and a site of democratic education.

Aftermath of Korea

Following the highly corrupt and ineffective decade of Korea's authoritarian President Syngman Rhee and a short-lived chaotic democratic administration, General Park Chung-Hee led a military coup that put him into power from 1961 until his assassination in 1979 by an old colleague, then the director of the Korean Central Intelligence Agency. Over the next eight years, the agitation for democratic governance culminated in the competitive elections of 1987 and democratic governance thereafter. In the meantime, the Korean economy grew at a remarkably high, sustained rate, converting one of the poorest countries to a wealthy member of the OECD and a significant provider of foreign assistance. Yet the North Korean threat has kept the South Korean military as a formidable force, with roughly 650,000 active troops and a US$30 billion military budget taking up more than 14 percent of the government budget and roughly 2.5 percent of GDP (Global Security 2015).

US development assistance and military assistance approaches have varied dramatically in the face of both the dramatic changes and the constant military threat from the North.

Following the cessation of the Korean War (i.e., the 1953 uneasy truce that still prevails without a formal treaty ending the war), the US government poured huge volumes of economic and military assistance into South Korea: US$5.2 billion from 1953 through 1976 in economic assistance; US$6.9 billion in military assistance, with the proportions of military assistance steadily increasing (38% in 1953–1961, 60% in 1962–1969, 75% in 1970–1976) (Korean Ministry of Finance and Strategy 2012, 38). Thereafter, as the Korean economy grew, US official development assistance amounted to only US$166 million through 1981; by 1982 loan repayments exceeded the flow of loans and grants (OECD ODA database).

The military assistance declined for a different reason. In 1977 Secretary of State Henry Kissinger was successful in reducing US military assistance to the South Korean government, due to the driving concern that Park might grow confident enough to attack North Korea (Kim 2011, 36). Presumably US policymakers, following the shock and huge US troop losses when China entered the Korean War, had been reconciled to the stalemate in the Korean Peninsula. The challenge for the United States was to draw down US troops in Korea without making South Korea seem vulnerable or provoking the South Korean military into transforming the country into a garrison state at the expense of democratic governance. At the time of the truce, US armed forces had nearly over 325,000 troops in Korea. It is true that the US support for General Park paid dividends—Park committed 320,000 Korean troops to join US forces in Vietnam—but pressures within the United States to withdraw troops have been strong from the beginning. In 1954 US troops were reduced to just over 225,000, and to just over 75,000 in 1955, with gradual reductions thereafter to the current level of less than 30,000 troops. The development of democratic governance in South Korea reflects interlocking dynamics that harken back to many of the theories of economic development and governance, but with more nuance than the original theories. The premise of modernization theory that prosperity would create the conditions for democratic governance seems to be consistent with post-1987 South Korea, but the theory did not anticipate the tutelage of the United States. On the other hand, South Korea's democratization defies the presumption that acute security threats dramatically increase the likelihood of military dominance. The continuing close collaboration between the South Korean and US armed forces reinforced the ethos of accepting civilian authority. Moreover, the South Korean military has not had to grovel for funding or to seek additional political power to gain respect—the military defends the country against an obvious enemy and provides technical and leadership training magnified by universal male conscription.

South Korea's remarkable economic growth is consistent with the Big Push implication that massive development assistance could trigger a take-off. For reasons clearly other than the standard sociopolitical underdevelopment that the Big Push theory attributed to developing countries, South Korea fulfilled the conditions Rostow (1956, 25) specified as necessary—"a society prepared to respond actively to new possibilities for productive enterprise; [requiring] political, social and institutional changes." The social conditions were strengthened by the fact that a mobilized civil society was in evidence immediately after the Japanese occupation, in the efforts to end the immediate postwar American occupation (Lee 1997), as well as during the authoritarian period under General Park (Richardson 2010, 166). The government's capacity to make tough economic decisions is supported by the North Korean threat, which, for example, has impelled (and permitted) governments to be less tolerant of business leaders taking advantage of protected industries to capture rents rather than take appropriate constructive risks.

South Korea also avoided the major institutional pitfall (see chapter 9) of the military's entrenchment in business. The 1973 government initiative to replace US-origin materiel, beginning with arms and ammunition and progressing to aircraft, ships, and missiles, awarded the production contracts to the private *chaebol* (business conglomerates) such as Daewoo and Hyundai (Klare 1997, 59), leaving little space for the armed forces to penetrate the business sector.

Vietnam

US foreign aid spending reached a peak in real terms in the 1960s, when fears were at their height that without US military and economic assistance, developing countries around the world would fall like dominoes to either Soviet- or Chinese-style communism. The US Congress authorized $1 billion in foreign aid in 1960, $3.6 billion in 1963, down to $2 billion in 1967 and $1.4 billion in 1968. In 1965 Vietnam received a larger portion of US foreign aid than any other country. Nearly 90 percent of this aid went to military appropriations, including support for the training of indigenous armed forces and intelligence services. Only a minute fraction of this money went to traditional development programs supporting industry and agriculture (Picard and Buss 2009, 96, 110–111). Out of this gross imbalance emerged perhaps the most creative, albeit most politically explosive, program in Vietnam, designed to combine security with social reform and change: the Strategic Hamlet program of civilian pacification led by John Paul Vann, a USAID official in the office of Civil Operations and Revolutionary Development Support (Herring 1979; Sheehan 1989).

Although the enormous military assistance and economic aid to Vietnam failed to prevent or win the war, the legacy of US foreign aid to Vietnam has to be measured, as with Chile and Korea, against the ability of the local system to absorb the demands of modernization, to cleanse itself of historic elements of power and greed, and to achieve some modicum of economic and political independence under conditions of abject dependency (Picard and Buss 2009, 112–117). Aid to Vietnam and the extreme conditions that motivated it pose an important test for the doctrine that development assistance, organized around the twin goals of security and stability, could be an effective deterrent to civil unrest.

South Vietnam was in a poor position to accomplish either of these ends with or without foreign assistance. It suffered from a debilitating history of dependency under French rule; a much older history of Mandarin elitism and Confucianism that relegated material progress to the demands of moral and spiritual uplift; and traditional anti-Western, antiindustrial attitudes. All these made it difficult for the country to absorb the much touted lessons of modernization: the desire for socioeconomic mobility, a technological orientation to change, and individual efficacy in the face of impersonal social forces (Dacy 1986, 94–96). These orientations made it unlikely that most South Vietnamese would respond effectively to the economic opportunities and democratic initiatives that some US policymakers hoped could be sustained.

Faced in 1974 with the imminent collapse of South Vietnam, the Senate Foreign Relations Committee found the blame not in the vagaries of US foreign aid, the growing strength of the North, or any other external forces, but in Vietnam itself: "Vietnam is rich in agricultural resources but cannot feed herself, has absorbed western technology but cannot afford the imports to operate it, has a well-trained labor force but cannot employ it, and provides a wide range of government services but does not have the means to pay for them" (Dacy 1986, 19).

The assumption of Vietnamese intractability in the face of US foreign economic aid and assistance is only partially true. Certainly the prospects for democracy were slim under the oligarchic, paramilitary governments of both Ngo Dinh Diem and Nguyen Van Thieu. It is practically a commonplace that "The unpopularity and ineffectuality of the Vietnamese government conditioned the effectiveness of assistance and ultimately neutralized any impact it might have had" (Picard and Buss 2009, 113). However, competing conceptions held by American economic and military advisors of the role of US foreign aid played no less a determining part.

These objectives—one pursued through economic programs designed to win legitimacy of the government, the other through military assistance programs designed to increase security—ran along more or less parallel lines at the beginning of US involvement in Vietnam, converging increasingly in the strategic hamlets and in other pacification programs. The legitimacy school argued for a Point Four approach to social and economic development, including technical assistance and capital input into public health and education, land reform, agricultural productivity, and small-scale industry. The security school sought very different kinds of results, "body counts and village stability ratings," that would lend credibility to the overriding American mission to prevent an infiltration of communist forces, which came to be called the Viet Cong Infrastructure (Montgomery 1986, 68). As the security threat grew (both cause and effect of the failure to achieve genuine legitimacy), the role of aid began to shift almost exclusively to counterinsurgency in what one USAID official described as "'Militarization' of Program" (USAID 1969, 408). Of course, to assume otherwise under the circumstances would have been naïve.

There is some evidence that USAID resisted this view. A full four years before the first major spike in military aid and US troop size in September 1962, a confidential USAID addendum to its 1963 Operating Year Budget claimed that the Vietnam program, due to "changing forces of the war," had been "drastically reoriented from the traditional economic development and technical assistance theme... to one primarily concerned with counter-insurgency" (USAID 1963, 1–2). This meant the allocation of $12.4 million in nonproject commodities (i.e., military and police equipment) for use in the villages and hamlets and $11.6 million in direct support of counterinsurgency activities to repair roads and improve provincial hospitals.

Much of this aid was coordinated by the USAID Public Safety Division until in May 1967 it was incorporated into CORDS under the Military Assistance Command, Vietnam. Headed by General William C. Westmoreland yet directed by a civilian, Robert W. Komer, then Ambassador to Vietnam, CORDS was charged with coordinating all civil and military pacification programs in Vietnam (USAID 1969, 481–482). Montgomery (1986, 76) concluded that this boiled down to "counter-insurgency through civic action, the theoretical hope that an army can contribute to nation-building by providing both security and an infrastructure of small public works constructed by troops and village volunteers."

As the war in Vietnam escalated, its original Cold War objectives dissolved into increasing numbers of friendly and enemy body counts: the struggle for legitimacy was overruled and finally corrupted by a perverse evaluation of progress based on spurious quantitative measures (Montgomery 1986, 71). The casualties for a soft power approach to peace and development were felt up and down the line within USAID. The whole relationship between theory and practice, between program planning, implementation, and review, was upended. "There was never time in the heat of war," lamented a USAID official, "to fully dissect theories and prove them in pilot operations before launching into large-scale operation." In the transition from peace time to war time, traditional programming areas within the Agency—public health, education, agriculture, refugee relief, and supply management—were no longer recognizable. Where public health had been labeled "advisory services on preventative medicine" it now meant "an unprecedented program of direct medical care for sick and wounded Vietnamese." Where education programs once entailed "long-range development of an integrated educational system," they now called for "a crash program of providing new elementary schools in the rural areas in order to gain political impact." Agricultural programs, elsewhere organized on an experiment station extension plan, in Vietnam supported "large operations with immediate impact" (USAID 1969, 379–380). Fitzgerald (1972, 346) has lampooned the waste with historical accuracy: "thousands of tons of bulgar wheat, thousands of gallons of cooking oil, tons of pharmaceuticals, enough seed to plant New Jersey with miracle rice, enough fertilizer for the same." Supply management, once a function of local intermediaries, grew to Total War proportions: "a modern logistics operation," reads an internal administrative history of USAID, "with hundreds of Americans in port management, coastal sealift, airlift, warehouse construction and operation, repair and maintenance" (USAID 1969, 380).

The USAID program in Vietnam was unique in one other important respect. In conventional settings, AID advisors worked collaboratively with their local counterparts. In Vietnam "major operations were sometimes planned, supervised and partially performed by American personnel," a dramatic reversal of the rhetoric of self-help and of the advising function that had driven development theory since the beginning. The prosecution of aid in Vietnam, the vast majority of it dedicated to counterinsurgency, had become a big business with aid workers serving no longer as advisors but as operators. In 1967 USAID established a separate Vietnam Bureau with a staff of over 400 to expedite the provision of aid and enable "quicker procurement of goods and services" than would be normal under traditional procedures requiring prior host country agreement before loans or shipments could be made (USAID 1969, 390, 395–396). The Vietnam experience elevated the doctrine that more uncertain conditions require more heavy-handed intervention by foreign assistance providers.

Economic and Political Renewal after the Second Indochina War

The evolution of Vietnam's economic policies reveals the pull of the private market. Like China, Vietnam today demonstrates what would probably have been unthinkable to many in both the West and the Communist states: a Communist party, still monopolizing governance, but shorn of what were thought to be the essential elements of a Communist economic system. When the Soviet Union collapsed, its support for Vietnam disappeared, and several crucial weaknesses thus revealed compelled the Vietnamese government to permit private entrepreneurship. Another unanticipated outcome has been how the conflict between Vietnam and China that flared into open war in 1978–1979 gave lie to the fear of a World Communist Movement. Policymakers easily confuse deeply ingrained nationalism for ideology. Chinese military and economic assistance in the amount of roughly $300 million per year came to an end in 1978, which forced the Vietnamese government to seek increased assistance from the Soviet Union that resulted in effective Soviet control over many of the industrial sectors of the economy (Steinfeld and Thai 2013, 19; Vo 1990, 90–98).

The first decade after the end of the Second Indochina War (1954–1976) produced slight economy recovery or growth.[15] No wonder. The country had been at war virtually nonstop since WWII and had suffered more than three million deaths and the exodus of something like two million refugees. The industrialized north had been essentially destroyed by relentless air attacks, while the agricultural south was crippled by battle-scarred terrain and monumental environmental devastation from defoliants. Flush with victory, the Communist Party set very ambitious goals for its Second Five-Year Plan (1976–1980) of double-digit rates of economic growth, major reconstruction of its industrial base, modernization of traditional agriculture, and integration of North and South Vietnam into a single nation (Beresford and Phong

2000). Success of the Plan was predicated on continuing and large foreign investments and assistance, which in the end were not forthcoming in amounts equal to the ambitions (Boothroyd and Pham 2000, 14). The Third Five-Year Plan (1981–1985) acknowledged these shortcomings with fewer, less ambitious goals that were doled out every year, rather than at once, during the full run of the plan. Tacit acceptance of an entrepreneurial private sector, particularly in the South for wholesale and retail trade, marked another realistic accommodation. Decentralization of industrial and agricultural activities paired with the sloughing off of inefficient state enterprises. State subsidies for a full range of goods and services were also sharply curtailed (Luong 2003).

All of these changes finally came together in 1986 with the *Doi Moi* economic renovation plan that eventually extended until 2006. During this remarkable period Vietnam's economy grew at an average 7–8 percent per year, foreign direct investment (including from the United States as part of the 1992 US-Vietnam Trade Bilateral Agreement) soared to over $60 billion per year, and numerous international, political, and economic partnerships, events, and agreements marked this era (Vuong 2014).

The country has hit rougher waters since the global economic downturn of 2008–2009 and is currently dealing with problems in the finance, banking, real estate, and many of the remaining state-owned or operated enterprises. Substantial corruption has also surfaced. On top of this lengthy list of domestic and economic challenges, the long-standing competition between Vietnam and China is once more coming to the fore—in economics (Malesky and London 2014) and in geopolitical matters in disputed territory and resources in the South China Sea (Steinfeld and Thai 2013, 70–72).

Conclusion

In the administration of aid in our three case studies, the goal of political and economic stabilization almost always took precedence over the social reforms associated with democratization—free elections, constitutional governance, broad economic opportunity, and equitable distribution. Not that these goals had any greater prospect of success in lieu of US assistance. The point is not that Chile, South Korea, or Vietnam would have been better off without US aid, but that opportunities were lost based on the kind and degree of that support. US containment doctrine in Latin American and Southeast Asia restricted the field of action for reform-minded USAID officials; it made the Agency a reluctant partner with the Defense Department. Joining USAID and MAP funding uncomfortably at the hip in effect subordinated the peace time uses of aid to the martial necessity of combatting Communism. In Chile this meant a rejection of Christian democratic liberalism on the one hand and an exaggerated and ultimately baseless fear of Allende's Marxist radicalism on the other, leaving US aid officials in an unconstructive limbo between policies that supported industrialization (at the expense of land reform and greater agricultural productivity) and new levels of potentially destabilizing anticommunist police preparation that discouraged foreign investment in the country's growth.

In Vietnam, almost wholly dependent on US military support as the prospects for both democracy and development grew dimmer, military solutions to problems of development sabotaged the good work of aid officials even in strategic hamlet and pacification programs, where the value of their contributions was often underappreciated. Instead the strategic hamlet became a kind of government decoy for local insurgents in search of asylum in the Vietnamese countryside. Visiting one such hamlet in March of 1962, the head of USAID's Public Safety Division reported back to Washington the remarks of their guide, Presidential Advisor Ngo Dinh Nhu: "Under my plan," said Nhu, "we force the Communists to attack the Strategic Hamlet; thus it is no longer guerrilla warfare" (Walton 1972, 2). Achieving the right mix of economic aid and MAP funding, so much a part of development assistance in countries like Chile and Korea, simply did not apply in Vietnam. Military and civilian operations were indistinguishable, with USAID workers forming counterinsurgency teams and US soldiers directing relief work under one unified command structure. Not even the best economic planning by the best minds under the most promising conditions could compensate for the hypermilitarized social and political landscape. This was the one overriding lesson of the US development experience in Vietnam.

CHAPTER 11

Conclusion: Linkages and Challenges

Taking Stock

We have seen that some aspects of the evolution of development thinking and practice have been quite positive. First, the development agenda has been greatly expanded, bringing development efforts more broadly in line with the full pursuit of human dignity by going beyond economic growth to include equitable distribution, responsive governance, environmental protection, gender equality, and minimization of violence. Second, the expertise in the field of development has deepened and broadened. Development economics, development administration, human ecology, institutional analysis, gender studies, anthropology, and a host of other fields have been deployed to address development challenges. A host of institutions has accumulated and organized vast amounts of information, leading to broader and more nuanced understanding of the dynamics and challenges of development. Third, with experience has come greater pragmatism; in many circles the extreme ideological positions have given way to asking what works in particular contexts. And, as the monitoring of the MDGs has indicated, significant progress has been made in reducing poverty, illiteracy, discrimination against women, and some diseases in many countries, though these problems remain severe in numerous nations as well.

However, other trends have been highly problematic. Some promising development strategies have been rejected prematurely or are applied only superficially. Some counterproductive strategies have persisted. Superficially definitive methods hide issues in need of deliberation. Often inappropriate objectives reign, despite the broadened agenda. As a consequence, the progress in addressing growth, poverty, conflict, the environment, health, and other aspects of human dignity has certainly not advanced as much as one would hope.

This chapter focuses on the remaining challenges of theory, doctrine, and practice by exploring the factors that still hold back progress in economic growth, poverty alleviation, reduction of violence, and environmental stewardship. These factors are found not just within each of the foci of this book (economic strategies, governance reform, and the roles of civil society, the military, and foreign assistance) but also at the intersections of these aspects of development. At the heart of the challenges is the pervasive weakness of development efforts to integrate the multiple facets of development. Here we are referring not just to the frequently criticized lack of coordination among government agencies, the clashes among international development agencies, the turf battles among NGOs, or the conflicts across these sectors. There is a deeper lack of integration of institutional development and economic policies, of foreign assistance doctrines and strategic considerations of international assistance agencies, and of development strategies and the role of the armed forces.

Economic Growth and Distribution

What Technical Analysis Can and Cannot Do

As chapters 2 and 3 recount, the apparent rigor of economic analytical tools can mask the considerable technical uncertainty that remains in evaluating projects, programs, and policies, both before and after they are implemented. Some analytic techniques developed through hard labor by first-rate economists—explicit rate-of-return analysis being the most obvious example—are applied more rarely or less seriously. This reflects the implicit recognition that consequences are simply more difficult to predict than the methods require. The space for ideology and political calculations is, therefore, expanded beyond what would prevail if the outcomes of particular actions were more fully known. However, technical certainty would not eliminate political concerns—who gets what, when, how,[1] and, in fact, should not—because distributional questions, and the question of what society ought to produce in terms of material and nonmaterial benefits, are beyond purely technical analysis. On the other hand, technical analysis can clarify the range of the plausible and help to make policy analysis more comprehensive and systematic.

Continuity and Adaptability

When the hopes for predictive certainty are dashed, policymakers must consider adopting policies that are robust in terms of the implications of uncertainty and often require the capacity to adapt. As important as adaptive management is to correct for false steps, consistency is also important. To address the potential tradeoffs, it is useful to note that uncertainty has two distinct aspects. One aspect is uncertainty as to whether objectives of an economic policy would be accomplished, such as reducing inflation or stimulating investment. In such cases, assuming that enough time has been given to adequately determine that the policy has truly failed compared to expected outcomes of policy alternatives, terminating the policy is a sensible option. Yet sometimes core policies include fixed parameters, such as exchange rates or tax rates, to provide certainty, to gain the acquiescence of stakeholders who would see risk in easily adjusted parameters, or both. Government officials may try to enhance the certainty, and their political support, by conveying that the commitment is irrevocable. However, it may reach the point that the fixed parameter creates such distortions that the economy seriously falters. In Argentina, for example, the peso-dollar parity, which for a while was stunningly successful in reducing inflation, was maintained far too long, resulting in the collapse of the Argentine economy. Argentine political institutions have been too fragile to permit governments to easily adapt policies introduced as important commitments.

The other aspect of unpredictability is that basically sound policies can have unexpectedly adverse impacts on particular groups or unexpectedly disruptive reactions of groups that regard the policy as contrary to their interests. Compensatory measures must be enacted—and publicized—both to cushion the possible adverse effects on groups made vulnerable by the new policies and to preempt disruptive opposition. In other words, if a policy is relatively sound for the nation as a whole, the adverse effects on particular groups ought to be addressed through social safety nets, tax structures that mitigate big losses, or other measures that leave a sound economic structure operating while compensating unexpected losses to a greater or lesser degree. The fundamental principle is to address efficiency and equity through different policies so as to avoid distorting the efficient operation of the economy.

The question, then, is how stabilization can be accomplished. One possibility is to rely on foreign assistance agencies, whether bilateral or multilateral. Agreements of developing country governments with such entities are often credibly binding, because of the potential costs that governments would incur if they reneged on an agreement. Yet here, too, the risk of being locked in too long arises, if the agreement ties the hands of the government beyond the period in which the commitment makes sense. One approach to mitigate this risk is for the agreement to have mutually agreed-upon thresholds for relevant parameters, such that exceeding these thresholds would void the agreement without penalizing the government and country. Another approach is for the foreign assistance agency to have the flexibility and monitoring capacity to recognize when the agreement is no longer optimal and to be able to terminate or adapt the agreement without harming the country. For example, although the IMF was roundly criticized for the stringency of the agreements reached with Indonesia, South Korea, and Thailand in the wake of the 1997 East Asian financial crisis, what the critics fail to appreciate is that the IMF backed off on the fiscal restrictions when it became clear in 1998 that they were too stringent.[2]

Matching Economic Policies and Institutions

As chapter 2 describes, one insight of development economics in the 1980s was that "institutions matter." This was greeted with some amusement by political scientists, in that political scientists (e.g., Plato) have recognized this for millennia. A common instinct of economic theorists, with the exception of transactions cost theorists, is simply to urge stronger institutions and rely on questionable aggregate indicators of institutional strength, often without exploring what specifically would constitute stronger institutions and how to accomplish this. There is little consideration of how policies need to be altered to take account of institutional weaknesses; perhaps that would seem to be caving in to a system's weakness to acknowledge that "second best" policy options must be adopted. However, even if the sound advice of governance experts and economists is followed, institutional development is often a slow and inconsistent process, and it often comes at a high political cost that calls for prioritizing the institutional changes in terms of their contributions to sound economic and political development. This gives rise to two challenges under the rubric of matching economic development policies and institutions: how should policies be adapted to match institutional weaknesses, and what priorities should guide institutional reform, in light of the fact that such reform often comes at a high political cost.

Adapting Economic Policies to Match Institutions

Given the frailties of the institutions through which economic development policies are formulated and conducted, how should the policies accommodate these limitations? The need to adapt economic strategies to the reality of weak institutions remains, but has been a largely unattended challenge. To adapt the economic strategies and policies that are developed under the assumption of sound institutions raises the question of what economic policies are more resilient under low institutional strength.

To address this requires clarifying what institutional weakness entails. One aspect is the inability or unwillingness to enforce regulations in a consistent manner. Government leaders may also lack the incentive or capability to allocate government resources in a sensible way to maximize societal benefits. A third weakness in many contexts is a high level of corruption. It is also a common problem that governments under strong political pressures and thin support have trouble maintaining policy continuity. However, some commitments are locked in to the point that policies cannot be adapted soundly to meet new conditions and challenges.

The challenge of adapting to weak institutions has been addressed constructively in taxation theory and practice, but has been considerably weaker in many other aspects of economic policy. Tax reforms undertaken in both developed and developing countries have embraced the virtues of simplicity and uniformity through an emphasis on value-added tax. Value-added tax regimes typically reduce the otherwise multiple sales tax rates applied to different products and reduce the importance of income taxes, which are typically highly complex because of multiple rates and exemptions for both different economic activities and different situations of taxpayers. Whereas income tax regimes that have undergone thorough reforms to reduce exemptions are often highly vulnerable to the reintroduction of exemptions, value-added taxation is much less prone to this erosion.

Because liberalization in general entails simplifying economic policy by reducing nonmarket elements (regulations, subsidies, monopolies, and state enterprises), the continuity of liberalization gains requires preserving these simplifications in the face of efforts by special interests to erode them. Liberalization is highly contested because of the redistributive implications, generating pressures to restore the privileges that liberalization has eliminated. Whereas tax reform can simplify in ways that do not have significant short-term redistributive impacts, the elimination of price controls and other subsidies and changes in budgetary allocations typically have highly visible, immediate redistributive impacts. The weakness of institutions with respect to maintaining the simplification comes in the inability or unwillingness of leaders to resist the pressures to restore subsidies and price controls.

Adapting Institutions to Match Economic Policies

In the long run, of course, the optimal situation would be that sounder institutions would permit technically optimal policies to be faithfully enacted. This is one of the most important topics covered in chapters 4 and 5. As mentioned above, in addition to complex challenges of policy inconsistency and policy rigidity, institutional weaknesses include weak regulatory enforcement, weak capacity to allocate budgetary resources, and corruption.

With respect to weak regulatory enforcement, although it is easy to lament the inability or weak capacity to enforce regulations, the scope of regulation and the magnitude of resources devoted to enforcing them are policy choices. One approach to make regulation more effective with a given level of resources is to reduce the scope of regulation to what is truly constructive rather than what is creating restrictions to competition in order to favor privileged firms or individuals. In other words, it is essential to distinguish between regulations that reduce damaging behavior and those that primarily restrict competition. The institutional weakness for reducing predominantly competition-restricting regulations is often the inability or unwillingness of leaders to fend off rent-seeking efforts. For constructive regulations, here again we see the interplay between governance and economic policy, in that one tack is to strengthen governance by devoting both budgetary and political resources to enforcement, but an alternative tack is to maintain the regulatory scope but streamline the regulations such that they are more easily monitored and enforced.

Bureaucratic Limitations

How can governance reforms minimize the damage potentially caused by intragovernmental rivalries, inadequate intra- and interorganizational communication, and lack of coordination? Economic growth requires coordinated calibration and timing of investment, goods, and services, but the conventional ministerial structure of governments requires more coordination and goodwill than many developing country governments can muster. Chapters 4 and 5 address the possible solution of shifting the source of demand from the multiple centralized agencies to subnational agencies. These agencies can order inputs according to the level of resources they generate or are allocated and the preferences that the local people express. However, although many governments have claimed to embrace decentralization and have lauded it as a democracy-enhancing governance doctrine, the financial wherewithal for effective decentralized governance is often lacking. Fiscal decentralization is required to avoid unfunded mandates.

Corruption

The major, though often highly controversial, economic policy instrument to address corruption is to increase the administrative budget to provide adequate pay for civil servants. This requires overcoming the public distaste for granting higher salaries to government and state officials, ranging from bureaucrats to police officers, when they are widely believed to be prospering from bribes or self-dealing practices.

Reforming economic policy measures, such as existing regulations that induce rent-seeking, is another imperative at the intersection of economic policy and governance. The intertwining of well-connected businesspeople and the state is a risk to competitive politics as well as a drag on the economy. However, economic policy reforms are clearly not enough. Credible deterrence, requiring adequate monitoring and the political will to punish high-profile violators, is necessary. Yet anticorruption campaigns must be insulated as much as possible from the reality and perception of political vendetta. Anticorruption measures that come down on the rivals of the top government leaders often contribute to the cynicism regarding the motives and thoroughness of the anticorruption initiative.

Armed Forces and the Economy

Economic efficiency is also challenged by the role of the military in many developing countries. The military often claims the prerogative to control those aspects of the economy and society that it deems to be relevant to national security. Yet national security, as argued in chapter 9, is a concept with no obvious boundary; few economic policies, programs, or projects cannot be justified as relevant to strengthening national defense, maintaining internal order, or simply strengthening the nation so that it can stand up to foreign or domestic threats. When the military is in control of the government, the option of leaving economic policy in the hands of technical experts often gives way to security strategies, such as launching expensive or inefficient programs to extend the reach of the state or of the military itself, or to populate a potentially contested area (e.g., the Brazilian Amazonian expansion). Sometimes the military insists on maintaining inefficient industries deemed as "commanding heights," or requires that these industries be held domestically even if international firms would be more efficient. In other cases, the military establishes and jealously maintains a large industrial complex, invoking national security, even if these industries are inefficient.

The military may also be implicated in economic problems through the patterns of conflict between the military and others. To be sure, in some circumstances the strength of the armed forces deters violence. Yet in other circumstances the military reacts with coercion to perceived threats to domestic order or to its own perquisites, provoking armed resistance. The cycle of violence, which is often exacerbated if the military has taken over governance, often delays the establishment of restoration of sustainable democracy. In addition, this cycle of coup d'état, resistance, greater coercion, and human rights violations is often a major deterrent to investment and disrupts economic activity in general.

Poverty Alleviation

Poverty alleviation, though certainly not the sum total of the efforts to further human dignity, is clearly essential. The obstacles to poverty alleviation begin with the inertia of discriminatory policies, exacerbated by misguided opposition to the reforms to eliminate these policies. Beyond these disadvantages, the government investments that could help the poor are often neglected because the benefits and costs of projects and programs are underexamined.

Persistence of Discriminatory Policies

As noted in chapter 3, many of the privileges of the wealthy come through the *illiberal* aspects of economic policy: restricted import licenses, import duties that protect manufacturers and the modern sector "labor elite," tax exemptions that reduce the government's capacity to spend on social programs, energy pricing subsidies that favor higher income people who consume more energy, and so on. Note that our argument is not that all regulations of market forces are bad but rather that many of them are, in terms of both equity and economic growth.[3]

It is therefore highly problematic that even these policies, unwarranted on both growth and equity grounds, have resisted reform. In part, as we argue in chapters 2 and 3, the ideological opposition to the dismantling of policies that favor the wealthy and the powerful rests on a serious misinterpretation of why income inequality often increases as development proceeds. The premise of an unavoidable tradeoff between efficiency and equity has defined much of the ideological confrontation. The most negative interpretation of growing inequality is that ongoing policy reforms, particularly liberalization initiatives, are prejudicial to the poor. Indeed, there are instances in which the enrichment of the already privileged comes at the expense of the poor. And it is true that economic growth typically enriches some people, particularly those with greater assets in the first place, much more than others, increasing inequality of wealth, education, access to health care, and environmental amenities. Growing inequality provokes skepticism about the trajectory of development as well as intergroup conflict. What is too often ignored is that growing inequality often results from the expansion and increasing productivity of the modern sector, elevating the profits and salaries of owners and workers within this sector, but without depriving the relatively lower income owners and workers outside of the modern sector. The radical critique against liberalization, that growing inequality is proof of unfairness to the poor, has often served as a convenient pretext for resisting liberalization reforms. Because growing inequality sometimes can have innocuous causes, a paradigm shift is required: from a preoccupation with distribution—always a politically fraught topic—to a concern over poverty alleviation.

Yet in practice the liberalization efforts themselves often fall short of protecting the poor while eliminating the special advantages of the rent-seekers. Although many governments have claimed that they were adhering to the suite of the Washington Consensus mentioned in several chapters (and thereby seeming to accept the development policy leadership of the United States), in fact crucial aspects of the doctrine frequently have been omitted in practice. Most importantly, of the ten elements of the Washington Consensus reforms (see chapter 2), the two most important for safeguarding the incomes and well-being of low income families have often been neglected. One is to redirect public spending to sectors with both high economic returns *and* "the potential to improve income distribution." The other is to secure property rights. Without these elements, invoking and applying the Washington Consensus can be highly regressive in terms of income distribution and can undermine income opportunities of local people living off natural resources such as forests. John Williamson (1999) has expressed regret about how the term has been hijacked by advocates of narrow neoliberal, market fundamentalist policies to portray a false image of what Washington-based development institutions such as the World Bank were really espousing. He argues that the truncated conception ("restriction by partial incorporation," one of the dynamics that we describe in chapter 1) is not an effective framework for combating poverty but rather a regressive rationalization for "laissez faire Reagonomics."

In short, because core economic policies often do not directly address poverty alleviation, the risks of both neglecting the poor and provoking disruption need to be addressed by policies specifically targeting the poor. However, economic theory calls for policies that minimally distort the market signals required for economic efficiency. The orthodox approach is to rely on government revenues to provide direct payments to the poor; a less preferable alternative—to provide goods and services to the poor—limits the capacity of the poor to purchase what they need and opens up the possibility of benefit leakage insofar as the nonpoor can manipulate eligibility criteria or the poor feel compelled to resell the goods. One of the most promising approaches, then, is the direct cash transfer, which can have additional long-term benefits, as well as a more compelling appeal to tax payers, if they are conditioned on families keeping the children in school and providing for their health care. As mentioned in chapter 2, conditional cash transfers have strong potential for improving human capital, as families become healthier and better educated; they can also ease the process of economic reform by reducing the pain of structural adjustment for the most vulnerable, and often most potentially disruptive, populations.

However, whether the compensatory benefits for the poor are regarded as worthwhile government expenditures depends on whether their advantages can be adequately appreciated. Sound economic policymaking requires budget decisions that take into account, as accurately as possible, the full range of benefits and costs of government spending, including the less tangible gains in education, health care, and environmental protection, and such costs as environmental degradation, population displacements, and so on. Incorporating the long-term, often intangible benefits of meeting the poor's needs and wants requires intensive analysis that many planning agencies neglect. The failure to do so is as much a governance issue as an economic methodology issue. The declining use of benefit-cost analysis to guide government investment decisions (see chapters 2 and 3) means that the investment decisions can be made without going through the difficult, imprecise, but necessary step of trying to inventory and value the consequences. Moreover, because the poor generally face greater environmental vulnerability, the neglect of environmental costs poses risks to the poor's income from degraded forestry, farming, and fishing, as well as threats to health because of pollution that the poor are less able to avoid.

The stability of social safety net programs is frequently essential for poverty alleviation and the general well-being of the most vulnerable populations. However, this need for continuity clashes with the conventional approaches to public finance prevalent in developed nations. Conventional public finance theory dictates that revenues from taxation and natural resource royalties ought to go directly into the central treasury, to be allocated across the whole range of possible programs and projects in order to maximize the overall societal benefit. However, this flexibility, despite its virtue of permitting planners to adapt to changing problems and opportunities, presumes enough societal consensus that the poverty alleviation and social safety nets would not be jettisoned. In a developed country with a stable set of practices and expectations regarding the broad outlines of income distribution and social safety net support for the poor, the resources devoted to social safety nets are likely, in many instances, to be fairly stable. Yet there is evidence that propoor social safety net programs are vulnerable to severe cuts when the economy falters and taxpayers become more self-protective.[4] Therefore, stabilizing the financing of these programs ought to be a high priority.

The logic for stabilizing social safety net programs conforms to the general logic outlined earlier regarding the advantages of simple, broad, and uniform measures. The political considerations come to the fore as well: they begin with the need to convince the nonpoor that social programs should benefit those who contribute minimally to taxes. A broad social safety net program is likely to be more resilient insofar as efforts to terminate or emasculate such a program would arouse more opposition. This does not mean that all eligible beneficiaries would receive the same magnitude of benefits, but they must be sufficiently motivated to organize against severe cutbacks. Another important requirement is simplicity of the compensatory program, in terms of eligibility criteria, lean administration, and insulation from egregious corruption. However, a social safety net program that depends on recipients' contributions (e.g., a conventional pension plan) may seem broad and uniform but would exclude workers outside of the formal sector and families too poor to contribute.[5]

Chapters 4 and 5 on governance reveal some of the obstacles to addressing the problems of the poor. Understanding the obstacles to fair and effective governance has advanced across the social sciences. And the doctrines of greater and fairer participation in governance have been considerably refined, evolving from traditional, centralized, top-down public administration to democratic governance, but in many countries have not been fully embraced. The typically far greater access to revenues enjoyed by central governments (whether through taxation or raw material export revenues) has reinforced the power of central governments to control subnational authorities, and in some instances, deconcentration as a variant of decentralization has imposed even greater central government control over local affairs.

Conflict-Sensitive Development

The fragmented nature of most government policymaking rarely brings sufficient knowledge of the potential for intergroup conflict into the selection of development policies. Strategies that entail population movements (e.g., resettlement programs, physical infrastructure projects that displace people, regional development policies that induce spontaneous migrations to favored areas) often create flashpoints involving the preexisting populations and the "intruders." Greater state presence may be required to keep the peace, but the state's expansion into hitherto lightly governed frontiers frequently provokes resistance from the indigenous population. Strategies that involve direct or indirect redistribution of income, job opportunities, educational access, and so on, often create hostility between favored and disfavored groups. They can also reinforce rigid group identities and stereotypes, as in the case of affirmative action programs that add economic and political salience of such identities as Afro-Brazilian or "backward castes" in India. Liberalization initiatives that hold the long-term promise of increasing economic efficiency and eliminating the special privileges of the wealthy and politically powerful have had a frightful history of riots, other forms of disruption, and military coups, especially when introduced alongside painful austerity measures.

Conflict-sensitive development policies include those that encourage broader sharing of economic roles to reduce the perception that ostensibly exploitive roles are held predominantly by one ethnicity (e.g., Chinese merchants in Southeast Asia; Lebanese shopkeepers in Senegal). They can also involve policies that create mutually beneficial economic interdependence among groups, deterring aggressive confrontations that might threaten the gains. They may entail compensatory programs that soften the impacts on the groups otherwise disadvantaged by the main thrust of the strategy. They may reduce the salience of group differences through integrated schools or military service.

However, given the lack of awareness of conflict potentials by analysts in economic policymaking institutions, the conflict-sensitive policies may seem suboptimal, even if taking the risk of violence and consequent economic disruption into account would reveal that conflict-inducing policies may be worse from an economic perspective as well. Just as the selection of development policies must heed the institutional constraints, it must also note the fragility of peace in many developing countries.

If order is to be maintained by the armed forces, the evolution of its own institutional roles becomes crucial. A positive trend has been that in many countries the armed forces seem to have accepted—in principle—the doctrine of military subservience and political neutrality. Yet the practice often falls short. In one variant, the military leaders claim the role of "safeguarding" democracy by dictating processes or policies, or even by ousting a civilian government they deem as bent on undermining the very democracy that brought that government to power.[6] In another variant, the government cannot control disruption, and military leaders conclude that maintaining order justifies intervention. In this context, antigovernment opposition leaders have an incentive to provoke confrontation. Yet another variant is an alliance between military leaders and particular segments of civilian political forces, as in Turkey and Thailand, that may or may not result in military intervention but provides the armed forces with more control than full subservience. Sometimes all of these circumstances converge, as in the case of Egypt currently. As chapter 9 clarifies, for many countries the crucial unfinished business is making civilian government robust enough to shepherd the military to a constructive and politically set of roles. If the armed forces are to be shifted to a less militarized, less fully mobilized form, whether a national guard or a proportionally greater reserve force, the civilian government institutions, and the robustness of the economy, must be stable enough.

Environment

All of the foci of this book bear on the weaknesses of environmental protection and conservation in developing countries, accounting for the largely dismal record in many countries.

Governance structures in many developing countries put the agencies overseeing environmental protection into a weak intragovernmental position vis-à-vis the ministries mandated to expand industry, agriculture, resettlement, and other initiatives that pose environmental and conservation threats. Even decentralization, often posed as a corrective for national governments' low environmental priorities, can undermine environmental protection. For example, in China the decentralization of environmental regulation enforcement to the provincial level permits the national government to claim environmental responsibility by having what appear to be stringent environmental regulations, but the provincial authorities have little incentive to enforce them at the cost of economic growth (Lo, Fryxell, and Wong 2006). The devolution of responsibility for forest exploitation has often left underfunded local governments or community groups to grapple with already degraded forestlands, and sometimes provides perverse incentives for local authorities to overexploit the forests (Andersson, Gibson, and Lehoucq 2004).

Conflicts over the distribution of revenues from resource extraction, exacerbated by lack of transparency in both contracting the extraction effort and channeling revenues, have led to the waste and misallocation of resource wealth and are root causes of some of the violence in resource-dependent countries. Vigorous international efforts by coalitions of international NGOs and bilateral and intergovernmental development agencies have been launched to promote improved resource governance for greater transparency and sound policies.[7] The challenge is to induce governments to comply with the guidelines despite the possibilities of huge gains through corrupt practices.

The relationship between the military and the environment is often driven by the dominance of national security as integral to the military's doctrine. The Brazilian expansion into the Amazon, initiated under a military government, has caused considerable forest degradation.

Environmental NGOs have played a highly constructive role in many countries, yet two types of conflicts frequently arise. One is that NGO leaders in some countries view their relationship with government as intrinsically adversarial, with the result that government leaders disregard NGO input. The other conflict is between environmental NGOs, particularly international ones, and local grassroots organizations when the NGOs call for conservation that threatens livelihoods.

Regarding economic methodology and doctrines, in the use of standard economic approaches to select programs and projects, the long-term, less easily identified or monetized environmental impacts are often given short shrift. While some international organizations, such as the World Bank, have adopted the approach of holding up projects that have open-ended, difficult-to-assess environmental risks.[8]

Final Thoughts

The problems facing developing countries in economic policy, governance, and international interactions, as played out among so many types of institutions, can easily be seen by pessimists as too entangled to hold much hope for sustained progress of inclusive development. Weak institutions preclude the optimal policies for growth and poverty alleviation; corruption has corrosive impacts on both growth and equity; political instability and limits to honest incomes make corruption all the more attractive; conflicting objectives among civilian government leaders, civil society, and the military reduce the stability and soundness of policies even beyond the uncertainties intrinsic to economically fragile nations.

Yet, as our chapters have conveyed, progress can have its own momentum. In many respects, economic policies have shed some of the more regressive and growth-restricting elements formerly propped up by now-discredited doctrines. The paths to stronger institutions have been blazed by enough nations to point to the social, political, and economic advantages of sounder, more inclusive governance. As nations gain political stability, especially if based on broader political participation through regular democratic channels that preempt the disruption to gain voice, the military's role is likely to be constrained as well. The institutions through which developed nations impinge upon developing nations, whether bilateral, multilateral, or nongovernmental, have also been learning how to play more constructive roles.

Notes

1 Introduction

1. We embrace the policy sciences framework (Lasswell 1971) to anchor our examination in the task of exploring how real-world problems—poverty, infringements on human rights, authoritarian and unresponsive governance, physical insecurity, and sluggish economic growth—in many developing countries can be understood and overcome.

2 Evolution of Economic Development Theories and Doctrines since World War II

1. The First Development Decade (1961–1970) specified a 5 percent sustained annual growth rate by 1970; actual was 4 percent. The Second Development Decade (1971–1980) specified a 6 percent annual growth rate by 1980; actual was 4.8 percent. The Third Development Decade specified a 7 percent growth rate during that decade; the actual was 2.8 percent. Similar shortfalls pertain to agricultural and industrial growth rates and savings rates (Roberts 2005, 114).
2. For example, see Chenery (1975), Hirschman (1981), Hoadley (1981), Killick (1976), Krugman (1994), Meier (1968, 1987), Meier and Seers (1984), Rostow (1956), Schultz (1986), Stern and Ferreira (1997), and Tignor (2006). Kapur, Lewis, and Webb (1997), with an exhaustive history and set of commentaries on the World Bank, also have highly relevant material on the evolution of development thinking.
3. Like for many broad labels, considerable confusion reigns as to whether a particular theorist ought to be classified as being a neoclassical economist. However, Remenyi (1979, 56–59) offers a helpful characterization of the hardcore propositions of the essential commitments of neoclassical economics: (1) Consumers and producers can be assumed to be rational decision makers who know their wants. (2) Economic activity is motivated by individual self-interest. (3) More is better than less. (4) Given perfect knowledge and good government, economic welfare is maximized by free competition. (5) Although welfare and economic welfare are not synonymous, the latter is a good approximation for the former. (6) Stable Pareto-efficient equilibrium solutions can be defined for any and all markets relevant to economic research and analysis. (7) Everything has its opportunity cost. (8) Abstract, reduced-form models and simplifying assumptions are valid tools of economic analysis.
4. The GDP growth rate is estimated on the basis of the incremental capital-output ratio—the change in capital required to generate a unit change in GDP. The domestic savings rate and the effectiveness of capital to generate economic growth determine the growth rate.

5. Kuznets (1955, 24–25) warned that "if and when industrialization begins, the dislocating effects on these societies, in which there is often an old hardened crust of economic and social institutions, are likely to be quite sharp—so sharp as to destroy the positions of some of the lower groups more rapidly than opportunities elsewhere in the economy may be created for them."
6. Columbia University economist Ragnar Nurkse (1953, 21) argued,

   > Why do [poor] countries not push their exports of primary products according to the rules of international specialization, and import the goods they need for a "balanced diet"?... For fairly obvious reasons, expansion of primary production for export is apt to encounter adverse price conditions on the world market, unless the industrial countries' demand is steadily expanding... In the present century... [t]here has been some sluggishness in the industrial countries' demand for primary products, and despite the recent raw material boom there is no certainty that this sluggishness is gone for good.

7. Tax reform, as endorsed by international financial institutions, entails marginal rate reduction, tax base broadening, and special tax treatment elimination. See Gillis (1988).
8. Pioneered by Wassily Leontief, who won the 1973 Nobel Prize in economics.
9. As Wing (2004, 2) notes,

   > Computable general equilibrium (CGE) models are simulations that combine the abstract general equilibrium structure... with realistic economic data to solve numerically for the levels of supply, demand and price that support equilibrium across a specified set of markets.

10. Chenery (1961, 21) defines balanced growth as "simultaneous expansion of a number of sectors of production," crediting Nurkse (1953) and Rosenstein-Rodan (1943) for establishing that market signals regarding comparative advantage are unreliable and that the synergies among sectors should not be neglected.
11. Anderson (1994, 16) concludes that "[b]roadly, developing countries tend to underprice farm products whereas industrial countries tend to overprice them."
12. The prominent 1981 World Bank Berg Report documented the policy biases and concluded,

   > There is a fairly widespread consensus as to the main factors behind the present rural crisis... disruptions caused by wars and civil strife, drought and poor rainfall patterns during the 1970s, and rapid population growth, which pushed cultivation into less productive areas. Agriculture was also neglected for a long time by government and donors, as it was by development theorists. (48–49)

   Bates (1981) also enumerates the bases of the bias in Sub-Saharan Africa and provides multiple political explanations for it.
13. So-called "tied aid."
14. Technically, it is the discount rate that would render the net present value of the project equal to zero.
15. Sometimes labeled the two-sector model, but this can be confused with the two-sector model of household and firm; this confusion can be avoided by using the term dual-sector.
16. See Zeller and Meyer (2002) and Flynn (2007) for useful histories and assessments.
17. Streeten and Burki (1978, 413) argue that "BN gives high priority (attaches considerable weight) to meeting specified needs of the poorest people, not primarily in order to raise productivity (though additional production is necessary), but as an end in itself."
18. Streeten and Burki (1978, 413) assert that "societies can define their own basket of basic goods and services. The list of goods included and the quantities in which they are to be consumed would differ according to the society's objective."

19. Streeten and Burki (1978, 415) had to rely on the following rough estimates:

> Well over 50% of the population has inadequate caloric intake; over 60% receive less than the minimum daily requirement of proteins and the entire population has vitamin deficiencies…this translates into about 2.5 million tons of foodgrains. At least 20–25% of the population does not have minimum clothing. This is equivalent to about 100 million square meters of cloth…The share of the absolutely poor in total income is estimated at only 30% or $2.6 billion, giving them an income per head of only $56…to satisfy the three core needs of these people, they must receive on average at least $43 in additional income. However, since perhaps only two-thirds of the extra income is spent on core basic needs, the absolute increase in incomes may amount to $65 *per capita.*

20. The World Bank lends for projects submitted by developing country governments along with their own assessments of the projects' costs and benefits; the World Bank staff then conducts its own appraisal. In practice, the Bank staff has significant prior input in the selection and design of submitted proposals.
21. See Ascher (1983) and Leff (1985b) for accounts of the experience of trying to introduce this method; see Cleaver (1980) for a prominent example of its use.
22. The broadest definition of the term institution is a regular pattern of interaction and expectations, whether formal or informal.
23. See Wolf (1979) for a prominent articulation of the policy (i.e., nonmarket) failure explanation for poor performance.
24. Ronald Coase, James Buchanan, Jr., Herbert Simon, Douglass North, Oliver Williamson, and Elinor Ostrom are generally regarded as fitting one or both criteria.
25. Neoclassical economics and institutional economics do not conceive of transaction costs in the same way. Allen (1999, 898, 901) notes that the neoclassical definition of transaction costs is "costs resulting from the transfer of property rights," whereas the broader conception from the property rights approach of the new institutional economics is "costs establishing and maintaining property rights." This latter conception entails a more central role of the state as protector of the institutions that secure property rights.
26. For example, see Buchanan, Tollison, and Tullock (1980).

3 Economic Policy and Program Practice

1. Lora (2001) developed indicators of structural reforms in trade, finance, taxation, privatization, and labor market regulation for 19 Latin American countries from 1985 to 1999.
2. Except for Chile, they are small nations: Bahamas, Belize, Costa Rica, and Panama.
3. On deregulation, 140th, 135th, 147th, and 151st, respectively; on trade openness, 150th, 67th, 81st, and 151st, respectively.
4. This includes both North Africa and Sub-Saharan Africa; the Economic Freedom assessment includes 40 African nations.
5. On deregulation, Tunisia, Morocco, Algeria, and Egypt were ranked 107th, 137th, and 146th, 148th, respectively; on trade openness, 79th, 75th, 141st, and 126th, respectively.
6. Similarly, the World Bank Group's ease of "trading across borders" rankings place nearly 80 percent of the African countries below the median; half are in the bottom quartile.
7. Calculated from the World Bank Databank site, http://www.doingbusiness.org/data/exploretopics/trading-across-borders
8. World Bank Group (2014). Djibouti, Morocco, and Tunisia are above the median.
9. Calculated from the IMF database (2014), using IMF country designations.
10. Ortiz and Cummins (2011). Apparently, given their reliance on World Bank data, the 36 low-income countries are 27 Sub-Saharan African countries, plus Afghanistan, Bangladesh, Cambodia, Haiti, Kyrgyzstan, Nepal, North Korea, and Tajikistan.

11. The neoclassical perspective is that protective tariffs protect the profits and wages of inefficient industry to the detriment of the rest of society (Little 1982; Lal 2000); the neoclassical position on credit is that market-sensitive credit mechanisms are superior because administered credit allocation favors the well-connected (World Bank 1975). Regarding minimum wages, the neoclassical argument is that the cost of labor in the privileged modern sector of the labor elite discourages hiring. Regarding tax exemptions designed to induce investment, the neoclassical position is that they reduce revenues that the government could invest in social services, divert investment into activities unjustified by before-tax rates of return, and favor the wealthy who typically have more capability to avoid or evade taxation (Harberger 1984, 432–434; Gillis 1988).
12. Cited in Nellis (2012, 2).
13. A better measure of tax effort than the ratio of tax revenues to GDP is the volume of tax revenues to the tax capacity, which takes into account factors such as per capita income, the size of the agricultural sector, and/or income distribution to determine how much revenue could feasibly be collected.
14. Pessino and Fenochietto (2010, 74) present different metrics of tax effort and choose the truncated normal heterogeneous model as their criterion for ranking, which takes into account the impact of corruption and inflation on the inefficiency of tax collection.
15. Useful summaries are found in Grosh et al. (2008) and Fiszbein and Schady (2009).
16. For Asia, real per capita GDP would increase by 0.53 percent; 0.44 percent for Eastern European and Central Asian nations; 0.56 for Latin America and the Caribbean; 0.47 percent for the Middle East and North Africa; 0.55 percent for Sub-Saharan Africa.
17. Reported in World Bank (2012c). Twenty-six African nations have reported data on public education expenditure by level. Unfortunately, only one North African country, Tunisia, reports this information; the latest figure of the Tunisian proportion of expenditure targeted to higher education is 23 percent. Educational data have been quite slow to update.
18. The three that achieved that level were Botswana, Rwanda, and Zambia.
19. This ranking excludes the city-states of Hong Kong and Singapore. World Bank (2011b).
20. In an updated publication, Alston (2010) still finds high rates of return.
21. For detailed information on the Chinese infrastructure strategy, see Kim and Nangia (2010).
22. Some impacts, such as the health and education improvements due to the greater ease of people to reach health clinics and schools, cannot be systematically assessed.
23. Calculated from Tables 6 and 7 in Canning and Bennathan (2007). For the roads sample, eleven of the countries are Latin American, eight are Sub-Saharan, two are South Asian, two are Southeast Asian, and two are Middle Eastern. For electricity sample, sixteen are Latin American, eleven are Sub-Saharan, six are Southeast Asian, five are Middle Eastern, five are South Asian, four are Middle Eastern, and the remaining two are Fiji and China.
24. Government policies unavoidably promote or discourage investment and overall spending in different sectors through the budget allocations to sectoral ministries (e.g., ministries of agriculture or education), tax policies, subsidies of input such as energy, the stringency or leniency of regulations, and so on.
25. Confusion over education returns arises because social returns are not returns to society as a whole, but rather only the private economic returns *minus* societal costs of providing the education. Patrinos and Psacharopoulos (2011, 10) note that "[e]stimates of social returns to education, as commonly found in the literature, ignore non-income benefits of education (e.g., improved health) and the possibility of positive externalities from education, such as productivity spillovers, lower crime, reduced use of social services, increased civic participation, and so on." Yet even when social includes impacts on the rest of society, the findings have been highly inconsistent (Venniker 2001).

26. See Bloom and Canning (2003) and Jack and Lewis (2009).
27. The World Bank (2013) provides a useful definition of the computable general equilibrium model: "Computable General Equilibrium (CGE) models offer a comprehensive way of modeling the overall impact of policy changes on the economy. They are completely-specified models of an economy or a region, including all production activities, factors and institutions... These models incorporate many economic linkages and can be used to try to explain medium- to long-term trends and structural responses to changes in development policy."
28. Van de Laar (1980, 221) argued that "[m]any Bank staff... feel that the calculation of project rates of return is principally intended to put the 'icing on the cake' after all the interesting decisions have been made. Insofar as Executive Directors need to be able to show that they have approved 'sound'... projects, it is necessary to have a yardstick."
29. The same problems hold for the regional development banks: the Asian Development Bank (2013, 1) reported, "Through a series of annual retrospectives from 2003 to 2008, ERD [the Asian Development Bank's Economic Research Department] reviewed the quality of economic analysis in the ADB and noted significant scope for improvement, particularly in the articulation of the projects, economic rationale, demand analysis, and alternatives analysis."
30. See Mok (2008) for a summary of the efforts and obstacles to extend education in Vietnam.

4 Evolution of Governance and Development Administration Theory

1. Cawson (1985: 38) defines corporatism as the "process in which organizations representing monopolistic functional interests engage in political exchange with state agencies over public policy outputs which involves those organizations in a role which combines interest representation and policy implementation through delegated self-enforcement."
2. As a Princeton professor, Woodrow Wilson (1887) wrote the landmark article "The Study of Administration."
3. For example, the University of Southern California's School of Citizenship and Public Administration became the School of Policy, Planning and Development, and is now the Sol Price School of Public Policy. See also Lynn (1996) and Allison (2008).
4. See especially Foucault et al. (1991) for a particularly philosophical treatment.
5. For example, the so-called Istanbul Consensus emerging from the 2011 Fourth UN Conference on the Least Developed Countries.
6. Of course, other institutions, from religious institutions to multinational banks dating back to the Renaissance, have always played a role.

5 Evolution of Governance and Development Administration Practice

1. Tosun and Yilmaz (2008, 8, 28–29) report that in the Middle East and North African region the deconcentrated units of the central government provide most public services, including health and education, while the role of decentralized units (municipalities) is limited to construction of local roads, street maintenance, and other mundane functions.
2. In political science, the most useful definition of legitimacy is behavioral: the belief on the part of citizens that the form of government is appropriate and the top government officials have been selected through appropriate processes. This is in contrast to conceptions

of legitimacy as an attribute apart from people's attitudes. The latter type of definitions is used in political discourse as a form of demand; for example, "This government is legitimate because the King/God/ the Party has blessed it." Using the behavioral definition, it is clear that undemocratic governments may be considered legitimate.

6 Evolving Roles of NGOs in Developing Countries

1. Article 71 of the UN Charter formalized NGOs and included them as an integral part of UN activities.
2. Davis (2013) notes the existence of religious orders, missionary groups, merchant combines, and scientific societies beginning at least in the seventeenth century as performing functions one could readily associate with NGOs in more recent times. Pressing back even farther in time, there is clear evidence of humanitarian societies providing nongovernmental public services along China's fabled river systems as early as the thirteenth century.
3. http://www.ushistory.org/documents/ask-not.htm
4. http://www.un.org/geninfo/bp/enviro.htm
5. For the reader's convenience, these are fiscal discipline, redirection of public expenditure priorities toward sectors with both high economic returns and potential to improve income distribution, tax reform, interest rate liberalization, competitive exchange rates, trade liberalization, liberalization of inflows of foreign direct investment, privatization of state-owned enterprises, deregulation to abolish barriers to entry and exit, and secure property rights.
6. Interview: www.democracynow.org/2004/8/23
7. Aggressive attempts to define and assess measures of effectiveness may actually impede performance because of burdensome record keeping and reporting. Small NGOs are especially vulnerable to this because of limited personnel.
8. Pelaez (2012) recites a list of confrontations in the human rights and civil liberties realms between NGOs and Venezuela, Nicaragua, Bolivia, Ecuador, Cuba, and Belarus. Russia continues to crackdown on a wide range of NGOs (Weir 2013). The various issues and tensions were hardly confined to socialist and former Communist states (Mayr 2010; Hubbard 2013; Murphy 2013).
9. It shows up in contemporary times in theories and practical activities related to social enterprises, a topic we return to later in this chapter.
10. The commercial possibilities of the poorest of the poor underlie and motivate theories to exploit "The Fortune at the Bottom of the Pyramid" (Prahalad and Hart 2002).
11. Dees and Anderson (2006, 39) conclude, "The construct of social entrepreneurship... reflects a breakdown in the boundaries between business and the nonprofit sector in the search for new approaches to social problems and needs. It is a development that is potentially promising, but also risky."

    The seminal Dees and Anderson survey has been augmented by Dacin, Dacin, and Matear (2010), where disdain for theorizing is signaled in the survey's subtitle: "Why We Don't Need a New Theory and How We Move Forward from Here."
12. Several organizations on this list are not well known outside of the social enterprise realm. Kiva Microfunds, Acumen, SELCO, and Seventh Generation are all interesting in their own right.
13. The late John Sawhill, former president of the Nature Conservancy.
14. Citizens United is a tax-exempt 501(c) 3 organization. The Supreme Court decision, *Citizens United vs. Federal Election Commission* (No. 08–205, decided January 21, 2010), opened up the full range of nonprofit, public benefit, nongovernmental institutional

options to heretofore prohibited political and commercial activities(http://www.law.cornell.edu/supct/html/08; http://www.followthemoney.org/Press).

15. It was overtaken in value by the Bill and Melinda Gates Foundation.
16. http://www.history.com/this-day-in-history/henry-ford-ii-leaves-post-at-ford-foundation
17. The *combined* State Department and USAID budget requested in FY 2014 was $47.8 billion (http://www.usaid.gov/results-and-data).
18. Four to five billion dollars of the proposed USAID budget is earmarked for Afghanistan and Pakistan.
19. Tetlock and Gardner (2013: 81) report,

    In the late 1980s one of us (Philip Tetlock) launched a [forecasting] tournament. It involved 284 economists, political scientists, intelligence analysts and journalists and collected almost 28,000 predictions. The results were startling. The average expert did only slightly better than random guessing. Even more disconcerting, experts with the most inflated views of their own batting averages tended to attract the most media attention. Their more self-effacing colleagues, the ones we should be heeding, often don't get on to our radar screens.

20. It is unusual to find thoughtful efforts to combine ecological and human systems directly and in terms of stability and resilience. The giant reinsurance organization, Munich Re, has done excellent work in the last ten years or so by sponsoring an annual Summer Academy on Social Vulnerability (Hamza et al. 2012).
21. Hurricane Mitch struck Central America in 1998 with such terrific force as to call into question the very legitimacy of Honduras and Nicaragua. The lessons learned in this disaster are still being assembled, although they indict national, international, and non-governmental organizations with nearly equal opprobrium (Bradshaw, Linneker, and Zúniga 2001; Allianz 2012: 112–128). The 2010 Haitian earthquake is likewise filled with important lessons.

7 Evolution of Foreign Assistance Theories and Doctrines

1. For incisive background works, see Montgomery (1962), Montgomery (1967), Shin (1969), Tendler (1975), Montgomery (1986), Riddell (1987), and Tarnoff and Lawson (2012).
2. US official development assistance exceeded US$30 billion in 2012, two-and-a-half times greater than that of the United Kingdom, the second largest donor by absolute volume (OECD 2013b).
3. Roughly ten percent of foreign assistance during the Cold War came from the Soviet Union (Dreier, Fuchs, and Nunnenkamp 2012, 2).
4. These included prominent political scientists and economists such as Gabriel Almond, Daniel Lerner, Max Millikan, Lucian Pye, Paul Rosenstein-Rodan, and Walt Rostow.
5. An excellent survey of early post-WWII thinking about modernization theory and the role of the United States can be found in Latham (2011, especially Chapter 2).
6. That is, its successor, the Mutual Security Agency (MSA) (1951–1953), and the Foreign Operations Agency (1953–1955), which superseded the MSA.
7. See Arase (1995, 2005) and Ascher (2007) for more details about the contrasting foreign assistance approaches.
8. A US Congressional Research Service notes that "For FY2000 and thereafter, Plan Colombia funds are assigned to the State Department's International Narcotics and Law Enforcement Bureau (INL) or the Andean Counterdrug Initiative (ACI). The State Department transfers funds to other agencies carrying out programs in Colombia, of which USAID has received the largest portion" (Beittel 2012, 38).

9. Comecon consisted of Bulgaria, Cuba, Czechoslovakia, East Germany, Hungary, Mongolia, Poland, Romania, the Soviet Union, and Vietnam.
10. As noted in chapter 2, the World Bank Independent Evaluation Group (2010) reported that even the World Bank, long a champion of systematic analysis, has used benefit-cost analysis less and less in its project evaluations, with the lowest utilization for education and health projects.
11. McCleary and Barro (2008, 512) report that "[i]n recent years, an estimated 41% of U.S. overseas development funds are channeled through PVOs, whereas in Japan, only 2%, and in the United Kingdom, 12% of development funds are estimated to flow through nonprofits."
12. Its functions were absorbed by the UN Food and Agriculture Organization and the World Food Programme, both predating the World Food Council's formation in 1974.
13. The definitional distinction is that the members of NGOs ostensibly act in the public interest, for people other than themselves, or both; whereas members of grassroots organizations explicitly act in their own interest.
14. Sachs (2003, 38) notes, "Rather than focus on improving institutions in sub-Saharan Africa, it would be wise to devote more effort to fighting AIDS, tuberculosis, and malaria; addressing the depletion of soil nutrients; and building more roads to connect remote populations to regional markets and coastal ports."
15. Kapur and Webb (2000, 2) argue,

    > Governance was thrust into prominence with the end of the cold war and the resulting need to recreate civil societies in former communist states. Even more broadly, international political culture has changed, making state sovereignty far less sacrosanct in international discourse. One aspect of globalization has been to change international rules and norms in ways that weaken the "sovereignty" defence against intervention. The setting up of the International Criminal Court and international conventions that allow crimes against humanity to be tried in countries where they were not committed are examples of this shift.

8 International Development in the American Grain: From Point Four to the Present

1. Quoted in *USA Today*, February 2, 2009, http://www.usatoday.com/news/washington/2009-02-01-aid-inside_N.htm
2. For an account of the internal machinations behind Kennedy's historic declaration, see Heffron (2009).
3. The relative percentages of US bilateral and multilateral ODA were calculated from figures at the US ODA Database, http://usoda.eads.usaidallnet.gov/data/summary_reports.html
4. In a 2010 survey conducted by the polling firm Polimetrix/YouGov, participants were asked "Would you prefer that the US give economic aid directly to a country or give aid to an international organization (such as the World Bank or International Monetary Fund) which then would give it to the country?" Only 22 percent of the respondents chose multilateral delivery, 78 percent preferring the bilateral option (Milner and Tingley 2013, 322–324).
5. Dean Achison, Secretary of State under Truman, stated, "We are willing to help people who believe the way we do, to continue to live the way they want to live" (cited in Williams 1973, 474).
6. As McVety (2012, 105) has recently written, "While many policymakers recognized the intrinsic humanitarianism of promoting economic development for its own sake, the realities of international affairs and domestic politics dictated that self-interest…had to be the foundation of U.S. foreign policy." See also Ruttan (1996, 1–4), where he paints the contradiction underlying American exceptionalism as one between our idealism and our realism.

7. Under current law, 100 percent of Food for Peace Act (FPA) disbursements must be produced in the United States and 50 percent shipped overseas on US-flagged ships, including a requirement for monetization of at least 15 percent (http://borgenproject.org/food-aid-reform-act-1/). This although "food aid currently compris[es] just 0.86 percent of total U.S. agricultural exports and 0.56 percent of net farm income" (http://www.washingtonpost.com/blogs/wonkblog/wp/2013/05/17/obamas-plan-to-overhaul-food-aid-is-running-into-trouble-in-congress/).
8. This last point is so essential to the orientations and limitations of US foreign assistance as well as the evolution of militaries in developing countries that chapter 10 is devoted to the security-economic assistance nexus in the crucial cases of US intervention in Korea, Vietnam, and Chile.
9. One candidate for notable exception to this generalization is Ekbladh (2010).
10. "No economic or technical assistance shall be supplied to any other nation unless," the Act stated, "the President finds that the supplying of such assistance will strengthen the security of the United States and promote world peace, and unless the recipient country has agreed to join in promoting international understanding and good will, and in maintaining world peace, and to take such action as may be mutually agreed upon to eliminate causes of international tension" (Mutual Security Act of 1951, sec. 511b).
11. The loss of China did not precipitate communist takeovers in Indonesia, Thailand, or other large Southeast Asian countries; the wars in Laos, Cambodia, and Vietnam were wars of national liberation from colonial and neocolonial governments, exploited perhaps by Moscow but never under its direct control. The Sino-Soviet split would make sure of that (Herring 1979, 149–150, 242; Hobsbawm 1994, 346–347). Eisenhower (1954) introduced his domino theory at a press conference.
12. Judging from the relative scope and magnitude of the US and Soviet aid programs, there was little real competition in the first place. In 1964, US foreign aid outstripped the Soviet Union by ten times. In FY 1968, US aid commitments equaled the total Soviet expenditures between 1954 and 1968.
13. See also USAID (1982).
14. See Prosterman (2007); Raup (1968, especially 574); and Gaud (1968). For a useful overview, see Prosterman and Riedinger (1987).
15. Rennack, Mages, and Chesser (2011, 1) note, "Over time, as enactment of foreign aid authorizations waned, the general provisions of foreign operations appropriations measures increasingly have become a legislative option for Congress to assert its views on the role and use of U.S. foreign policy, put limits or conditions on assistance, or even authorize new programs. As a result, some contend, general provisions have become more important."
16. For supplemental emergency funding estimates, see Lawson, Epstein, and Nakamura (2010).
17. For a treatment of McNamara's global liberalism and its critics, those who viewed the Bank "as a self-indulgent institution that forced unsuccessful, neo-socialist programs on nations that would otherwise have successfully embraced private enterprise," see Milobsky and Galambos (1995, 173, 169).
18. One of the earliest expositions of this concept is Watkins (1998).
19. See the US Foreignassistance.gov site, http://www.foreignassistance.gov/web/AgencyLanding.aspx

9 Evolving Roles of the Military in Developing Countries

1. Simple classifications of nations in this era as Modern, Traditional, or Modernizing, as Pye and others did in the late 1950s and early 1960s, facilitated the joining of military with

development objectives—typically to no good end in general and especially for development ends.

2. One overlooked and consistent outcome of trying to force military and development objectives during this period was failure, often to accomplish either goal. For instance, few African officers had even basic military training; courses in economics or business management skills were likewise unknown in military academies around the world—including those in developed countries.
3. Coast guards and border patrols actually qualify as *internal* agencies more than heavy-duty *external* agencies having tanks, fighters, bombers, and large naval vessels. At least they do for this discussion.
4. Egypt, Pakistan, Myanmar, Thailand, and many other countries are illustrative.
5. Huntington and others following his lead assumed the existence of a stable constitutional order, with competent civilian policy elites. Such a condition may exist in the United States and other developed or modern countries; it seldom does in developing countries.
6. Fitch (1998, Chapter 3) points out that for Latin America, at least, military professionalism in a democratic system implies that armed forces training should not go beyond the conventional military roles and the policymaking roles should be confined narrowly to the few top-ranking officers who must interact with national security councils and their equivalents.
7. The title of Jones's article, "The Nationbuilder, Soldier of the Sixties," says it all.
8. Transition to civilian control and away from the military took notable strides in 2005 with the creation of the Council of Civil Aviation and in 2011 with the creation of Secretariat of Civil Aviation. Airports and related infrastructure were privatized (Infraero) during the run-up to the 2014 World Cup, largely because of charges of ineffectiveness. The control of all airspace remains the responsibility of the Air Force's Department of Air Space Control. http://www.abag.org.br.
9. So-called military tourism, where actual military sites and equipment are featured attractions, is one form. Russia and Ukraine are notable for this type. Another form occurs when military interests actually own regular tourist sites and infrastructure. Thailand, Indonesia, Ecuador, Egypt, and Pakistan, among others, are cases in point.
10. Realistic, as opposed to perceived, security threats are subject to interpretation—and manipulation.
11. Sudan, Ethiopia, Bangladesh, Thailand, Indonesia, India, Nicaragua, Guatemala, Liberia, Sierra Leone, and so on.
12. Conscription historically has severe coercive connotations, for example, cannon fodder, press gangs. Our use of the label goes well beyond this to include national service in which the military is only one of several options. It also includes voluntary service through contractual enlistment for different periods of time.
13. Mercenaries date back to antiquity. The modern-day equivalent for outsourcing and privatizing the use of deadly force is private military contractors, a notorious example of which is Blackwater.
14. The literature is extensive—especially that generated in the 1960s to the 1980s, with the 1970s being most productive. Contributions by Schmitter (1972, 1974), O'Donnell (1973, 1986), Przeworski et al. (1986) and Stepan (1971) are illustrative.
15. Peru (1975), Honduras (1975), Argentina (1976), El Salvador (1979), Bolivia (1980), Haiti (1988), Panama (1989), Paraguay (1989), Haiti (1991), Venezuela (1992), Peru (1992), Guatemala (1993), Ecuador (2000), Venezuela (2002), Haiti (2004), Honduras (2009), and Ecuador (2010).
16. We are indebted to our colleague Sam Fitch for this sharp insight and for many other improvements in this chapter.
17. Military entrepreneurs challenge civil control over means of production and political institutions and processes. Khaki capitalism is a recent label bestowed by a wag at *The*

Economist (2011), in a revealing summary of military enterprises in Egypt, Pakistan, China, Iran, Indonesia, and Thailand. Lengthier, more scholarly sources include Brömmelhörster and Paes (2004) and Mani (2007). Mani (2011) is an excellent review of Latin American experiences.

18. The Iran-Iraq War (1980–1988) provided a long-lived and profitable market. It also flooded the Middle East with weapons of war in use to the present. Saudi Arabia and Pakistan were also good customers for the Chinese PLA firms.
19. For instance: The China Poly Group, linked to the PLA's General Staff Department; China Xinxing, controlled by the PLA General Logistics Department; China Carrie and the PLA General Political Department; NORINCO, an arm of the Northern Army Group; and China Songhai, operated by the PLA Navy.
20. PLA companies listed in Hong Kong are referred to as red chips. The British Virgin Islands, Liechtenstein, and Panama each has PLA-affiliated financial offices.
21. Unclassified sources are not as numerous as one would hope. Business sources are seldom forthcoming for proprietary and competitive reasons. Chinese language sources, including translations, are also challenging—primarily because those sources are located in and controlled by China.
22. Goldman Sachs published the book *BRICS and Beyond in 2007*, when things were going well. A careful reading of the book is reminder of the cloudiness of economic crystal balls much beyond a year or two. http://www.goldmansachs.com/our-thinking/archive/BRICs-and-Beyond.html
23. The relationship here to the Theory of Universal Conscription is evident.
24. At 50 million human beings, this is the largest number of refugees since WWII according to the UN High Commissioner for Refugees.
25. The mega-cities scenario and forecast presented in the NGO chapter describe this.
26. Economic improvement explains most of the history of migration north out of Mexican poverty. Other conflicts in Central America, as well as in Mexico, now contribute to the exodus of women and children and the humanitarian crisis this presents to the United States.
27. Costa Rica abolished its military in 1948. In addition some 20 nations around the world do not have formally constituted militaries. Unsurprisingly, these nations are small, even tiny. Surprisingly, 11 of them are island nations at high risk of increasing sea levels. http://en.wikipedia.org/wiki/List_of_countries_without_armed-forces
28. Codelco, the state-owned copper company, is the result of nationalization of the privately held copper industry in 1971 by Salvador Allende and then its reorganization during the military reign of Augusto Pinochet in 1976. http://upi,com/Business_News/Security-Industry/2014/01/22Chilean-defense-spending
29. Hurricane Mitch, in 1998, struck Central America with terrific force and also called into question the very legitimacy of Honduras and Nicaragua. The lessons learned in this disaster are still being assembled, although they basically implicate, even indict, national, international, and nongovernmental organizations with nearly equal opprobrium (Bradshaw, Linneker, and Zúniga, 2003; Allianz 2012, 112–128). The 2010 Haitian earthquake is likewise filled with important lessons, some of which informed recovery efforts in the Philippines in the wake of Typhoon Haiyan (Holtz 2014).

10 Complementarity of Security and Development Doctrines: Historical Cases and Aftermaths

1. "The security and stability of nations half a globe away," Secretary of Defense Robert S. McNamara made imperative in 1966, is directly linked to the security of the United States, "whether communists are involved or not" (Nelson 1968, 19, 20).

2. Described on Rabe (2010, 38).
3. For the meaning and usage at the time of the term political development, see Pye (1960). See also Butler (1968).
4. CORFO refers to the Production Development Corporation (in Spanish: Corporación de Fomento de la Producción de Chile), an organization founded in 1939, by President Pedro Aguirre Cerda, to promote domestic economic development.
5. See, for example, Loveman (1976, 197, 189–220, 293); Lower (1968, 283–287); and Kaufman (1972, 39–44).
6. For example, USAID workers expressed alarm that in Santiago, a city of three million people, less than half of the 78 patrol cars in its inventory were operational.
7. USAID/Chile (1980, 28); USSC (1975, 34).
8. See Lowden (1996) and Meyer (2013).
9. Public Library of US Diplomacy n.d., https://wikileaks.org/plusd/cables/06CARACAS3356_a.html
10. http://www.usaid.gov/who-we-are/organization/bureaus/bureau-democracy-conflict-and-humanitarian-assistance/office-1.
11. The Open Society Foundations' website mentions, "We give priority to civil society groups and individuals that work to: monitor and report on human rights abuses and advocate for rights-respecting policies and practices in select countries (Mexico and Venezuela)." http://www.opensocietyfoundations.org/about/programs/latin-america-program. The Open Society Institute & Soros Foundations Network also supports Human Rights Watch, which issued scathing reports on the Maduro government in the *World Reports 2014* and *2015*.
12. For example, Human Rights Watch 2015. http://www.hrw.org/world-report/2015/country-chapters/venezuela?page=1
13. For the details of this assassination attempt and the US response, see Lee (2006, 56–57).
14. For the full study, analyzed here in critical detail, see Mason et al. (1980).
15. The First Indochina War was against the French (1945–1954), the second against the United States and its Allies (1954–1976), and the third refers to the short conflict with China over Vietits incursions into Cambodia in 1978–1979 on behalf of the Khmer Rouge, which had instigated an ethnic cleansing campaign against Vietnamese living there.

11 Conclusion: Linkages and Challenges

1. Lasswell (1936/1958).
2. See IMF (2000). The budget restrictions were loosened as early as 1998.
3. The debates over liberalization still rage among prominent economists, such as Edwards (2010), Lal (2012), Rodrik (2008), Stiglitz (2000, 2002, 2011).
4. For example, chapter 3 cites Ravallion (2002) on the greater declines in Argentine propoor safety net spending than overall spending.
5. Evidence of this bias is provided in depth in Grosh et al. (2008, 357–359).
6. Guillermo O'Donnell (1973, 166–199) called this the "Impossible Game" in reference to military intervention in Argentina beginning in the 1950s when Peronists, regarded by the military as likely to impose authoritarian governance, won elections that were then annulled by the military.
7. The Extractive Industries Transparency Initiative is the most prominent effort to promote improvements; the most prominent effort to establish governance guidelines is the Resource Charter. See Bebbington (2013).

8. Nielson and Tierney (2003, 264) note regarding the World Bank's project appraisal:

> Disbursement for any project categorized as environmentally sensitive A or B types now requires two distinct environmental assessments... [A]n environmental impact assessment (EIA) must be submitted and a pilot program must be designed to test for environmental degradation... After data is collected during the pilot phase, a second EIA must be submitted for the project as a whole Loan disbursements cannot commence until these administrative procedures are completed and the documents submitted to the board by authorized environmental staff.

References

ABAG. 2013. *ABAG yearbook 2012*. Brasilia: Brazilian Association for General Aviation.

Abdullahi, Najad. 2008 February 17. Pakistani army's "$20bn" business, *Al Jazeera*. http://www.aljazeera.com/focus/pakistanpowerandpolitics/2007/10/200852518451598412

Abu-Bader, Suleiman, and Aamer Abu-Qarn. 2003. Government expenditures, military spending and economic growth: Causality evidence from Egypt, Israel, and Syria. *Journal of Policy Modeling* 25(6): 567–83.

Acheson, Christopher. 2005. Review of Taylor's reconstructing macroeconomics: Structuralist proposals and critiques of the mainstream, *Journal of Economic Literature* 43(1): 141–43.

Adams, Dale. 1971. What can under-developed countries expect from foreign aid to agriculture? Case study: Brazil 1950–70, *Inter-American Economic Affairs* 25(1): 47–64.

Agarwal, Bina. 1997. "Bargaining" and gender relations: Within and beyond the household, *Feminist Economics* 3(1): 1–51.

Agrawal, Arun, and Jesse Ribot. 1999. Accountability in decentralization: A framework with South Asian and West African cases, *Journal of Developing Areas* 33: 473–502.

Agrawal, Arun, and Jesse Ribot. 2006. Recentralizing while decentralizing. In David Cameron, Gustav Ranis, and Annalisa Zinn, eds., *Globalization and self-determination: Is the nation-state under siege?* London: Routledge, pp. 301–32.

Ahluwalia, Montek, and Hollis Chenery. 1974. The economic framework. In Hollis Chenery, Monteek Ahluwalia, C. L. G. Bell, John Duloy, and Richard Jolly, eds., *Redistribution with growth*. Oxford: Oxford University Press, pp. 38–51.

Alagappa, Muthiah. 2004. *Civil society and political change in Asia: Expanding and contracting democratic space*. Palo Alto, CA: Stanford University Press.

Allen, Douglas. 1999. Transaction costs. In Boudewijn Bouckaert and Gerrit De Geest, eds., *Encyclopedia of law and economics*, Vol. 1. Surrey, UK: Edward Elgar.

Allianz, Ryan. 2012. Unsupervised recovery: Adaptation strategies by two NGOs in post-Mitch Honduras, *SOURCE* 16: 112–28.

Allison, Graham. 2008. Emergence of schools of public policy: Reflections by a founding dean. In Michael Moran and Martin Rien, eds., *The Oxford handbook of public policy*. New York: Oxford University Press, pp. 58–79.

Alston, Julian. 2010. The benefits from agricultural research and development, innovation, and productivity growth. OECD Food, Agriculture and Fisheries Paper No. 31. Paris: OECD Publishing.

Alston, Julian, Connie Chan-Kang, Michele C. Marra, Philip G. Pardey, and Tim J. Wyatt. 2000. *A meta-analysis of rates of return to agricultural R&D: Ex pede herculem?* Washington, DC: International Food Policy Research Institute.

Alvarado, Facundo, and Leonardo Gasparini. 2013 November. Recent trends in inequality and poverty in developing countries. Rio de la Plata, Argentina: Centro de Estudios Distributivos, Laborales y Sociales, Universidad Nacional de la Plata Documento de Trabajo Nr0. 151.

Anderson, Kym. 1994. Food price policy in East Asia, *Asian-Pacific Economic Literature* 8 (2): 15–30.

Anderson, Kim, John Cockburn, and Will Martin, eds. 2010. Introduction and summary. In Kym Anderson, John Cockburn, and Will Martin, eds., *Agricultural price distortions, inequality, and poverty.* Washington, DC: World Bank, pp. 3–46.

Anderson, Mary. 1999. *Do no harm: How aid can support peace—or war.* Boulder, CO: Lynne Rienner Publishers.

Andersson, Krister, Clark Gibson, and Fabrice Lehoucq. 2004. The politics of decentralized natural resource governance, *Political Science and Politics* 37(3): 421–26.

Andrews, Stanley. 1970 October 31. Oral history interview. Sec. 14, 12. Oral Histories. Kansas City, MO: Harry S. Truman Public Library.

Appiah, Elizabeth, and Walter McMahon. 2002. The social outcomes of education and feedbacks on growth in Africa, *Journal of Development Studies* 38(4): 27–68.

Arase, David. 1995. *Buying power: The political economy of Japan's foreign aid.* Boulder, CO: Lynne Rienner.

Arase, David. 2005. *Japan's foreign development, not foreign aid: Old continuities and new directions.* London, UK: Routledge.

Ascher, William. 1983. New development approaches and the adaptability of international agencies: The case of the World Bank, *International Organization* 37(3): 415–39.

Ascher, William. 1998. From oil to timber: The political economy of off-budget development financing in Indonesia," *Indonesia* 65: 37–61.

Ascher, William. 2007. Foreign aid and alternative approaches to leadership: Japanese and U.S. contrasts. In Dennis Rondinelli and Jay Heffron, eds., *Leadership for development in a globalizing society.* Sterling, VA: Kumarian Press.

Asian-African Conference. 1955 April 24. Final communique of the Asian-African conference of Bandung.

Asian Development Bank. 1997. Governance bank policies. *Operations manual.* Section 54. Manila: Asian Development Bank.

Asian Development Bank. 2010. *Strategy 2020.* Manila: Asian Development Bank.

Asian Development Bank. 2013. *Cost-benefit analysis: A practical guide.* Mandaluyong City, The Philippines: Asian Development Bank.

Asociación Latinoamericana de Organizaciones de Promoción al Desarrollo. 2010. Open forum for CSO effectiveness and development, Mexico City.

Auer, Matthew R. 2007. More aid, better institutions, or both? *Sustainability Science* 2(2): 179–87.

Austin, James E., and John C. Ickis. 1986. Management, managers, and revolution, *World Development* 14(7): 775–90.

Baldacci, Emanuele, Benedict Clements, Sanjeev Gupta, and Qiang Cui. 2008. Social spending, human capital, and growth in developing countries, *World Development* 36(8): 1317–41.

Ball, George. 1964 November 10. Briefing paper on USAID Public Safety Program with Carabineros de Chile. Office of Public Safety. USAID Record Group 286. Entry 26– Chile. Box 23. Folder IPS-1 File material/Chile 1 of 2. NACP.

Bang, Henrik, and Anders Esmark. 2009. Good governance in network society: Reconfiguring the political from politics to policy, *Administrative Theory & Praxis* 31(1): 7–37.

Barash, David. 2013 December15. Costa Rica's peace dividend: How abolishing the military paid off, *Los Angeles Times.* http://www.articles.latimes.com/print/2013/dec/15/opinion/la-oe-barash-costa-rica-demilitarization

Barbier, E. B. 2014. Disaster management. A global strategy for protecting vulnerable coastal populations, *Science* 345(6202): 1250–51.

Bates, Robert. 1981. *Markets and states in Tropical Africa*. Berkeley, CA: University of California Press.

Bates, Robert, and Da-Hsiang D. Lien. 1985. A note on taxation, development and representative government, *Politics and Society* 14(1): 53–70.

Bauer, Peter. 1957/1965. *Economic analysis and policy in underdeveloped countries*. Cambridge: Cambridge University Press.

Bauer, Peter, and Basil Yamey. 1957. *The economics of underdeveloped countries*. Cambridge, UK: Cambridge University Press.

Bebbington, Anthony. 2013 June. Natural resource extraction and the possibilities of inclusive development: Politics across space and time. ESID Working Paper No. 21. Manchester: University of Manchester.

Beeson, Mark. 2008. Civil-military relations in Indonesia and the Philippines: Will the Thai coup prove contagious? *Armed Forces & Society* 34(3): 474–90.

Beittel, June S. 2012. *Colombia: Background, US relations, and congressional interest*. Washington, DC: Congressional Research Service.

Benmaamar, Mustapha. 2006. Financing of road maintenance in Sub-Saharan Africa. Sub-Saharan Africa Transport Policy Program, Discussion Paper 6. Washington, DC: World Bank.

Benoit, Emile. 1968. The monetary and real costs of national defense, *American Economic Review* 58(2): 398–416.

Benoit, Emile. 1973. *Defense and economic growth in developing countries*. Lexington, MA: Lexington/D.C. Heath.

Beresford, M., and D. Phong. 2000. *Economic transition in Vietnam: Trade and aid in the demise of the centrally planned economy*. Cheltenham: Edward Elgar.

Bernstein, Thomas, and Xiaobo Lü. 2008. Taxation and coercion in rural China, *Taxation and State-building in Developing Countries*: 89–108.

Bevir, Mark, Rod Rhodes, and Patrick Weller. 2003. Traditions of governance: Interpreting the changing role of the public sector, *Public Administration* 81(1): 1–17.

Bezemer, Dirk, and Derek Headey. 2008. Agriculture, development, and urban bias, *World Development* 36(8): 1342–364.

Bickel, Lennard. 1974. *Facing starvation: Norman Borlaug and the fight against hunger*. New York: Reader's Digest Press.

Bickford, Thomas. 1994. The Chinese military and its business operations: The PLA as entrepreneur, *Asian Survey* 34(5): 460–74.

Bickford, Thomas. 1999. Sunrise or sunset for the Chinese military? The business operations of the Chinese People's Liberation Army, *Problems of Post-Communism* 46 (4): 35–46.

Bird, Richard, Jorge Martinez-Vazquez, and Benno Torgler. 2008. Tax effort in developing countries and high income countries: The impact of corruption, voice and accountability, *Economic Analysis and Policy* 38(1): 55–71.

Black, C. E. 1967. *The dynamics of modernization: A study in comparative history*. New York: Harper & Row.

Blind, Peride. 2009. *Democratic institutions of undemocratic individuals*. New York: Palgrave Macmillan.

Bloom, David, and David Canning. 2003. The health and poverty of nations: From theory to practice, *Journal of Human Development* 4(1): 47–71.

Bloom, David, David Canning, and Jaypee Sevilla. 2004. The effect of health on economic growth: A production function approach, *World Development* 32(1): 1–13.

Blunt, Peter, and Merrick Jones. 1992. *Managing organizations in Africa*. Berlin and New York: Walter de Gruyter.

Bockh, Laszlo, and Mary Blakeslee. n.d. Budget analysis—Federal agency money matters. http://www.budgetanalyst.com/The%20CR.htm

Boothroyd, P., and Pham Xuan Nam, eds. 2000. *Socioeconomic renovation in Vietnam: The origin, evolution, and impact of DoiMoi*. Ottawa: International Development Research Centre.

Bornhorst, Fabian, Sanjeev Gupta, and John Thornton. 2009. Natural resource endowments and the domestic revenue effort, *European Journal of Political Economy* 25(4): 439–46.

Bossert, T. J., and J. C. Beauvais. 2002 March. Decentralization of health systems in Ghana, Zambia, Uganda and the Philippines: A comparative analysis of decision space, *Health Policy and Planning* 17(1): 14–31.

Boughton, James. 2003. *Who's in charge? Ownership and conditionality in IMF-supported programs*. Washington, DC: International Monetary Fund.

Bradshaw, Sarah, Brian Linneker, and Rebecca Zúniga. 2001. Social roles and spatial relations of NGOs and civil society: Participation and effectiveness in Central America post Hurricane "Mitch," *The Nicaraguan Academic Journal* 2(1): 73–113.

Brazinsky, Gregg. 2007. *Nation building in South Korea: Koreans, Americans, and the making of a democracy*. Chapel Hill: University of North Carolina Press.

Brewer, Garry D. 2007. Inventing the future: Scenarios, imagination, mastery and control, *Sustainability Science* 2(2): 159–77.

Briceño-Garmendia, Cecilia, Antonio Estache, and Nemat Shafik. 2004. *Infrastructure services in developing countries: Access, quality, costs, and policy reform*. Washington, DC: World Bank.

Brodzinsky, Sibylla. 2013 July 22. Foreign funding dries up for Latin American NGOs, *Christian Science Monitor*. http://www.csmonitor.com/World/Americas/2013/0722

Brömmelhörster, Jörn, and Wolf-Christian Paes. 2004. *The military as an economic actor: Soldiers in business*. New York: Palgrave Macmillan.

Brooks, Arthur. 2008. *Social entrepreneurship: A modern approach to social value creation*. Upper Saddle River, NJ: Pearson Prentice Hall.

Buchanan, James, Jr., Robert Tollison, and Gordon Tullock, eds. 1980. *Toward a theory of the rent-seeking society*. College Station, TX: Texas A&M University Press.

Buckley, Chris. 2014 July 1.China's antigraft push snares and ex-general, *New York Times*, A6.

Buffett, Peter. 2013 July 27. The charitable-industrial complex, *New York Times*, A-19.

Burnside, Craig, and David Dollar. 2000. Aid, policies, and growth, *American Economic Review* 90(4): 847–68.

Busch, Gary. 2008. The Chinese military-commercial complex. http://www.ocnus.net/artman2/publish/Editorial_10/

Butler, Brian E. 1968. Title IX of the foreign assistance act: Foreign aid and political development, *Law & Society Review* 3(1): 115–51.

Butterfield, Samuel. 2004. *Hale U.S. development aid—An historic first: Achievements and failures in the twentieth century*. Westport, CT: Praeger

Caffrey, Craig. 2014 October 2. South Korea announces 5.3% boost in defence spending. http://www.janes.com/article/44030/south-korea-announces-5-3-boost-in-defence-spending

Campbell, Joel, Leena Thacker Kumar, and Steve Slagle. 2010. Bargaining sovereignty: State power and networked governance in a globalizing world, *International Social Science Review* 85(3–4): 107–23.

Canning, David, and Esra Bennathan. 2007. The rate of return to transportation infrastructure. Report of the Round Table on Transport Economics, European Conference of Ministers of Transport.

Carothers, Thomas. 1999. *Civil society*. Washington, DC: Carnegie Endowment for International Peace.

Carson, Rachel. 1962. *Silent spring*. Boston, MA: Houghton Mifflin.

Castells, Manuel. 2006. The network society: From knowledge to policy. In Manuel Castells and Gustavo Cardoso, eds., *Societies in transition to the network society*. Baltimore, MD: The Johns Hopkins University Press, pp. 1–21.

Castillo, Jasen, Julia Lowell, Ashley Tellis, Jorge Munoz, and Benjamin Zycher. 2001. *Military expenditures and economic growth*. Santa Monica, CA: RAND Corporation.

Cawson, A. 1985 *The political economy of corporatism*. London: Sage.

Chalmers, Douglas. 1977. The politicized state in Latin America. In James Malloy, ed., *Authoritarianism and corporatism in Latin America*. Pittsburgh, PA: University of Pittsburgh Press, pp. 23–45.

Cheema, G. Shabbir. 2005. *Building democratic institutions: Governance reform in developing countries*. West Hartford, CT: Kumarian Press.

Cheema, G. Shabbir. 2010. Civil society engagement and democratic governance. In G. Shabbir Cheema and Vesselin Popovski, eds., *Engaging civil society: Emerging trends in democratic governance*. Tokyo: United Nations University, pp. 1–20.

Cheema, G. Shabbir, and Dennis Rondinelli. 2007. *Decentralizing governance: Emerging concepts and practices*. Washington, DC: Brookings Institution Press.

Cheema, G. Shabbir, and Vesselin Popovski. 2010. *Engaging civil society: Emerging trends in democratic governance*. Tokyo: United Nations University.

Chenery, Hollis. 1961. Comparative advantage and development policy, *The American Economic Review* 51(1): 18–51.

Chenery, Hollis. 1974. Introduction. In H. Chenery, M. S. Ahluwalia, C. L. G. Bell, J. H. Duloy, and R. Jolly, eds., *Redistribution with growth*. London: Oxford University Press.

Chenery, Hollis. 1975. The structuralist approach to development policy, *American Economic Review* 65(2): 310–16.

Cheung, Tai Ming. 2001. *China's entrepreneurial army*. New York: Oxford University Press.

Christian Science Monitor. 2010 August 5. Warren Buffett, Bill Gates, and the billionaire challenge. http://www.csmonitor.com/Commentary/the-monitors-view/2010/0805/

Christian Science Monitor. 2014 December 8. When loss is gain: A global move to end fuel subsidies, *Christian Science Monitor* 107(3): 34.

Clark, John. 1991. *Democratizing development: The role of voluntary organizations*. West Hartford, CT: Kumarian Press.

Clark, John. 2010. The role of transnational civil society in promoting transparency and accountability in global governance. In G. Shabbir Cheema and Vessilin Popovski, eds., *Engaging civil society: Emerging trends in democratic governance*. Tokyo: United Nations University Press, pp. 44–57.

Cleveland, Harlan. 1972. *The future executive: A guide for tomorrow's manager*. New York: Harper & Row.

Cleaver, Kevin M. 1980. Economic and social analysis of projects and of price policy: The Morocco fourth agricultural credit project. World Bank Staff Working Paper, International Bank for Reconstruction and Development.

Clinton-Gore Administration History Project. 2000. *USAID's role, 1993–2001*. Washington, DC: US Agency for International Development.

Cochrane, Joe. 2014 July 23. A child of the slums rises as President of Indonesia, *New York Times*, A-4.

Cohen, Jean, and Andrew Arato. 1994. *Civil society and political theory*. Cambridge, MA: MIT Press.

Colclough, Christopher. 1980. Primary schooling and economic development: A review of the evidence. World Bank Staff Working Paper. Washington, DC: World Bank.

Collier, David, and Ruth Collier. 1977. Who does what, to whom, and how: Toward a comparative analysis of Latin American corporatism. In James Malloy, ed., *Authoritarianism and corporatism in Latin America*. Pittsburgh, PA: University of Pittsburgh Press, pp. 489–512.

Collier, Simon, and William Sater. 1999. *A history of Chile, 1808–1994*. New York: Cambridge University Press.

Commission on Global Governance. 1995. *Our global neighbourhood*. Oxford: Oxford University Press.

Committee on International Relations and Committee on Foreign Relations. 2003. *Legislation on foreign relations through 2002*. Volume 1-A, Sec. 296 Foreign Assistance Act of 1961. Washington, DC: US Government Printing Office.

Congressional Quarterly Research. 1951 May 3. Future of foreign aid. Washington, DC: Congressional Quarterly. http://library.cqpress.com/cqresearcher/document.php?id=cqresrre1951050300

Constable, Pamela, and Arturo Valenzuela. 1989. Chile's return to democracy, *Foreign Affairs* 68(5): 169–86.

Cooley, Alexander, and James Ron. 2002. The NGO scramble: Organizational insecurity and the political economy of transnational action, *International Security* 27(1): 5–39.

Cornia, Giovanni Andrea, Richard Jolly, and Frances Stewart. 1987. *Adjustment with a human face: Protecting the vulnerable and promoting growth*. Oxford: Oxford University Press.

Cowen, Tyler 2015. Does China hitting the wall reflect a deeper reality about emerging economy growth? *Marginal Revolution* http://marginalrevolution.com/marginalrevolution/2015/09/does-china-hitting-the-wall-reflect-a-deeper-reality-about-emerging-economy-growth.html

Cullather, Nick. 2010. *The hungry world*. Cambridge, MA: Harvard University Press.

Cuny, Frederick C. 1989. *Use of the military in humanitarian relief*. Dallas, TX: Intertect Relief and Reconstruction Corporation. http://www.pbs.org/wgbh/pages/frontline/shows/cuny/laptop//humanrelief

Cutshall, Charles R., Dustin C. Emery, Daniel J. Fitzpatrick, Sarah J. Hammer, Leslie J. Kelley, and Kirill Meleshevich. 2009. Integrating USAID and DOS: The future of development and diplomacy, Maxwell School of Citizenship and Public Affairs. http://www.developmentgap.org/americas/US/the_effectiveness_development_assistance_programs_under_new_.pdf

Daalder, Hans. 1962/6969. *The role of the military in the emerging countries*. The Hague, The Netherlands: Institute of Social Studies.

Dacin, Pete, Tina Dacin, and Margaret Matear. 2010. Social entrepreneurship: Why we don't need a new theory and how we move forward from here, *The Academy of Management Perspectives* 24(3): 37–57.

Dacy, Douglas. 1986. *Foreign aid, war, and economic development: South Vietnam, 1955–1975*. Cambridge: Cambridge University Press.

Daniño, Roberto. 2006. The legal aspects of the World Bank's work on human rights in development outreach, *The International Lawyer* 41(1): 21–25.

Davis, Kevin, and Anna Gelpern. 2010. *Peer-to-peer financing for development: Regulating the intermediaries*. New York: NYU Law and Economics Working Papers.

Davis, Thomas. 2013 January 24. NGOs: A long and turbulent history, *The Global Journal*. http://theglobaljournal.net/article/view/981

De Janvry, Alain, and Jean-Jacques Dethier. 2012. The World Bank and governance: The Bank's efforts to help developing countries build state capacity. World Bank Policy Research Working Papers. Washington, DC: World Bank.

De la Rosa, Braulio. 2007. *Transforming armed forces to National Guard units in Latin America*. Carlisle Barracks, PA: US Army War College.

de Rugy, Veronique. 2012. Military Keynesians. http://reason.com/archives/2012/11/20/military-keynesians/print

de Waal, Alex. 1997. *Famine crimes: Politics and the disaster relief industry in Africa*. Bloomington: Indiana University Press.

Dees, J. Gregory. 2011. Social ventures as learning laboratories, *Tennessee's Business* 20(1): 3–5.

Dees, J. Gregory, and Beth Anderson. 2004. Scaling social impact, *Stanford Social Innovation Review* (Spring). http://www/ssireview.org/articles/entry/scaling_social_impact

Dees, J. Gregory, and Beth Anderson. 2006. Framing a theory of social entrepreneurship: Building on two schools of practice and thought, *Research on Social Entrepreneurship: Understanding and Contributing to an Emerging Field* 1(3): 39–66.

Devarajan, Shantayanan, Tuan Minh Le, and Gaël Raballand. 2010 April. Increasing public expenditure efficiency in oil-rich economies: A proposal. World Bank Policy Research Working Paper 5287. Washington, DC: World Bank.

Diamond, Larry. 1994. Toward democratic consolidation, *Journal of Democracy* 5(3): 4–17.

Dodge, Joseph M. 1954 November 22. Considerations relating to the problem and its solution. Report on the Development and Coordination of Foreign Economic Policy. White House Central Files Confidential. Box 18. Council on Foreign Economic Policy. Dwight D. Eisenhower Library, Abilene, Kansas.

Dollar, David, and Aart Kraay. 2001 April. Growth is good for the poor. World Bank Policy Research Working Paper 2587. Washington, DC: World Bank.

Dorfman, Robert. 1991. Review article: Economic development from the beginning to Rostow, *Journal of Economic Literature* 29(2): 573–91.

Doornbos, Martin. 2004. "Good Governance": The pliability of a policy concept, *Trames* 4: 372–87.

Dreier, Axel, Andreas Fuchs, and Peter Nunnenkamp. 2012. New donors. Kiel: Kiel Institute for the World Economy Working Paper.

Drèze, Jean, and Amartya Sen. 2013. *An uncertain glory: India and its contradictions.* Princeton, NJ: Princeton University Press.

Dryzek, John, and Patrick Dunleavy. 2009. *Theories of the democratic state.* New York: Palgrave Macmillan.

Duit, Andreas, and Victor Galaz. 2008. Governance and complexity: Emerging issues for governance theory, *Governance* 21(3): 311–35.

Dunleavy, Patrick, and Hood, Christopher. 1994. From old public administration to new public management, *Public Money and Management* 14(3): 9–16.

Duranton, Gilles. 2009. Are cities engines of growth and prosperity for developing countries? In Michael Spence, Patricia Annez, and Robert Buckley, eds., *Urbanization and growth.* Washington, DC: World Bank for the Commission on Growth and Development, pp. 67–113.

Ebrahim, Alnoor S., and V. Kasturi Rangan. 2010. The limits of nonprofit impact: A contingency framework for measuring social performance. Harvard Business School General Management Unit Working Paper 10–099.

Economist. 2011 December 3. Khaki capitalism. http://www.economist.com.node/21540985/print

Economist. 2013a July 27. The great deceleration, 10.

Economist. 2013b November 23. Let quite a few flowers bloom, 16.

Economist. 2013c June 1. Towards the end of poverty, 11.

Economist. 2013d July 27. When giants slow down, 20.

Economist. 2014a June 14. Sharif versus Sharif, 35–36.

Economist. 2014b July 5. Chairman of everything, 38.

Economist. 2014c September 20. Nosebags: The army in Pakistan, 78–79.

Edelman Trust Barometer. 2013. NGOs most trusted institution globally. http://www.trust.edelman.com/trusts/trust-in-institutions-2/ngos-remain-most-trusted

Edwards, Michael, and David Hulme, eds. 1996. *Beyond the magic bullet: NGO performance and accountability in the post-Cold War world.* West Hartford, CT: Kumarian Press.

Edwards, Sebastian. 2010. *Left behind: Latin America and the false promise of populism.* Chicago: University of Chicago Press.

Eggertsson, Thráinn. 1990. *Economic behavior and institutions: Principles of neoinstitutional economics.* Cambridge: Cambridge University Press.

Eisenhower, Dwight D. 1954. The President's News Conference. Online by Gerhard Peters and John T. Woolley, The American Presidency Project. http://www.presidency.ucsb.edu/ws/?pid=10202

Ekbladh, David. 2010. *The Great American Mission: Modernization and the construction of an American world order.* Princeton, NJ: Princeton University Press.

El Beblawi, Hazem. 2008. Economic growth in Egypt: Impediments and constraints (1974–2004). Washington, DC: World Bank Working Paper 14.

Elkington, John. 1994. Towards the sustainable corporation: Win-win-win business strategies for sustainable development, *California Management Review* 36: 90–100.

Elkington, John. 1997. Cannibals with forks. In *The triple bottom line of 21st century.* Oxford, UK: Capstone.

Encyclopedia Britannica. 2014. Corporatism. http://www.britannica.com/EBchecked/topic/138442/corporatism

Engle, Bryon. 1963 September 27. Military assistance plan—Korea. USAID Record Group 286. Entry 25. Box 69. Folder: IPS#1 General Policy Guidelines and Background, 1961–1967 Korea. NACP.

Engle, Bryon. 1968. Action memorandum for the administrator. USAID Record Group 286. Box 70. Folder: IPS#1 General Policy Guidelines, General Information Korea '68. NACP.

Esman, Milton J., and Norman T. Uphoff. 1984. *Local organizations: Intermediaries in rural development.* Ithaca, CA: Cornell University Press.

Eswaran, Mukesh, and Ashok Kotwal. 2006. The role of agriculture in development. In A. V. Banerjee, ed., *Understanding poverty.* Oxford: Oxford University Press, pp. 111–23.

European Bank for Reconstruction and Development. 2010. http://www.ebrd.com/downloads/research/economics/macrodata/sci.xls

Fair, C. Christine. 2014. *Fighting to the end: The Pakistan army's way of war.* Oxford: Oxford University Press.

Falconer, Peter. 1997. Public administration and the new public management: Lessons from the UK experience, *Línea.* http://www.Vus.Uni-Lj.si/Anglescina/FALPOR97.Doc

Farnsworth, Elizabeth. 1974 Autumn. More than admitted, *Foreign Policy* 16: 127–41.

Fay, Marianne, and Mary Morrison. 2007. *Infrastructure in Latin America and the Caribbean: Recent developments and key challenges.* Washington, DC: World Bank.

Feasel, Edward M. 2013. Official development assistance (ODA) and conflict: A case study on Japanese ODA to Vietnam. In William Ascher and Natalia Mirovitskaya, eds., *Development strategies, identities, and conflict in Asia.* New York: Palgrave Macmillan, pp. 209–40.

Feder, Ernest. 1965. Land reform under the alliance for progress, *Journal of Farm Economics* 47(3): 652–68.

Felix, David. 1961. An alternative view of the "monetarist"-"structuralist" controversy. In Albert Hirschman, ed., *Latin American issues.* New York: Twentieth Century Fund, pp. 81–93.

Fernholz, Rosemary Morales. 2010. Infrastructure and inclusive development through "free, prior, and informed consent" of indigenous peoples. In William Ascher and Corinne Krupp, eds., *Physical infrastructure development: Balancing the growth, equity, and environmental imperatives.* New York: Palgrave Macmillan, pp. 225–58.

Fesselmeyer, Eric, and Kien T. Le. 2010. Urban-biased policies and the increasing rural–urban expenditure gap in Vietnam in the 1990s, *Asian Economic Journal* 24(2): 161–78.

Field, C., V. Barros, T. Stocker, Q. Dahe, D. Dokken, K. Ebi, M. Mastrandrea, K. Mach, G. Plattner, and S. Allen. 2012. IPCC. Managing the risks of extreme events and disasters to advance climate change adaptation. Special Report of the Intergovernmental Panel on Climate Change. Cambridge: Cambridge University Press.

Finkelstein, Lawrence S. 1995. What is global governance? *Global Governance*: 367–72.

Fiszbein, Ariel, and Norbert Schady. 2009. *Conditional cash transfers: Reducing present and future poverty*. Washington, DC: World Bank.

Fitch, J. Samuel. 1998. *The armed forces and democracy in Latin America*. Baltimore: The Johns Hopkins University Press.

Fitzgerald, Frances. 1972. *Fire in the lake: The Vietnamese and the Americans in Vietnam*. New York: Vintage.

Fjeldstad, Odd-Helge, and Mick Moore. 2008. Tax reform and state building in a globalized world. In D. Brautigam, O.-H. Fjeldstad, and M. Moore, eds., *Taxation and state-building in developing countries: Capacity and consent*. Cambridge: Cambridge University Press, pp. 235–60.

Flynn, Patrice. 2007. Microfinance: The newest financial technology of the Washington consensus, *Challenge* 50(2): 110–21.

Flyvbjerg, Bent, Massimo Garbuio, and Dan Lovallo. 2009. Delusion and deception in large infrastructure projects: Two models for explaining and preventing executive disaster, *California Management Review* 51(2): 170–93.

Flyvbjerg, Bent, Nils Bruzelius, and Werner Rothengatter. 2003. *Megaprojects and risk: An anatomy of ambition*. Cambridge: Cambridge University Press.

Foucault, Michel, Graham Burchell, Colin Gordon, and Peter Miller. 1991. *The Foucault effect: Studies in governmentality*. Chicago: University of Chicago Press.

Frederickson, H. George. 2004. Whatever happened to public administration? Governance, governance, everywhere. Queens University Institute of Governance, Public Policy and Social Research Working Paper QU/GOV/3/2004. Belfast: Queens University.

Frei Montalva, Eduardo. 1967. The alliance that lost its way, *Foreign Affairs* 45: 437–48.

Fritz, Verena, Kai Kaiser, and Brian Levy. 2009. *Problem-driven governance and political economy analysis: Good practice framework*. Washington, DC: World Bank.

Fukuda-Parr, Sakiko. 2013 April. Global development goal setting as a policy tool for global governance: Intended and unintended consequences. International Policy Centre for Inclusive Growth Working Paper No. 108, Brasilia.

Gaud, William. 1968 March 8. The Green Revolution: Accomplishments and apprehensions. Administrator, Agency for International Development. Washington, DC: The Society for International Development. http://www.agbioworld.org/biotech-info/topics/borlaug/borlaug-green.html

Gerakas, E., D. S. Patterson, W. F. Sanford, and C. B. Yee, eds. 1995. National Security Council Record of Action No. 2447. Foreign Relations of the United States, 1961–1963, Vol. IX. Foreign Economic Policy, Section 7.

Gerschenkron, Alexander. 1957. Reflections on the concept of "prerequisites" of modern industrialization, *L'Industria* 2: 31–51.

Gerth, Hans, and C. Wright Mills, eds. and trans. 1946. *Politics as a vocation in Max Weber*. New York: Oxford University Press.

Gettleman, Jeffrey. 2014 December 7. As Ebola rages, poor planning thwarts efforts, *New York Times*, A-1, A-4.

Gibson, Bill, and Dirk Ernst Van Seventer. 2000. A tale of two models: Comparing structuralist and neoclassical computable general equilibrium models for South Africa, *International Review of Applied Economics* 14(2): 149–71.

Gillis, Justin. 2013 December 3. Panel says global warming carries risk of deep changes, *New York Times*, A-24.

Gillis, Malcolm, ed. 1988. *Lessons from fundamental tax reform in developing countries*. Durham, NC: Duke University Press.

Gimlin, Hoyt. 1988. Foreign aid: A declining commitment, *Editorial Research Reports*, Vol. II. Washington, DC: Congressional Quarterly Press.

Global Security. 2015. U.S.-Korean relations. http://www.globalsecurity.org/military/world/rok/forrel-us.htm

Goodman, David. 1994. *Corruption in the People's Liberation Army.* Murdoch, Australia: Asia Research Centre, Murdoch University.

Goodman, David, Gerald Segal, and Richard H. Yang. 1996. Corruption in the PLA, *Chinese Economic Reform: The Impact on Security*: 35–52.

Goodman, Louis. 1996. Military roles past and present, *Civil-Military Relations and Democracy*: 30–43.

Goodrich, Malinda. 1989. The Council for Mutual Economic Assistance—Soviet Union, Appendix B. In Raymond E. Zickel, ed., *Soviet Union: A country study.* Washington, DC: Library of Congress Congressional Research Service. http://rs6.loc.gov/frd/cs/soviet_union/su_appnb.html

Gordon, Lincoln. 1975 July 17. Oral History Interview. Sec. 162 and 158. Oral Histories. Harry S. Truman Public Library.

Gourevitch, Peter, David Lake, and Janice Gross Stein. 2012. *The credibility of transnational NGOs: When virtue is not enough.* New York: Cambridge University Press.

Graham, Carol. 2002. *Crafting sustainable social contracts in Latin America: Political economy, public attitudes, and social policy.* Washington, DC: Center on Social and Economic Dynamics.

Graham, Norman, ed. 1994. *Seeking security and development: The impact of military spending and arms transfers.* Boulder, CO: Lynne Rienner.

Gravel, Mike, ed. 1971. Establishment of the Special Group (Counterinsurgency). The Pentagon Papers, Gravel Edition, Vol. 2. Boston, MA: Beacon Press.

Grosh, Margaret E., Carlo Del Ninno, Emil Tesliuc, and Azedine Ouerghi. 2008. *For protection and promotion: The design and implementation of effective safety nets.* Washington, DC: World Bank.

Guest, L. Hayden. 1925 October. British labor and the empire, *Nation* 121: 392–94.

Gupta, Sanjeev, Alvar Kangur, Chris Papageorgiou, and Abdoul Wane. 2011. Efficiency-adjusted public capital and growth, *World Development* 57: 164–78.

Gupta, Sanjeev, Luiz De Mello, and Raju Sharan. 2001. Corruption and military spending, *European Journal of Political Economy* 17(4): 749–77.

Gwartney, James, Joshua Hall, and Robert Lawson. 2014. *2014 economic freedom dataset.* Vancouver: Fraser Institute. http://www.freetheworld.com/datasets_efw.html.

Hacking, Theo, and Peter Guthrie. 2008. A framework for clarifying the meaning of triple bottom-line, integrated, and sustainability assessment, *Environmental Impact Assessment Review* 28(2): 73–89.

Hagen, James, and Vernon Ruttan. 1987. Development policy under Eisenhower and Kennedy. Economic Development Center, University of Minnesota, Bulletin No. 87–10. Minneapolis: University of Minnesota.

Halim, Nafisa. 2008. Testing alternative theories of bureaucratic corruption in less developed countries, *Social Science Quarterly* 89(1): 236–57.

Halimi, Abdul-Latif. 2014 April 10. The regional implications of Indonesia's rise, *The Diplomat.* http://thediplomat.com/2014/04/the-regional-implications-of-indonesias-rise

Hamilton, Fowler. 1962 May 25. Memorandum for the President. Methods for Improving the Coordination of Economic and Military Aid Programs. USAID Record Group 286. Box 6. Folder: Program FY63. Entry P39: Classified Vietnam. NACP.

Hamza, Mohamed, Cosmin Corendea, Andrea Wendeler, and Katharina Brach. 2012. *Climate change and fragile states: Rethinking adaptation.* Tokyo: United Nations University.

Harberger, Arnold. 1984. Economic policy and economic growth. In Arnold Harberger, ed., *World economic growth.* San Francisco, CA: ICS Press, pp. 427–66.

Harmonization for Health in Africa. 2011. *Investing in health for Africa: The case for strengthening systems for better health outcomes.* Geneva: World Health Organization.

Hauslohner, Abigail. 2014. Egypt's military expands its control of the country's economy. *Washington Post.* http://www.washingtonpost.com/world/middle_east/egyptian-Military-Expands-its-Economic-control/2014/03/16/39508b52-a554-11e3-b865-38b254d92063_story.Html

Hechinger, John, and David Golden. 2006 July 8. The great giveaway, *Wall Street Journal*, A-1, 8–9.

Heffron, John M. 2009. Leadership for a "Decade of Development": John F. Kennedy and the Foreign Assistance Act of 1961. In Dennis Rondinelli and John M. Heffron, eds. *Leadership for development: What globalization demands of leaders fighting for change.* Sterling, VA: Kumarian Press.

Hellinger, Steven, Fred O'Regan, and Douglas Hellinger. 2009. The effectiveness of development assistance programs under New Directions legislation: Criteria for assessment. The Development Group for Alternative Policies. http://www.developmentgap.org/foriegn_aid/proposal_to_reform_us_development_assistance.pdf

Henry, Clement, and Robert Springborg. 2001. *Globalization and the politics of development in the Middle East.* Cambridge: Cambridge University Press.

Herring, George C. 1979. *America's longest war: The United States and Vietnam, 1950–1975.* New York: Willey.

Hirschman, Albert. 1958. *The strategy of development.* New Haven, CT: Yale University Press.

Hirschman, Albert. 1981. The rise and decline of development economics. In Albert Hirschman, ed., *Essays in trespassing: Economics to politics and beyond.* Cambridge: Cambridge University Press, pp. 1–24.

Hoadley, J. Stephen. 1981. The rise and fall of the basic needs approach, *Cooperation and Conflict* 16(3): 149–64.

Hobsbawm, Eric. 1994. *The age of extremes: A history of the world, 1914–1991.* New York: Vintage.

Hoebink, Paul. 2006. European donors and "good governance": Condition or goal? *The European Journal of Development Research* 18(1): 131–61.

Hofmann, Charles-Antoine, and Laura Hudson. 2009 September. Military responses to natural disasters: Last resort or inevitable trend? *Humanitarian Exchange Magazine.* http://www.odilhpn.org/humanitarian-exchange-magazine/issue-44

Holling, Crawford S. 1973. Resilience and stability of ecological systems, *Annual Review of Ecology and Systematics* 4: 1–23.

Holtz, Michael. 2014 December 1. In the Philippines, a year of recovery, *The Christian Science Monitor* 103(2): 18, 20.

Hout, Wil. 2010. *Governance and the rhetoric of international development.* Rotterdam: Erasmus University, International Institute of Social Studies.

Hubbard, Ben. 2013 June 5. Egypt convicts workers at foreign nonprofit groups, including 16 Americans, *New York Times*, A-4.

Hubert, Leo, ed. 2004. *The integrity of governance.* New York: Palgrave Macmillan.

Hulme, David, and Michael Edwards. 1996. *Making a difference: NGOs and development in a changing world.* London: Routledge.

Human Rights Watch. 2014a. Russia: Constitutional court upholds "Foreign Agents" law: Blow to freedom of association. http://www.hrw.org/news/2014/04/08/russia-constitutional-court-upholds-foreign-agents-law

Human Rights Watch. 2014b. World report 2014: Zimbabwe. http://www.hrw.org/world-report/2014/country-chapters/zimbabwe?page=2

Human Rights Watch 2015. Human Rights Watch world report: Venezuela. http://www.hrw.org/world-report/2015/country-chapters/venezuela?page=1

Huntington, Samuel. 1957. *The soldier and the state: The theory and politics of civil-military relations.* Cambridge, MA: Harvard University Press.

Hutchison Port Holdings. 2014. http://www.hhp.com

Hydén, Goran. 2007. The challenges of making governance assessments nationally owned. Paper presented at the 2007 Bergen Seminar on Governance Assessments and the Paris Declaration, September 24–25. Bergen: Mickelsen Institute.

Hydén, Goran. 2011. Rethinking governance theory and practice. In Göran Hydén and John Samuel, eds., *Making the state responsive: Experience with democratic governance assessments.* New York: UN Development Programme.

Hydén, Goran. 2012. *African politics in comparative perspective.* Cambridge: Cambridge University Press.

Hydén, Goran, and John Samuel, eds. 2011. *Making the state responsive: Experience with democratic governance assessments.* New York: UN Development Programme.

Hyman, Gerald. 2010. *Foreign policy and development: Structure, process, and the drip-by-drip erosion of USAID.* Washington, DC: Center for Strategic and International Studies.

Ickis, John, Ed C. de Jesus, and M. Maru Rushikesh. 1986. *Beyond bureaucracy: Strategic management of social development.* West Hartford, CT: Kumarian Press.

IISD. 2013. The rise and role of NGOs in sustainable development. In *Business and sustainable development: A global guide.* https://www.iisd.org/business/ngo/roles.aspx.

International Monetary Fund. 2000 June. Recovery from the Asian crisis and the role of the IMF. Washington, DC: International Monetary Fund. http://www.imf.org/external/np/exr/ib/2000/062300.htm#V

International Monetary Fund. 2002 August 31. The IMF and good governance, a factsheet. Washington, DC: International Monetary Fund.

International Monetary Fund. 2014. World economic outlook, October 2014. Washington, DC: International Monetary Fund. http://www.imf.org/external/pubs/ft/weo/2014/02/weodata/weorept.aspx?pr.x=53&pr.y=7&sy=2000&ey=2015&scsm=1&ssd=1&sort=country&ds=.&br=1&c=001%2C110%2C200%2C901%2C505%2C511%2C205%2C440%2C406%2C603&s=PPPPC&grp=1&a=1

IPCC Working Group II. 2014. *Climate change 2014: Impacts, adaptation, and vulnerability.* Philadelphia, PA: Saunders. http://www.ipcc.ch/report/ar5/wg2/

Isham, Jonathan, and Daniel Kaufmann. 1999. The forgotten rationale for policy reform: The productivity of investment projects, *Quarterly Journal of Economics* 114(1): 149–84.

Ishikawa, Shigeru. 1967. *Economic development in Asian perspective.* Tokyo: Kinofuniya Bookstore Co.

Jack, William, and Maureen Lewis. 2009. Health investments and economic growth: Macroeconomic evidence and microeconomic foundations. World Bank Policy Research Working Paper Series No. 4877.

Jain, H. K. 2010. *Green revolution: History, impact and future.* New Delhi: Studium Press LLC.

Jalal, Ayesha. 2014. *The struggle for Pakistan.* Cambridge, MA: Belknap Press.

Janowitz, Morris. 1960. *The professional soldier: A social and political portrait.* New York: Free Press.

Janowitz, Morris. 1964. *The military in the political development of new nations: An essay in comparative analysis.* Chicago: University of Chicago Press.

Jaskoski, Maiah. 2013. *Military politics and democracy in the Andes.* Baltimore: The Johns Hopkins University Press.

Jayasooria, Denison. 2010. Civil society engagement in Malaysia. In G. Shabbir Cheema and Vesselin Popovski, eds., *Engaging civil society: Emerging trends in democratic governance.* Tokyo: United Nations University, pp. 214–31.

Jepma, Catrinus. 1991. *The tying of aid.* Paris: Organisation for Economic Cooperation and Development.

Jimenez, Emmanuel, and Harry Anthony Patrinos. 2008 March. Can cost-benefit analysis guide education policy in developing countries? World Bank Policy Research Working Paper 4568. Washington, DC: World Bank.

JLT Specialty Limited. 2014. *JLT expropriation database*. London: JLT Specialty Limited. http://expropriation.jltgroup.com/home/listview?Country=&IndustryDataObject[]=3&EventTypeDataObject[]=&FromDate=2004–01&ToDate=2013–12&sortby=

Johnson, John J. ed. 1962. *The role of the military in underdeveloped countries*. Princeton, NJ: Princeton University Press.

Jones, Richard. 1965 January. The nationbuilder, soldier of the sixties, *Military Review* 45: 63–67.

Kahn, Herman. 1962. *Thinking the unthinkable*. New York: Horizon.

Kalman, Matthew. 2013 July 11. Israel's military entrepreneurial complex owns big data, *MIT Technology Review*. http://www.technologyreview.com/news/516511/israels-military-entrepreneurial-complex

Kaplan, Jacob. 1950. United States foreign aid programs: Past perspectives and future needs, *World Politics* 3(1): 55–71.

Kapur, Devesh, John Lewis, and Richard Webb. 1997. *The World Bank: Its first half century*. Washington: The Brookings Institution Press.

Kapur, Devesh, and Richard Webb. 2000 August. Governance-related conditionalities of the international financial institutions. G-24 Discussion Paper Series No. 6. New York: United Nations.

Katz, Jonathan. 2013. *The big truck that went by: How the world came to save Haiti and left behind a disaster*. New York: Macmillan.

Kaufman, Robert. 1972. *The politics of land reform in Chile, 1950–1970: Public policy, political institutions, and social change*. Cambridge: Harvard University Press.

Kaufmann, Daniel. 1992. *How macroeconomic policies affect project performance in the social sectors*. Washington, DC: World Bank.

Kaufmann, Daniel. 2009. Aid effectiveness and governance: The good, the bad and the ugly, *Development Outreach* 11(1): 26–29.

Kaufmann, Daniel and Yan Wang. 1995. Macroeconomic policies and project performance in the social sectors: A model of human capital production and evidence from LDCs, *World Development* 23(5): 751–765.

Kaysen, Carl. n.d. Secret–Korea–Field Plan, National Security Files. Foreign Aid Subjects. Military Assistance Programs. Box 373. John F. Kennedy Library.

Kennan, George. 1947. The sources of Soviet conduct, *Foreign Affairs* 25(4): 566–82.

Kennedy, John F. 1960. *The strategy of peace*. New York: Harper.

Kennedy, John F. 1961a. Remarks at the Eighth National Conference on International Economic and Social Development. http://www.presidency.ucsb.edu

Kennedy, John F. 1961b. Special message to the congress on foreign aid. http://www.presidency.ucsb.edu

Kennedy, John F. 1963 September 17. Address before the White House Conference on Exports. http://www.presidency.ucsb.edu/ws/?pid=9411

Kerr, Richard. 2013 December 3. Abrupt climate change: Still looming. *Science*.

Kettle, D. F. 2000. *The global public management revolution: A report on the transformation of governance*. Washington, DC: Brookings Institution Press.

Khan, Shahrukh. 2007. Pakistan's economy since 1999 has there been real progress? *South Asia Economic Journal* 8(2): 317–34.

Khan, Shahrukh. 2012 October. The military and economic development in Pakistan. Amherst, Political Economy Research Institute, Working Paper Series, No. 291.

Khator, Renu. 1998. The new paradigm: From development administration to sustainable development administration, *International Journal of Public Administration* 21(12): 1777–801.

Killick, Tony. 1976. The possibilities of development planning, *Oxford Economic Papers* 28(2): 161–84.

Kim, Hyung-A. 2011. Heavy and chemical industrialization, 1973–1979: South Korea's homeland security measures. In Hyung-A Kim and Clark Sorenson, eds., *Reassessing the Park Chung Hee era 1961–1979*. Seattle, WA: University of Washington Press, pp. 19–42.

Kim, M. Julie, and Rita Nangia. 2010. Infrastructure development in India and China—A comparative analysis. In William Ascher and Corinne Krupp, eds., *Physical infrastructure development: Balancing the growth, equity, and environmental imperatives.* New York: Palgrave Macmillan, pp. 97–140.

Kimmelman, Michael. 2014 July 5. Refugee camp evolves as a do-it-yourself city, *New York Times*, A-1, A-8.

Klare, Michael. 1997. East Asia's militaries muscle up, *The Bulletin of Atomic Scientists* 53(1): 136–152.

Knio, Karim. 2010. Investigating the two faces of governance: The case of the Euro-Mediterranean Development Bank, *Third World Quarterly* 31(1): 105–21.

Kohli, Harpaul, and Phillip Basil. 2010. Requirements for infrastructure investment in Latin America under alternate growth scenarios: 2011–2040, *Global Journal of Emerging Market Economies* 3(1): 59–110.

Kooiman, Jan. 1993. *Modern governance: New government-society interactions.* Thousand Oaks, CA: Sage.

Korean Ministry of Strategy and Finance. 2012. *Impact of foreign aid on Korea's development. Seoul: Ministry of Strategy and Finance.* Seattle, WA: University of Washington Center for Korean Studies, pp. 19–42.

Korten, David. 1987. Third generation NGO strategies: A key to people-centered development, *World Development* 15: 145–59.

Korten, David. 1990. *Getting to the twenty-first century: Voluntary action and the global agenda.* West Hartford, CT: Kumarian Press.

Krueger, Anne, Maurice Schiff, and Alberto Valdés. 1991. *The political economy of agricultural pricing policy, Volume 2.* Baltimore, MD: The Johns Hopkins University Press.

Krugman, Paul. 1994. The fall and rise of development economics. In Lloyd Rodwin and Donald Schon, eds., *Rethinking the development experience: Essays provoked by the work of Albert O. Hirschman.* Washington, DC: Brookings Institution Press, pp. 39–58.

Kuczynski, Pedro-Pablo, and John Williamson, eds. 2003. *After the Washington Consensus: Restarting growth and reform in Latin America.* Washington, DC: Peterson Institute.

Kuznets, Simon. 1955. Economic growth and income inequality, *American Economic Review*: 1–28.

Kwong, Julia. 1997. *The political economy of corruption in China.* Armonk, NY: ME Sharpe.

Labini, Paolo. 2000. Growth models and the explanation of forces behind development processes. In Roger Blackhouse and Andrea Salanti, eds., *Macroeconomics and the real world.* New York: Oxford University Press, pp. 263–74.

Lal, Deepak. 2000. *The poverty of development economics.* Cambridge, MA: MIT Press.

Lal, Deepak. 2012. Is the Washington Consensus dead? *Cato Journal* 32(3): 493–512.

Lanjouw, Peter, and Gershon Feder. 2001. Rural non-farm activities and rural development: From experience towards strategy. World Bank Rural Development Strategy Background Paper No. 4. Washington, DC: World Bank.

Lansdale, Edward. 1962. Civic action helps counter the guerrilla threat, *Army Information Digest* 17: 50–53.

LaPalombara, Joseph. 1967a. Bureaucracy and political development: Notes, queries, and dilemmas. In Joseph LaPalombara, ed., *Bureaucracy and political development.* Princeton: Princeton University Press, pp. 34–61.

LaPalombara, Joseph. 1967b. Preface and acknowledgements. In Joseph LaPalombara, ed., *Bureaucracy and political development.* Princeton: Princeton University Press, pp. ix–xi.

Laquian, A. A. 1981. Review and evaluation of urban accommodationist policies in population redistribution. In United Nations Department of International Economic and Social Affairs, eds., *Population distribution policies in development planning.* New York: United Nations, pp. 101–12.

Lasswell, Harold D. 1936/1958. *Politics: Who gets what, when, how*. New York: Meridian.

Lasswell, Harold D. 1965. The policy sciences of development, *World Politics* 17(2): 286–309.

Lasswell, Harold D. 1971. *A pre-view of policy sciences*. New York: Elsevier.

Lasswell, Harold D., and Abraham Kaplan. 1950. *Power and society*. New Haven: Yale University Press.

Lasswell, Harold D., and Myres McDougal. 1992. *Jurisprudence for a free society*. Dordecht: Kluwer.

Lastarria-Cornhiel, Susana. 2006. Feminization of agriculture: Trends and driving forces. Background paper for the World Development Report 2008. http://siteresources.worldbank.org/INTWDR2008/Resources/2795087-1191427986785/LastarriaCornhiel_FeminizationOfAgri.pdf

Latham, Michael E. 2011. *The right kind of revolution: Modernization, development, and US foreign policy from the Cold War to the present*. Ithaca: Cornell University Press.

Lawson, Marian, Susan Epstein, and Kennon Nakamura. 2010. *State, foreign operations, and related programs: FY2011 budget and appropriations*. Washington, DC: Congressional Research Service.

Lawson, Marian, Susan Epstein, and Tamara Resler. 2011. *State, foreign operations, and related programs: FY2012 budget and appropriations*. Washington, DC: Congressional Research Service.

Lee, Chae-Jin. 2006. *A troubled peace: U.S. policy and the two Koreas*. Baltimore: John Hopkins University Press.

Lee, Hyesook. 1997. State formation and civil society under American occupation: The case of South Korea, *Korea Journal of Population and Development* 26(2): 15–32.

Leff, Nathaniel. 1985a. Optimal investment choice for developing countries: Rational theory and rational decision-making, *Journal of Development Economics* 18(2): 335–60.

Leff, Nathaniel. 1985b. The use of policy-science tools in public-sector decision making: Social benefit-cost analysis in the World Bank, *Kyklos* 38(1): 60–76.

Lehmann, Tim. 2013 March 6. Embodying the new era: Models for NGOs in the 21st century, *Huffington Post*. www.huffingtonpost.com/tim-lehmann/wef-ngos

Levy, Brian. 2010. *Development trajectories: An evolutionary approach to integrating governance and growth*. Washington, DC: World Bank. https://openknowledge.worldbank.org/handle/10986/1018

Lewis, David, and Nazneen Kanji. 2009. *Non-governmental organizations and development*. New York: Routledge.

Lewis, W. Arthur. 1954. Economic development with unlimited supplies of labour, *The Manchester School* 22(2): 139–91.

Lindert, Kathy, Emmanuel Skoufias, and Joseph Shapiro. 2006. Redistributing income to the poor and the rich: Public transfers in Latin America and the Caribbean. Social Safety Nets Primer Series Discussion Paper 605. Washington, DC: World Bank.

Lipton, Michael. 2009. *Land reform in developing countries: Property rights and property wrongs*. New York: Routledge.

Little, Ian. 1982. *Economic development: Theory, policy, and international relations*. New York: Basic Books.

Little, Ian, and James Mirrlees. 1969. *Manual of industrial project analysis in developing countries*. Paris: Development Centre of the Organisation for Economic Cooperation and Development.

Little, Ian, and James Mirrlees. 1990. Project appraisal and planning twenty years on. Paper presented at the World Bank Annual Conference of Development Economics.

Lo, Carlos Wing-Hung, Gerald E. Fryxell, and Wilson Wai-Ho Wong. 2006. Effective regulations with little effect? The antecedents of the perceptions of environmental officials on enforcement effectiveness in china, *Environmental Management* 38(3): 388–410.

Looney, Robert. 1989. Military Keynesianism in the third world, *Journal of Political and Military Sociology* 17(Spring): 43–64.

Lora, Eduardo. 2001 December. *Structural reforms in Latin America: What has been reformed and how to measure it.* Washington, DC: Inter-American Development Bank.

Lora, Eduardo. 2007. *The state of state reform in Latin America.* Palo Alto, CA: Stanford University Press.

Loveman, Brian. 1976. *Struggle in the countryside: Politics and rural labor in Chile, 1919–1973.* Bloomington: Indiana University Press.

Lowden, Pamela. 1996. *Moral opposition to authoritarian rule in Chile, 1973–90.* New York: St. Martins Press.

Lower, Milton D. 1968. Institutional bases of economic stagnation in Chile, *Journal of Economic Issues* 2(3): 283–97.

Lucero, José. 2008. *Struggles of voice: The politics of indigenous representation in the Andes.* Pittsburgh: University of Pittsburgh Press.

Luong, H. V. 2003. Wealth, power, and inequality: Global market, the state, and local sociocultural dynamics. In H.V. Luong, ed. *Postwar Vietnam: Dynamics of a changing society.* Boulder: Rowman and Littlefield, pp. 81–106.

Lynn, Lawrence. 1996. *Public management as art, science, and profession.* Chatham, NJ: Chatham House Publishers.

MacAuslan, John, and Mark Addison. 2010. Governance in central government: Reconciling accountability and capability, *The Political Quarterly* 81(2): 243–52.

Maldonado, Nicole. 2010. The World Bank's evolving concept of good governance and its impact on human rights. Paper presented at doctoral workshop on development and international organizations, Stockholm.

Malesky, Edmund, and Jonathan London. 2014. The political economy of development in China and Vietnam, *Annual Review of Political Science* 17: 395–419.

Mani, Kristina. 2007. Militaries in business state-making and entrepreneurship in the developing world, *Armed Forces & Society* 33(4): 591–611.

Mani, Kristina. 2011. Militares empresarios: Approaches to studying the military as an economic actor, *Bulletin of Latin American Research* 30(2): 183–97.

Manyin, Mark E., Emma Chanlett-Avery, Mary Beth D. Nikitin, Ian E. Rinehart, and Brock R. Williams. 2015 June 11. *U.S.-South Korea relations* Washington, DC: Congressional Research Service.

Manzano, Osmel, Francisco Monaldi, and Federico Sturznegger. 2008. The political economy of oil production in Latin America, *Economía* 9(1): 59–98.

Maren, Michael. 1997. *The road to hell: The ravaging effects of foreign aid and international charity.* New York: Simon and Schuster.

Marino, Andrea. 2011. Economic factors and political turmoil in North Africa. In Franco Praussello, ed., *Euro-Mediterranean partnership in the aftermath of the Arab Spring* Milan: Franco Angeli, pp. 21–36.

Marquand, Robert. 2014 November 17. China unveils its own hard-line Putin, *Christian Science Monitor* 106(52): 21–23.

Martin, Roger, and Sally Osberg. 2007. Social entrepreneurship: The case for definition, *Stanford Social Innovation Review* 5(2): 28–39.

Mason, Edward, Mahn Je Kim, Dwight H. Perkins, and Kwang Suk Kim. 1980. *The economic and social modernization of the Republic of Korea.* Cambridge, MA: Harvard University Press.

Mauro, Paolo. 1995. Corruption and growth, *The Quarterly Journal of Economics*: 681–712.

Mavroudeas, Stavros, and Fanis Papadatos. 2007. Reform, reform the reforms or simply regression? The "Washington consensus" and its critics, *Bulletin of Political Economy* 1(1): 43–66.

Mayama, Katsuhiko. 2000 March. Chinese People's Liberation Army: Reduction in force by 500,000 and trend of modernization, *NIDS Security Reports* 1: 116–34.

Mayr, Walter. 2010 November 15. Exotic birds in a cage—Criticism grows of Afghanistan's bloated NGO industry, *Speigel Online.* http://www.spiegel.de/international/world/exotic-birds-in-a-cage-criticism-grows-of-afghanistan-s-bloated-ngo-industry-a-718656.html

McCleary, Rachel, and Robert Barro. 2008. Private voluntary organizations engaged in international assistance 1939–2004, *Nonprofit and Voluntary Sector Quarterly* 37: 512–36.

McVety, Amanda. 2012. *Enlightened aid: US development as foreign policy in Ethiopia.* New York: Oxford University Press.

Meier, Gerald. 1968. *The international economics of development.* New York: Harper & Row.

Meier, Gerald. 1987. *Pioneers in development*, Vol. 2. New York: Oxford University Press.

Meier, Gerald. 2004. *Biography of a subject: An evolution of development economics.* New York: Oxford University Press.

Meier, Gerald, and Dudley Seers. 1984. *Pioneers in development*, Vol. 1. New York: Oxford University Press.

Meltzer, Alan. 2000. *Report of the International Financial Institutions Advisory Commission.* Washington, DC: US Government Printing Office. http://www.house.gov/jec/imf/meltzer.pdf

Meyer, Peter. 2013. *Chile: Political and economic conditions and U.S. relations.* Washington, DC: Congressional Research Service.

Michaels, Albert. 1976. The Alliance for Progress and Chile's "Revolution in Liberty," 1964–1970, *Journal of Interamerican Studies and World Affairs* 18(1): 74–99.

Mickelwait, Donald, Charles Sweet, and Elliott Morss. 1979. *New directions in development: A study of US AID.* Boulder, CO: Westview Press.

Mikesell, Raymond. 1983. Appraising IMF conditionality: Too loose, too tight, or just right? In John Williamson, ed., *IMF conditionality.* Cambridge, MA: MIT Press, pp. 47–62.

Millikan, Max. 1955 November. Economic policy as an instrument of political and psychological policy. Psychological Aspects of United States Strategy. Panel Report. White House Central Files. Confidential. Box 61. Rockefeller. Dwight D. Eisenhower Library, Abilene, Kansas.

Milner, Helen V., and Dustin Tingley. 2013. The choice for multilateralism: Foreign aid and American foreign policy, *Review of International Organizations* 8(3): 313–41.

Milobsky, David, and Louis Galambos. 1995. The McNamara Bank and its legacy, 1968–1987, *Business and Economic History* 24(2): 173, 169.

Minford, Paul. 2011. Review of Lance Taylor, Maynard's revenge: the collapse of free market macroeconomics, *Journal of Economics* 103: 289–91.

Mishra, Pankaj. 2014 August 4. The places in between, *The New Yorker*: 64–69.

Mitra-Kahn, Benjamin. 2008. Debunking the myths of computable general equilibrium models. Schwartz Center for Economic Policy Analysis Working Paper 2008–1. New York: New School.

Modernizing Foreign Assistance Network. 2008 June 1. New day, new way: U.S. foreign assistance for the 21st century. http://www.cgdev.org/content/publications/detail/16210/

Mok, Ka Ho. 2008. When socialism meets market capitalism: Challenges for privatizing and marketizing education in China and Vietnam, *Policy Futures in Education* 6(5): 601–15.

Montgomery, John. 1962. *The politics of foreign aid: American experience in Southeast Asia.* New York: Council on Foreign Relations.

Montgomery, John. 1967. *Foreign aid in international politics.* Englewood Cliffs, NJ: Prentice-Hall.

Montgomery, John. 1986. *Aftermath: Tarnished outcomes of American foreign policy.* Dover: Auburn House Publishing Company.

Moore, Mick. 2004. Revenues, state formation, and the quality of governance in developing countries, *International Political Science Review* 25(3): 297–319.

Moore, Mick. 2007. *How does taxation affect the quality of governance?* Brighton, UK: Essex University Institute of Development Studies.

Morgenthau, Hans. 1978. *Politics among nations: The struggle for power and peace*. New York: Knopf.

Mosher, A. 1976. *Thinking about development*. New York: Agricultural Development Council.

Moss, R. H., G. A. Meehl, M. C. Lemos, J. B. Smith, J. R. Arnold, D. Behar, G. P. Brasseur, S. B. Broomell, A. J. Busalacchi, and S. Dessai. 2013. Climate change-hell and high water: Practice-relevant adaptation science, *Science* 342: 696–98.

Mulvenon, James. 1998. Military corruption in china: A conceptual examination, *Problems of Post-Communism* 45(2): 12–21.

Mulvenon, James. 1999. Soldiers of fortune: The rise and fall of PLA, Inc. Paper presented at American Political Science Association Meeting, Toronto.

Mulvenon, James. 2001a December. PLA divestiture and civil-military relations: Implications for the sixteenth Party Congress leadership, *China Leadership Monitor* 1(2).

Mulvenon, James. 2001b. *Soldiers of fortune: The rise and fall of the Chinese military-business complex, 1978–1998*. Armonk, NY: ME Sharpe.

Murphy, Dan. 2013 May 29. Why do they hate our NGO funding? Well, because it's threatening to foreign governments, *Christian Science Monitor*. http://www.csmonitor.com/World/Security-Watch/Backchannels/2013/0529

Murty, K. N., and A. Soumya. 2006. Macroeconomic effects of public investment in infrastructure in India. Working Paper series No. WP-2006–003. Mumbai: Indira Gandhi Institute of Development Research.

Muscat, Robert J. 2002. *Investing in peace: How development aid can prevent or promote conflict*. Armonk, NY: ME Sharpe.

Mutual Security Act of 1951. Title 5, sec. 511b. Washington, DC: US Government Printing Office.

National Research Council. 2002. *Abrupt climate change: Inevitable surprises*. Washington, DC: National Academies Press.

National Research Council. 2009. *Informing decisions in a changing climate*. Washington, DC: National Academies Press.

National Research Council. 2010a. *America's climate choices: Adapting to the impacts of climate change*. Washington, DC: National Academies Press.

National Research Council. 2010b. *America's climate choices: Advancing the science of climate change*. Washington, DC: National Academies Press.

Nellis, John. 2012 January. The international experience with privatization: Its rapid rise, partial fall and uncertain future. Calgary: University of Calgary School of Public Policy Research Paper No. 12–3.

Nelson, Joan. 1968. *Aid, influence, and foreign policy*. New York: Macmillan.

Nielson, Daniel, and Michael Tierney. 2003. Delegation to international organizations: Agency theory and World Bank environmental reform, *International Organization* 57(2): 241–76.

Nurkse, Ragnar. 1953. *Problems of capital formation in developing countries*. New York: Columbia University Press.

O'Donnell, Guillermo. 1973. *Modernization and bureaucratic-authoritarianism: Studies in South American politics*. Berkeley, CA: Institute of International Studies, University of California.

O'Donnell, Guillermo, Philippe Schmitter, and Laurence Whitehead. 1986a. *Transitions from authoritarian rule: Southern Europe*, Vol. 1. Baltimore, MD: The Johns Hopkins University Press.

O'Donnell, Guillermo, Philippe Schmitter, and Laurence Whitehead, eds. 1986b. *Transitions from authoritarian rule*, Vol. 2. Baltimore, MD: Johns Hopkins University Press.

Organisation for Economic Cooperation and Development. 1995. *Participatory Development and Good Governance.* Paris: Organisation for Economic Cooperation and Development.

Organisation for Economic Cooperation and Development, Development Assistance Committee. 2011. Peer review on the United States. http://www.oecd.org/development/peer-reviews/48434536.pdf

Organisation for Economic Cooperation and Development. 2013a. *Aid untying: 2012 report review of the implementation of the 2001 recommendation and the Accra and Busan untying commitments.* Paris: Organisation for Economic Cooperation and Development.

Organisation for Economic Cooperation and Development. 2013b. *Official Development Assistance 2013 update.* Paris: Organisation for Economic Cooperation and Development. http://search.oecd.org/officialdocuments/publicdisplaydocumentpdf/?cote=DCD/DAC(2012)39/FINAL&docLanguage=

Orme, John. The original megapolicy: America's Marshall Plan. In John D. Montgomery and Dennis Rondinelli, eds. *Great policies: Strategic innovations in the Pacific Basin.* Westport, CT: Praeger.

Ortiz, Isabel, and Matthew Cummins. 2011 April. Global inequality: Beyond the bottom billion. UNICEF Social and Economic Policy Working Paper. New York: UNICEF.

Osborne, David, and Ted Gaebler. 1993. *Reinventing government: How the entrepreneurial spirit is transforming the public sector.* New York: Plume.

Osborne, Stephen, ed. 2010. *The new public governance? Emerging perspectives on the theory and practice of public governance.* London: Routledge.

Owusu, Francis. 2003. Pragmatism and the gradual shift from dependency to neoliberalism: The World Bank, African leaders and development policy in Africa, *World Development* 31(10): 1655–72.

Pansuwan, Apisek, and Jayant K. Routray. 2011. Policies and pattern of industrial development in Thailand, *Geojournal* 76(1): 25–46.

Parker, Laura. 2015. Treading water, *National Geographic* 227(2): 106–27.

Partners in Progress. 1951. A report to the President by the International Advisory Board, ca. March 1951. Official File, Truman Papers in the Harry S. Truman Presidential Library, Independence, Missouri.

Patrinos, Harry, and George Psacharopoulos. 2011 March. Education: Past, present and future global challenges. Washington, DC: World Bank Human Development Network Education Team Policy Research Working Paper 5616.

Patterson, Thomas G., ed. 1989. *Kennedy's quest for victory: American foreign policy, 1961–1963.* New York: Oxford University Press.

Pelaez, Vicky. 2012 July 13. The insidious role of NGOs in the 21st century, *The Moscow News.* www.moscownews.ru/international/20120713/189957078

Perkins, Dwight H. 2001. Industrial and financial policy in China and Vietnam: A new model or a replay of the East Asian experience? In Joseph Stiglitz and Shahid Yusuf, eds., *Rethinking the East Asian economic miracle.* New York: Oxford University Press, pp. 247–94.

Pessino, Carola, and Ricardo Fenochietto. 2010. Determining countries' tax effort, *Revista de Economía Pública* 195: 65–87.

Petermann, Jan-Henrik. 2013. *Between export promotion and poverty reduction: The foreign economic policy of untying official development assistance.* Dordrecht: Springer.

Peters, B. Guy. 2001. *The future of governing,* 2nd ed. Lawrence, KS: University Press of Kansas.

Pfiffner, James. 2004. Traditional public administration versus the new public management: Accountability versus efficiency. In A. Benz, H. Siedentopf, and K. P. Sommermann, eds., *Institutionenbildung in Regierung und Verwaltung: Festschrift fur Klaus Konig.* Berlin: Duncker & Humbolt, pp. 443–54.

Picard, Louis, and Terry Buss. 2009. *A fragile peace: Re-examining the history of foreign aid, security, and diplomacy*. Sterling: Kumarian Press.

Popovski, Vesselin. 2010. The role of civil society in global governance. In G. Shabbir Cheema and Vesselin Popovski, eds., *Engaging civil society: Emerging trends in democratic governance.* Tokyo: United Nations University, pp. 23–43.

Powell, Walter, and Paul DiMaggio. 1991. *The new institutionalism in organizational analysis.* Chicago: University of Chicago Press.

Prahalad, Coimbatore, and Stuart Hart. 2002. The fortune at the bottom of the pyramid, *Strategy + Business* 26(Spring).

Prebisch, Raul. 1949. *The economic development of Latin America and its principal problems.* New York: United Nations.

President's Advisory Committee on Government Organization. 1954 July 12. Memorandum for President Eisenhower. Organization for the Development and Coordination of Foreign Economic Policy. Executive Office of the President. White House Central Files Confidential. Box 18. Council on Foreign Economic Policy. Dwight D. Eisenhower Library, Abilene, Kansas.

President's Committee to Study the United States Military Assistance Program. 1959 August 17. USAID DEC (PC-AAA-444).

Prince, Erik. 2013. *Civilian warriors: The inside story of blackwater and the unsung heroes of the war on terror.* New York: Penguin.

Prosterman, Roy. 2007. Land reform. In *International encyclopedia of the social sciences*, 2nd edn. New York: Macmillan, pp. 341–45.

Prosterman, Roy, and Jeffrey Riedinger. 1987. *Land reform and democratic development.* Baltimore: Johns Hopkins University Press.

Przeworski, Adam, Guillermo O'Donnell, Philippe C. Schmitter, and Laurence Whitehead. *Transitions from authoritarian rule: Comparative perspectives.* Baltimore, MD: The Johns Hopkins University Press.

Psacharopoulos, George, and Harry Patrinos. 2004. Education: Past, present and future global challenges. World Bank Policy Research Working Paper 5616. Washington, DC: World Bank.

Pye, Lucian. 1960. Political development and foreign aid, USAID Development Experience Clearinghouse. http://dec.usaid.gov/

Pye, Lucian. 1961. Armies in the process of political modernization, *European Journal of Sociology* 2: 82–92.

Pye, Lucian. 1964. Military training and political and economic development. In Don Piper and Taylor Cole, eds., *Post-primary education and economic development.* Durham, NC: Duke University Press, pp. 75–94.

Rabe, Stephen. 2010. *John F. Kennedy, world leader.* Dulles, VA: Potomac Books.

Radelet, Steven. 2006. A primer on foreign aid. Working Paper 92. Washington, DC: Center for Global Development.

Rai, Kul. 1980. Foreign aid and voting in the UN general assembly, 1967–1976, *Journal of Peace Research* 17(3): 269–77.

Ramphersad, Hubert, and Saleh Hussain. 2014. *Authentic governance: Aligning personal governance with corporate governance.* Dordrecht: Springer.

Rana, Raj. 2004. Contemporary challenges in the civil-military relationship: Complementarity or incompatibility? *International Review of the Red Cross* 86(855): 565–92.

Raup, Philip. 1968. Land reform. In *International encyclopedia of the social sciences.* New York: Macmillan, pp. 562–75, esp. 574.

Ravallion, Martin. 2002. Are the poor protected from budget cuts? Evidence for Argentina, *Journal of Applied Economics* 5(1): 95–121.

Reimann, Kim. 2005. Up to no good? Recent critics and critiques of NGOs. In Oliver Richmond and Henry Carey, eds., *Subcontracting peace: The challenges of NGO peacebuilding*, Aldershot, UK: Ashgate Publishing, pp. 37–54.

Remenyi, Joseph. 1979. Core demi-core interaction: Toward a general theory of disciplinary and subdisciplinary growth, *History of Political Economy* 11(1): 30–63.

Rennack, Dianne, Lisa Mages, and Susan Chesser. 2011 August. Foreign operations appropriations: General provisions. CRS Report for Congress. Washington, DC: Congressional Research Service.

Revkin, Andrew. 2013 December 3. An update on risks of abrupt jolts from global warming, *DOT Earth/The New York Times*. http://dotearth.blogs.nytimes.com/2013/12/03/an-update-on-risks-of-abrupt-jolts-from-global-warming/

Richardson, Michael. 2010. Civil society and the state in South Korea, *SAIS U.S.-Korea yearbook*. Washington, DC: Johns Hopkins School of Advanced International Studies, pp. 165–176.

Riddell, Roger. 1987. *Foreign aid reconsidered*. Baltimore: Johns Hopkins University Press.

Rigg, Jonathan. 2003. *South East Asia: The human landscape of modernization and development*. London: Routledge.

Rioja, Felix, 2003. The penalties of inefficient infrastructure, *Review of Development Economics* 7(1): 127–37.

Ritzer, George. 2000. *The MacDonaldization of society*. Newbury Park, CA: Pine Forge Press.

Roberts, John. 2005. Millennium development goals: Are international targets now more credible? *Journal of International Development* 17: 113–29.

Robichau, Robbie. 2011. The mosaic of governance: creating a picture with definitions, theories, and debates, *Policy Studies Journal* 39: 113–31.

Rodrik, Dani. 2008. *One economics, many recipes: Globalization, institutions, and economic growth*. Princeton, NJ: Princeton University Press.

Rondinelli, Dennis. 1987. *Development administration and US foreign aid policy*. Boulder, CO: Lynne Rienner.

Rondinelli, Dennis. 1990. Financing the decentralization of urban services in developing countries: Administrative requirements for fiscal improvements, *Studies in Comparative International Development* 25(2): 43–59.

Rondinelli, Dennis. 2007. Parallel and partnership approaches in decentralized governance: Experience in weak states. In G. Shabbir Cheema and Dennis Rondinelli, eds., *Decentralizing governance: Emerging concepts and practices*. Washington, DC: Brookings Institution Press, pp. 21–42.

Rosenstein-Rodan, Paul. 1943. Problems of industrialization of eastern and south-eastern Europe, *Economic Journal* 53: 202–11.

Rosenstein-Rodan, Paul. 1944. The international development of economically backward areas, *International Affairs* 20(2): 157–65.

Rostow, Walt. 1952. *The process of economic growth*. New York: W.W. Norton.

Rostow, Walt. 1956. The take-off into self-sustained growth, *The Economic Journal* 66(261): 25–48.

Rostow, Walt. 1960. *The stages of economic growth: A non-communist manifesto*. Cambridge: Cambridge University Press.

Roy, Arundhati. 2002. *Power politics*, 2nd edn. Cambridge, MA: South End Press.

Roy, Denny. 2009. China's dilemma over civil society organizations. Paper presented to the Regional Workshop on Engaging Civil Society, Honolulu, East-West Center.

Rusk, Dean. 1962 July 19. Methods for improving the coordination of economic and military aid programs. AID/Washington. USAID Record Group 286. Box 6. Folder: Program FY 63. Entry P39: Classified Vietnam. NACP.

Ruttan, Vernon. 1984. Integrated rural development programmes: A historical perspective, *World Development* 12(4): 393–401.

Ruttan, Vernon. 1996. *United States development assistance policy: The domestic politics of foreign economic aid*. Baltimore, MD: Johns Hopkins University Press.

Sachs, Jeffery. 2003. Institutions matter, but not for everything, *Finance and Development* 40: 38–41.

Salamon, Lester, and Helmut Anheier. 1992a. In search of the non-profit sector I: The problem of classification. *Voluntas: International Journal of Voluntary and Nonprofit Organizations* 3(2): 125–51.

Salamon, Lester, and Helmut Anheier. 1992b. In search of the non-profit sector II: The problem of classification. *Voluntas: International Journal of Voluntary and Nonprofit Organizations* 3(3): 267–309.

Salamon, Lester, S. Wojciech Sokolowski, and Associates. 2004. *Global civil society: Dimensions of the nonprofit sector*, Vol. II. Bloomfield, CT: Kumarian Press.

Sally, Razeen, and Rahul Sen. 2011. Trade policies in Southeast Asia in the wider Asian perspective, *The World Economy* 34(4): 568–601.

Sankey, Mara. 2014. The role of transnational non-state actors in Chilean democracy: The NED, WOLA, and Freedom House and the 1988 Plebiscite. http://www.academia.edu/6663614/

Sarfaty, Galit. 2012. *Values in translation: Human rights and the culture of the World Bank.* Stanford: Stanford University Press.

Savitz, Andrew. 2013. *The triple bottom line: How today's best-run companies are achieving economic, social and environmental success—and how you can too.* New York: John Wiley & Sons.

Scahill, Jeremy. 2007. *Blackwater: The rise of the world's most powerful mercenary army.* New York: Nation Books.

Schaub Jr., Gary, and Volker Franke. 2009. Contractors as military professionals? *Parameters* 39(4): 88–104.

Scheffran, Jürgen, Michael Brzoska, Jasmin Kominek, Michael Link, and Janpeter Schilling. 2012. Climate change and violent conflict, *Science* 336(6083): 869–71.

Schmidhuber, J., J. Bruinsma, and G. Boedeker. 2009. Capital requirements for agriculture in developing countries to 2050. Paper presented at the FAO Expert Meeting on How to Feed the World in 2050, 24–26 June, Rome.

Schmitter, Philippe. 1972. Paths to political development in Latin America, *Proceedings of the Academy of Political Science*: 83–105.

Schmitter, Phillipe. 1974. Still the century of corporatism? *Review of Politics* 36(1): 85–131.

Schmitter, Philippe 1996. The influence of international context upon the choice of national institutions and policies in neo-democracies. In Laurence Whitehead, ed., *The international dimensions of democratization: Europe and the Americas.* Oxford: Oxford University Press, pp. 26–54.

Schou, Arild, and Marit Haug. 2005. Decentralisation in conflict and post-conflict situations. Working Paper No. 139. Oslo: Norwegian Institute for Urban and Regional Research.

Schultz, T. Paul. 1986. Economic demography and development: New directions in an old field. In Gustav Ranis and T. Paul Schultz, eds., *The state of development economics.* London: Basil Blackwell.

Scott, James C. 1998. *Seeing like a state: How certain schemes to improve the human condition have failed.* New Haven: Yale University Press.

Secret Seoul 4499. 1968. Assistance to Korean National Police (KNP). Office of Public Safety: USAID Record Group 286. Entry 25. Box 5. Folder: General Correspondence Relating to Geographical Areas, 1965–1971 February 26. Jamaica to Korea. NACP. Box 70. Folder: IPS#1 General Policy Guidelines, General Information Korea '68. NACP.

Sen, Amartya. 1985. *Commodities and capabilities.* New York: North-Holland.

Sen, Amartya. 2000. *Development as freedom.* New York: Random House.

Sheehan, Neil. 1989. *A bright shining lie: John Paul Vann and America in Vietnam.* New York: Vintage Books.

Shihata, I. F. J. 1997. Corruption—General review and the role of the World Bank. In B. Rider, ed., *Corruption, the enemy within*. The Hague: Kluwer Law International.

Shin, Roy. 1969. *The politics of foreign aid: A study of the impact of United States aid in Korea from 1945 to 1966*. Doctoral dissertation, Department of Political Science, University of Minnesota, Minneapolis.

Shlaim, Avi. 1983. *The United States and the Berlin blockade, 1948–1949: A study of crisis decision making*. Berkeley, CA: University of California Press.

Siddiqa, Ayesha. 2007. *Military Inc: Inside Pakistan's military economy*. London: Pluto.

Siegel, Adam. 1998. *Mission creep: An examination of an overused and misunderstood term*. Alexandria, VA: Center for Naval Analysis.

Silicon Valley Business Journal Staff. 2012. Strategies for success—nonprofits: Nonprofit execs talk donor fatigue, recruiting, mission creep. http://www.bizjournals.com/sanjose/print-edition/2012/06/29/

Singer, Hans. 1950. The distribution of gains between investing and borrowing countries, *American Economic Review* 40(20): 473–85.

Singer, Hans. 1968. Targets, commitments and realities, *Intereconomics* 3(2): 53–55.

Singh, Kavaljit. 2003. Aid and good governance. A discussion paper for the reality of aid. Delhi: Public Interest Research Group (India).

SIPRI. 2008. *The effectiveness of foreign military assets in natural disaster response*. Stockholm: SIPRI.

Skocpol, Theda. 1996. Unsolved mysteries: The Tocqueville files: Unravelling from above, *The American Prospect* 25(March–April): 20–25.

Slaper, Timothy, and Tanya Hall. 2011. The triple bottom line: What is it and how does it work? *Indiana Business Review* 86(1): 4–8.

Slocum-Bradley, Nikki, and Andrew Bradley. 2010. Is the EU's governance "good"? An assessment of EU governance in its partnership with ACP states, *Third World Quarterly* 31(1): 31–49.

Solomone, Stacey. 1995. The PLAs commercial activities in the economy-effects and consequences, *Issues & Studies* 31(3): 20–43.

Solt, Frederick. 2010. *Standardized world income inequality database, Version 3.0*. Carbondale, IL: Southern Illinois University.

Squire, Lyn, and Herman van der Tak. 1975. *Economic analysis of projects*. Baltimore, MD: Johns Hopkins University Press.

Steinberg, David. 1982. The economic development of Korea: Sui Generis or Generic? A.I.D. Evaluation Special Study 6. USAID DEC. PN-AAJ-177.

Steinfeld, Joshua, and Khi V. Thai. 2013. Political economy of Vietnam: Market reform, growth, and the state, *Maryland Series on Contemporary Asian Studies* 3: 1–72.

Stepan, Alfred. 1971. *The military in politics: Changing patterns in Brazil*. Princeton, NJ: Princeton University Press.

Stepan, Alfred. 1988. *Rethinking military politics: Brazil and the southern cone*. Princeton, NJ: Princeton University Press.

Stern, Nicholas, and Francisco Ferreira. 1997. *The World Bank as "Intellectual Actor."* Washington, DC: Brookings Institution Press.

Stier, Ken. 2011. Egypt's military-industrial complex. *The Times* 9(11).

Stiglitz, Joseph. 2000 April 17. What I learned at the world economic crisis, *The New Republic* 222(16/17): 56–60.

Stiglitz, Joseph. 2002. *Globalization and its discontents*. New York: W.W. Norton & Company.

Stiglitz, Joseph. 2011. Rethinking development economics, *The World Bank Research Observer* 26(2): 230–36.

Stoker, Gerry. 2006. Public value management: A new narrative for networked governance? *American Review of Public Administration* 36(1): 41–57.

Stokke, Olav. 2009. *The UN and development: From aid to cooperation.* Bloomington, IN: Indiana University Press.

Stolper, Wolfgang F. 1966. *Planning without facts: Lessons in resource allocation from Nigeria's development.* Cambridge, MA: Harvard University Press.

Streeten, Paul. 1959. Unbalanced growth, *Oxford Economic Papers* 11(2): 167–91.

Streeten, Paul, and Shahid Burki. 1978. Basic needs: Some issues, *World Development* 63(March): 411–21.

Streeten, Paul, Shahid Burki, Mahbub ul Haq, Norman Hicks, and Frances Stewart. 1981. *First things first: Meeting basic human needs in the developing countries.* New York: Oxford University Press.

Studwell, Joe. 2013. *How Asia works: Success and failure in the world's most dynamic region.* New York: Grove Press.

Svoboda, Eva. 2014 April. The interaction between humanitarian and military actors: Where do we go from here? Policy Brief 58. London: HPG/Overseas Development Institute.

Swyngedouw, Erik. 2005. Governance innovation and the citizen: The Janus face of governance-beyond-the-state, *Urban Studies* 42(11): 1991–2006.

Tadros, Sherine. 2012 February 15. Egypt military's economic empire, *Aljazeera.* http://www.aljazeera.com/indepth/features/2012/02/2012215195912519142.html

Taffet, Jeffrey. 2007. *Foreign aid as foreign policy: The Alliance for Progress in Latin America.* New York: Routledge.

Tandon, Rajesh. 1992. *NGOs and civil society.* Boston, MA: Institute for Development Research.

Tarnoff, Curt, and Marian Lawson. 2011. *Foreign aid: An introduction to U.S. programs and policy.* Washington, DC: Congressional Research Service.

Tarnoff, Curt, and Marian Lawson. 2012. *Foreign aid: An introductory overview of US programs and policies.* Washington, DC: Congressional Research Service.

Tarnoff, Curt, and Larry Nowels. 2004. *Foreign aid: An introductory overview of US programs and policies.* Washington, DC: Congressional Research Service.

Tarrow, Sidney. 1994. *Power in movements: Social movements, collective action, and politics.* Cambridge: Cambridge University Press.

Task Force on Foreign Economic Policy. 1960 December 31. Report to the Honorable John F. Kennedy. National Security Files. Foreign Economic Policy. Box 297, 10–11. Boston, MA: John F. Kennedy Library.

Task Force on International Development. 1970. U.S. foreign assistance in the 1970s: A new approach. Report to the President. Washington, DC: US Government Printing Office.

Taylor, Lance. 2011. *Maynard's revenge: The collapse of free market macroeconomics.* Cambridge, MA: Harvard University Press.

Tendler, Judith. 1975. *Inside foreign aid.* Baltimore, MD: Johns Hopkins University Press.

Tetlock, Philip, and Dan Gardner. 2013. Who's good at forecasts? *The Economist*, 81.

Thirkell-White, Ben. 2003. The IMF, good governance and middle-income countries, *The European Journal of Development Research 15*(1): 99–125.

Tignor, Robert. 2006. *W. Arthur Lewis and the birth of development economics.* Princeton, NJ: Princeton University Press.

Tosun, Mehmet, and Serdar Yilmaz. 2008. Centralization, decentralization and conflict in the Middle East and North Africa. Paper delivered at the Economic Research Forum Fifteenth Annual Conference, November 23–25, Cairo.

Trinkunas, Harold. 2005. *Crafting civilian control of the military in Venezuela.* Chapel Hill, NC: University of North Carolina Press.

Trofimenko, Henry. 1981. The third world and the US-soviet competition: A soviet view, *Foreign Affairs* (Summer): 1021–40.

Truman, Harry S. 1949 January 20. Inaugural address. http://www.presidency.ucsb.edu/ws/?pid=13282

Ul Haq, Mahbub. 1999. *Human development in South Asia 1999: The crisis of governance*. New York: Oxford University Press.

UN Conference on Trade and Development. 2007 July. Economic development in Africa: Reclaiming policy space: Domestic resource mobilization and developmental states. Overview by the UNCTAD Secretariat, Geneva.

UN Department of Economic and Social Affairs 2009. *The contribution of the United Nations to the improvement of public administration: A 60-year history*. New York: United Nations.

UNICEF. 1996. Military expenditure—The opportunity cost. http://www.unicef.org/sowc96/8mlitary.htm

United Nations. 1962. *The United Nations development decade: Proposals fora*. New York: United Nations.

United Nations. 1987. *Report of the World Commission on Environment and Development: Our common future*. Oxford: Oxford University Press.

United Nations. 1997. *Agenda for development*. New York: UN Department of Public Information.

United Nations. 2011. *Global issues: Governance*. New York: United Nations. http://www.un.org/en/globalissues/governance/

United Nations Capital Development Fund. 2005. *Delivering the goods—Building local governance capacity to achieve Millennium Development Goals*. New York: United Nations.

United Nations Development Programme. 1990. *Human development report 1990*. New York: United Nations Development Programme.

United Nations Development Programme. 1995. Public sector management, governance, and sustainable human development: A discussion paper. New York: Management Development and Governance Division, Bureau for Policy and Programme Support, United Nations Development Programme.

United Nations Development Programme. 1997. *Governance for sustainable human development*. New York: United Nations Development Programme.

United Nations Development Programme. 2002. *Human development report 2002*. New York: United Nations Development Programme.

United Nations Development Programme. 2011. *Human development report 2011. Sustainability and equity: A better future for all*. New York: United Nations Development Programme.

United Nations High-Level Panel of Eminent Persons on the Post-2015 Development Agenda. 2013. *Report of the UN high-level panel of eminent persons on the post-2015 development agenda*. New York: United Nations.

UNRRA. 1950. *The history of the United Nations relief and rehabilitation administration*. New York: Cornell University Press, Vols. 1, 3, 4.

USA Today. 2009 February 2. http://www.usatoday.com/news/washington/2009-02-01-aid-inside_N.htm

USAID. 1959 September 1. Committee to Strengthen the Frontiers of Freedom. "The time has come to face the facts." USAID DEC. PC-AAA-217.

USAID. 1962 December 13. Country Assistance Program–Korea. USAID DEC. PD-ACC-396.

USAID. 1963. Introduction. Classified Vietnam Subject Files. USAID Record Group 286. Entry #P 39. Box 1. Folder: AGR Agriculture FY 1962, 1963. 1–2. NACP.

USAID. 1964a November 23. Office of Public Safety. Airgram–Public Safety. USAID Record Group 286. Box 23. Folder IPS-1. File Material Chile 2 of 2. National Archive at College Park.

USAID. 1964b March 13. Office of Public Safety. Joint State/AID to Santiago AIDTO. USAID Record Group 286. Box 24, 1964–73. Folder IPS1 General Policy–Chile 1964. NACP.

USAID. 1969. *Administrative history—Agency for International Development.* Vol. 1. Chapter 13. Washington, DC: USAID.

USAID/Chile. 1980 September. A history: United States of America assistance to Chile, A.I.D. and predecessor agencies, 1943–1980. USAID DEC. PN-AAJ-651.

USAID. 1982. Population assistance. AID Policy Paper. Washington: US Government Printing Office.

USAID. 2005. U.S. overseas loans and grants: Obligations and loan authorizations. July 1, 1945-September 30. Cong-R-0105. http://pdf.usaid.gov/pdf_docs/PNADH500.pdf

USAID. 2013. *USAID strategy on democracy, human rights and governance.* Washington, DC: USAID.

USAID 2014. *Greenbook 2014.* Washington, DC: US Agency for International Development. https://eads.usaid.gov/gbk/data/program_report.cfm

USAID Alumni Association. 2008 December. Foreign assistance reform and strengthening USAID: Talking points for the President-elect's transition team for foreign assistance. Washington, DC: USAID Alumni Association. http://pdf.usaid.gov/pdf_docs/PCAAB936.pdf

US Department of State. 1950 November 29. Report on American opinion. Public discussion of the Gordon Gray Report. Washington, DC: US Department of State.

US Department of State. 2014 May. Congressional budget justification: Department of State, foreign operations, and related programs. Washington, DC: Government Printing Office.

US Foreign Assistance Act of 1961, Section 610. 2002. Legislation on foreign relations through 2002. Washington, DC: US Government Printing Office.

US ODA 2014. *U.S. ODA database, summary reports.* Washington, DC: US ODA.

USSC (US Senate Select Committee to Study Governmental Operations with Respect to Intelligence Activities). 1975. *Covert action in Chile, 1963–1973.* Washington, DC: US Government Printing Office.

Valdés, Alberto. 1983. The role of agricultural exports in development, *Agriculture and the State: Growth, Employment, and Poverty in Developing Countries*: 84–115.

Valente, Michael. 2011 February 10. Egyptian democracy meets the military-industrial complex, *Sustainable Business Forum.* http://www.sustainablebusinessforum.com/mikevalente/49733/democracy-hands-corporations

Van de Laar, Aart. 1980. *The World Bank and the poor.* Dordrecht: Springer.

Venniker, Richard. 2001. Social returns to education: A survey of recent literature on human capital externalities. The Hague: CPB Netherlands Bureau for Economic Policy Analysis Report 00/1.

Vo, 1990. *Vietnam's economic policy since 1975.* Singapore: ASEAN Economic Research Unit.

Vuong, Quan Hoang. 2014. Vietnam's political economy: A discussion on the 1986–2016 period. Working Papers CEB 14–010, Universite Libre de Bruxelles.

Walt, Vivienne. 2014 November 10. Missing in action: Why the World Health Organization failed to stop Ebola, *Time* 184(18): 32–37.

Walterhouse, Harry. 1962. Civic action, a counter and cure for insurgency, *Military Review* 42(8): 47–54.

Walters, Robert S. 1970. *American and Soviet aid: A comparative perspective.* Pittsburg, PA: University of Pittsburgh Press.

Walton, Richard J. 1972. *Cold War and counterrevolution: The foreign policy of John F. Kennedy.* New York: Viking Press.

Watkins, Kevin. 1998. *Economic growth with equity: Lessons for East Asia.* Oxford, UK: Oxfam GB.

Weil, David. 2007. Accounting for the effect of health on economic growth, *The Quarterly Journal of Economics* 122(3): 1265–1306.

Weir, Fred. 2013 June 24. Russia widens its crackdown, *Christian Science Monitor* 105(31): 11–12.

Weiss, Thomas. 2000. Governance, good governance and global governance: Conceptual and actual challenges, *Third World Quarterly* 21(5): 795–815.

Wekwete, Kadmiel. 2007. Decentralization to promote effective and efficient pro-poor infrastructure and service delivery in the least developed countries. In G. Shabbir Cheema and Dennis Rondinelli, eds., *Decentralizing governance: Emerging concepts and practices.* Washington, DC: Brookings Institution Press, pp. 242–66.

Welker, D. 1997. The Chinese military industrial complex goes global, *The Multinational Monitor* 18(6). http://multinational monitor.org/hyper/mm0697.05.html

Westad, Odd Arne. 2003. *Decisive encounters: The Chinese civil war, 1946–1950.* Palo Alto, CA: Stanford University Press.

White, Gordon. 1994. Civil society, democratization and development (I): Clearing the analytical ground, *Democratization* 1(2): 375–90.

White House. 2002. *The national security strategy 2002.* Washington, DC: US Government Printing Office.

White House Office of the Press Secretary. 2010a. Fact sheet: US global development policy. http://www.whitehouse.gov/the-press-office/2010/09/22/fact-sheet-us-global-development-policy

White House Office of the Press Secretary. 2010b September 22. Remarks by the president at the Millennium Development Goals Summit in New York. http://www.whitehouse.gov/the-press-office/2010/09/22/remarks-president-millennium-development-goals-summit-new-york-new-york

Widianto, Bambang. 2007. Are budget support and cash transfer effective means of social protection. Paper presented at Forum on Inclusive growth and poverty reduction in the new Asia and the Pacific.

Williams, William. 1973. *The contours of American history.* New York: Verso Books.

Williamson, John. 1990. *Latin American adjustment: How much has happened?* Washington, DC: Institute for International Economics.

Williamson, John. 1999. What should the World Bank think about the Washington consensus? Paper prepared as a background to the World Bank's World Development Report 2000. Institute for International Economics Papers. http://scienzepolitiche.unipg.it/tutor/uploads/williamson_on_washington_consensus_002.pdf

Williamson, John. 2004. A short history of the Washington consensus, Conference on From the Washington Consensus towards a New Global Governance, September 24–25. Barcelona: Fundación CIDOB.

Williamson, John. 2009. Short history of the Washington consensus, *Law and Business Review of the Americas* 15: 7–26.

Williamson, Oliver. 2002. The theory of the firm as governance structure: From choice to contract, *Journal of Economic Perspectives*: 171–95.

Wills, Garry. 1987. *Reagan's America.* New York: Penguin.

Wilson, Margaret. 2011. *Does your nonprofit suffer from "mission creep"?* Frederick, MD: McLean, Koehler, Sparks & Hammond.

Wilson, Woodrow. 1887. The study of administration, *Political Science Quarterly* 2(2), 197–222.

Wing, Ian. 2004 September. Computable general equilibrium models and their use in economy-wide policy analysis. Cambridge, MA: MIT Joint Program on the Science and Policy of Global Change Technical Note No. 6.

Winn, Peter, and Cristobal Kay. 1974. Agrarian reform and rural revolution in Allende's Chile, *Journal of Latin American Studies* 6(1): 135–59.

Wintrop, Norman. 1983. *Liberal democratic theory and its critics.* London: Croom Helm.

Wolf, Jr., Charles. 1979. A theory of nonmarket failure: Framework for implementation analysis, *Journal of Law and Economics* 22(1): 107–39.

Wolf, Jr., Charles, Xiao Wang, and Eric Warner. 2013. *China's foreign aid and government-sponsored investment activities: Scale, content, destinations, and implications*. Santa Monica, CA: RAND Corporation.

World Bank. 1975. *Agricultural credit: Sector policy paper.* Washington, DC: World Bank.

World Bank. 1981. *Accelerated development in Sub-Saharan Africa: An agenda for action.* Washington, DC: World Bank.

World Bank. 1989. *Sub-Saharan Africa: From crisis to sustainable growth: A long-term perspective study.* Washington, DC: World Bank.

World Bank. 1992. *Governance and development.* Washington, DC: World Bank.

World Bank. 1994. *Governance: The World Bank's experience.* Washington, DC: World Bank.

World Bank. 1997. *World development report 1997: The state in a changing world.* New York: Oxford University Press.

World Bank. 2011a August. Breaking even or breaking through: Reaching financial sustainability while providing high quality standards in higher education in the Middle East and North Africa, Washington, DC. http://siteresources.worldbank.org/INTMENA/Resources/FinancingHigherEducation_MENA.pdf

World Bank. 2011b. *The changing wealth of nations: Measuring sustainable development in the new millennium.* Washington, DC: World Bank.

World Bank. 2012a. An update to the World Bank's estimates of consumption poverty in the developing world. Washington, DC: World Bank. http://siteresources.worldbank.org/INTPOVCALNET/Resources/Global_Poverty_Update_2012_02-29-12.pdf

World Bank. 2012b. World Bank data bank. Percentage distribution of public current expenditure by level. http://databank.worldbank.org/ddp/home.do?Step=3&id=4

World Bank. 2012c. World Bank databank. Secondary school enrollments. http://databank.worldbank.org/ddp/home.do?Step=3&id=4

World Bank. 2013. *Computable general equilibrium models, poverty and social impact analysis site.* Washington, DC: World Bank.

World Bank. 2014a. *The state of social safety nets 2014.* Washington, DC: World Bank.

World Bank. 2014b. *World development indicators 2013.* Washington, DC: World Bank.

World Bank Independent Evaluation Group. 2010. *Cost-benefit analysis in World Bank projects.* Washington, DC: World Bank.

World Bank Group. 2014. *Doing business 2015: Going beyond efficiency.* Washington, DC: World Bank.

World Health Organization. 2014. *World health statistics 2014.* Geneva, Switzerland: World Health Organization.

Yunus, Mohammed. 2010. *Building social business: The new kind of capitalism that serves humanity's most pressing needs.* New York: Public Affairs.

Zeller, Manfred, and Richard Meyer, eds. 2002. *The triangle of microfinance: Financial sustainability, outreach, and impact.* Baltimore: Johns Hopkins University Press

Index

CPSIA information can be obtained
at www.ICGtesting.com
Printed in the USA
LVOW01*2258210116
471729LV00011B/452/P